2016 11th International Conference on Advanced Semiconductor Devices & Microsystems (ASDAM 2016)

Smolenice, Slovakia
13 – 16 November 2016

IEEE Catalog Number: CFP16469-POD
ISBN: 978-1-5090-3084-2

Copyright © 2016 by the Institute of Electrical and Electronics Engineers, Inc
All Rights Reserved

Copyright and Reprint Permissions: Abstracting is permitted with credit to the source. Libraries are permitted to photocopy beyond the limit of U.S. copyright law for private use of patrons those articles in this volume that carry a code at the bottom of the first page, provided the per-copy fee indicated in the code is paid through Copyright Clearance Center, 222 Rosewood Drive, Danvers, MA 01923.

For other copying, reprint or republication permission, write to IEEE Copyrights Manager, IEEE Service Center, 445 Hoes Lane, Piscataway, NJ 08854. All rights reserved.

***This publication is a representation of what appears in the IEEE Digital Libraries. Some format issues inherent in the e-media version may also appear in this print version.**

IEEE Catalog Number: CFP16469-POD
ISBN (Print-On-Demand):   978-1-5090-3084-2
ISBN (Online):   978-1-5090-3083-5

Additional Copies of This Publication Are Available From:

Curran Associates, Inc
57 Morehouse Lane
Red Hook, NY 12571 USA
Phone: (845) 758-0400
Fax: (845) 758-2633
E-mail: curran@proceedings.com
Web: www.proceedings.com

2016 11th International Conference on Advanced Semiconductor Devices & Microsystems (ASDAM 2016)

Smolenice, Slovakia
13 – 16 November 2016

IEEE Catalog Number: CFP16469-POD
ISBN: 978-1-5090-3084-2

ASDAM 2016

Conference Proceedings

The 11th International Conference on

ADVANCED SEMICONDUCTOR DEVICES AND MICROSYSTEMS

Smolenice Castle, Slovakia

November 13-16, 2016

edited by

Štefan Haščík

Jaroslav Dzuba

Gabriel Vanko

ORGANIZERS

Dept. of Microelectronics and Sensors
Institute of Electrical Engineering
Slovak Academy of Sciences
Bratislava, Slovakia

and

Institute of Electronics and Photonics
Faculty of Electrical Engineering and Information Technology
Slovak University of Technology
Bratislava, Slovakia

COSPONSORSHIP

The IEEE Electron Devices Society

INTERNATIONAL PROGRAMME COMMITEE

J. Osvald	(IEE SAS Bratislava, Slovakia) – Chair
P. Kordoš	(STU Bratislava, Slovakia) – Vice-chair
J. Breza	(STU Bratislava, Slovakia)
D. Donoval	(STU Bratislava, Slovakia)
F. Dubecký	(IEE SAS Bratislava, Slovakia)
E. Gombia	(IMEM CNR Parma, Italy)
Zs. J. Horváth	(HAS Budapest, Hungary)
M. Husák	(CTU Prague, Czech Republic)
T. Lalinský	(IEE SAS Bratislava, Slovakia)
M. Mikulics	(PGI Jülich, Germany)
M. Missous	(University of Manchester, UK)
J. Novák	(IEE SAS Bratislava, Slovakia)
D. Pogany	(TU Wien, Austria)
M. Tlaczala	(University of Wroclaw, Poland)
G. Wachutka	(TU Munich, Germany)

ORGANIZING COMMITEE

G. Vanko	(IEE SAS Bratislava, Slovakia) – Chair
A. Laurenčíková	(IEE SAS Bratislava, Slovakia)
O. Babchenko	(IEE SAS Bratislava, Slovakia)
J. Dzuba	(IEE SAS Bratislava, Slovakia)
M. Grujbár	(IEE SAS Bratislava, Slovakia)
Š. Haščík	(IEE SAS Bratislava, Slovakia)
T. Lalinský	(IEE SAS Bratislava, Slovakia)

CONTENTS

Theme/Authors	Page

Threshold voltage instabilities in AlGaN/GaN MOS-HEMTs with ALD-grown Al_2O_3 gate dielectrics: Relation to distribution of oxide/semiconductor interface state density
M. Ťapajna, L. Válik, D. Gregušová, K. Fröhlich, F. Gucmann, T. Hashizume, and J. Kuzmík
1

Pseudo-vertical GaN-based trench gate metal oxide semiconductor field effect transistor
R. Hentschel, A. Wachowiak, A. Grosser, A. Jahn, U. Merkel, A. Wille, H. Kalisch, A. Vescan, S. Schmult, and T. Mikiolajick
5

DC and pulsed IV characterisation of AlGaN/GaN MOS-HEMT structures with Al_2O_3 gate dielectric prepared by various techniques
F. Gucmann, D. Gregušová, L. Válik, M. Ťapajna, Š. Haščik, K. Hušeková, K. Fröhlich, O. Pohorelec and J. Kuzmík
9

Analysis of Hybrid Pair Si IGBT-SiC diode for Very High Power Applications
M. Nawaz, F. Chimento
13

SiO2/4H-SiC Interface Traps Effects on the Input Capacitance of DMOSFET
G. D. Licciardo. L. Di Benedetto
19

Optimization of semi-insulating GaAs detector for thermal neutron detection
A. Šagátová, B. Zaťko, K. Sedlačková, V. Nečas, P. Boháček
23

Towards III-nitride nano-LED based single photon emitters: technology and applications
H. Hardtdegen, M. Mikulics
27

GaAs-based photodetector with applied PDMS membrane with photonic crystal in the surface
D. Pudiš, M. Tlaczala, L. Suslik, W. Dawidowski, B. Sciana, J. Kovač, J. Kovač jr., M. Goraus, P. Gašo, J. Ďurišová, I. Zborowska-Lindert
33

Improving Light Extraction Efficiency of PhC LED
P. Hronec, J. Kováč, J. Škriniarová, J. Kováč jr., F. Uherek
37

Photoluminescence of InGaN/GaN MQW structures – technological aspects
A. Hospodková, J. Pangrác, J. Oswald, K. Kuldová, M. Ziková, and E. Hulicius
41

Electrical Characterization of MIS Photoanodes Annealed Under Different Conditions for Solar Fuel Generation
M. Mikolášek, J. Racko, V. Řeháček, L. Harmatha, M. Ťapajna, and K. Fröhlich
45

Electro-Physical 2-D/3-D TCAD Analysis of Reduce The Temperature-related Negative Effects on Silicon Heterojunction Solar Cell for Concentrated Photovoltaic
P. Príbytný, M. Mikolášek, J. Marek, A. Chvála, and D. Donoval
49

Polymer 3D photonic crystals for LED   **53**
M. Goraus, D. Pudiš, P. Gaso, L. Suslik, D. Jandura

GaAsSb/InAs QDs structures for advanced telecom lasers   **57**
M. Zíková, A. Hospodková, J. Pangrác, J. Oswald, and E. Hulicius

Minimization of Self-Heating in SOI MOSFET Devices with SELBOX Structure   **61**
M. Narayanan, H. Al-Nashash

Highly reliable long-term operation of AlGaN/GaN/AlN HFETs grown on silver substrate   **65**
A. Fox, M. Mikulics, M. Marso, M. Kočan, Z. Sofer, P. Kordoš, H. Lüth, J. Schubert, D. Grützmacher and H. Hardtdegen

InGaN mesoscopic structures for low energy consumption nano-opto-electronics   **69**
H. Lüth, M. Mikulics, A. Winden, St. Trellenkamp, Z. Sofer, M. Marso, D. Grützmacher and H. Hardtdegen

Electrical and optical characterization of freestanding Ge$_1$Sb$_2$Te$_4$ nano-membranes integrated in coplanar strip lines   **73**
M. Mikulics, M. Marso, R. Adam, M. Schuck, A. Fox, R. Sobolewski, P. Kordoš, H. Lüth, D. Grützmacher and H. Hardtdegen

Hybrid Optoelectronics Based on a Nanocrystal/III-N Nano-LED Platform   **77**
M. Marso, M. Mikulics, H. Lüth, Z. Sofer, P. Kordoš and H. Hardtdegen

Fabrication of UV sources for novel lithographical techniques: development of nano-LED etching procedures   **81**
J. Moers, M. Mikulics, M. Marso, St. Trellenkamp, Z. Sofer, D. Grützmacher and H. Hardtdegen

The effect of process parameters and annealing on the properties of Ti/Pt films for miniature temperature sensors   **85**
I. Hotový, Š. Haščík, M. Predanocy, M. Mikolášek, V. Řehaček, I. Kostič, P. Nemec, A. Benčurová, D. Rossberg, L. Spiess

Multi-bit Adder Design using a ME-MTJ technology   **89**
N. Sharma, A. Marshall, J. Bird, P. Dowben

Technology of AlSb/GaSb based LED nanostructures for high temperature superlinear luminescence   **93**
E. Hulicius, A. Hospodková, J. Pangrác, M. Zíková

3D energy harvester with tunable resonant frequency   **97**
V. Janicek

Reducing Crosstalk in the Internal Structures of Integrated Circuits   **101**
J. Novak, J. Foit

Magnetically Levitated and Guided Systems   **105**
F. Puci

Model of the Triboelectric Generator   **109**
M. Husák, A. Bíly

Gas Analyzer for Quick Indicative Measurements   **113**
M. Husák, A. Laposa, J. Kroutil

Acoustic Method for Respiratory Monitoring   **117**
J. Kroutil, A. Laposa, M. Husák, R. Sio

Gas sensor based on sputtered NiO thin films   **121**
M. Predanocy, I. Hotový, V. Řehaček

Gold nanoisland arrays prepared by sputtering and bio-functionalization of their surfaces   **125**
O. Szabó, V. Tvarožek, S. Kováčová, I. Novotný

Study of Repetitive Avalanche Stress Invoked Degradation of Electrical Properties of DMOS and TrenchMOS Transistors   **129**
J. Marek, Ľ. Stuchlíková, M. Jagelka, A. Chvála, P. Príbytný, M. Donoval and D. Donoval

Study of negative electron beam nanoresist HSQ on GaAs substrate   **133**
R. Andok, A. Benčurová, I. Kostic, A. Ritomský, J. Škriniarová, K. Vutova

Simple patterning method of sub-micro- and nanometer structures for gas sensor   **137**
P. Ďurina, A. Benčurová, M. Truchlý, R. Andok, I. Kostič, B. Grančič, A. Plecenik, P. Kúš, K. Vutova, E. Koleva

Particle detectors based on 4H-SiC epitaxial layer and their properties   **141**
B. Zaťko, L. Hrubčín, A. Šagátová, P. Boháček, F. Dubecký, K. Sedlačková, M. Sekáčová, J. Arbet, V. Nečas, V. A. Skuratov

Investigation of AlGaN/GaN Schottky Structures by Deep Level Fourier Transient Spectroscopy with Optical Excitation   **145**
Ľ. Stuchliková, A. Kósa, R. Szobolovszky, M. Petrus, L. Harmatha, S. L. Delage, J. Kováč

DLTFS study of InGaAs/AlInAs heterostructures grown on n-InP:S substrates   **149**
A. Kósa, Ľ. Stuchliková, L. Harmatha, J. Kováč, M. Badura, K. Bielak, B. Sciana, M. Tlaczala

Impedance characterisation of NiO based gas sensor   **153**
M. Mikolášek, M. Predanocy, P. Jom, I. Hotový

Schottky contact metallization stability on AlGaN/GaN heterostructure during the diamond deposition process   **157**
O. Babchenko, G. Vanko, J. Dzuba, T. Ižák, M. Vojs, T. Lalinský and A. Kromka

Characterization and gas sensor testing of single and mixed metal oxides based on NiO and TiO_2 thin films   **161**
I. Hotový, M. Predanocy, M. Mikolášek, P. Benko and V. Řehaček

3-D Device and Circuit Electrothermal Simulations of Power Integrated Circuit Including Package — 165
A. Chvála, P. Benko, P. Príbytný, J. Marek and D. Donoval

Analysis and modeling of the electric field in the Gate oxide of 4H-SiC DMOSFET — 169
L. Di Benedetto, G. D. Licciardo and A. Rubino

Impact of repetitive UIS on modern GaN power devices — 173
J. Marek, Ľ. Stuchlíková, M. Jagelka, A. Chvála, P. Príbytný, M. Donoval, and D. Donoval

Post-deposition annealing and thermal stability of integrated self-aligned E/D-mode n++GaN/InAlN/AlN/GaN MOS HEMTs — 177
M. Blaho, D. Gregušová, Š. Haščík, A. Seifertová, M. Ťapajna, J. Šoltýs, A. Šatka, L. Nagy, A. Chvála, J. Marek, J. Priesol and J. Kuzmík

Scaling and Traps Induced Degradation of Cutoff Frequency in GaN HEMT — 181
B. C. Ubochi, S. Faramehr, K. Ahmeda, P. Igič, and K. Kalna

Trap analysis in GaN-based heterostructures using current transients measurements — 185
M. Florovič, J. Škriniarová, D. Gregušová, J. Kováč, P. Kordoš

Temperature-dependent of sub-threshold slope of AlGaN/GaN MOSHFETs with HfO$_2$ gate oxide prepared by ALD — 189
R. Stoklas, D. Gregušová, K. Fröhlich, and J. Kuzmík

Nanocrystalline diamond films for electronic monitoring of gas and organic molecules — 193
A. Kromka, T. Ižák, M. Davydova, B. Rezek

Modeling the I-V-T characteristics of 4H-SiC DMOSFET in presence of SiO$_2$/SiC interface traps and fixed oxide — 199
G. D. Licciardo, L. Di Benedetto, A. Rubino

The Effect of Self-Heating and Electrical Stress Induced Polarization in AlGaN/GaN Heterojunction Based Devices — 203
K. Ahmeda, S. Faramehr, P. Igic, K. Kalna, S. J. Duffy, A. Soltani, B. Benbakhti

Effect of HCl pretreatment on the oxide/semiconductor interface state density in AlGaN/GaN MOS-HEMT structures with MOCVD grown Al$_2$O$_3$ gate dielectric — 207
M. Ťapajna, K. Hušeková, O. Pohorelec, L. Válik, Š. Haščík, F. Gucmann, K. Fröhlich, D. Gregušová, and J. Kuzmík

RF performance optimization of SiGe:C transistors via Ge-B doping strategies and RTP — 211
A. M. Yarimbiyik, D. I. Öksüz, E. Cesur

Silicon carbide thin films deposited by PECVD technology for applications in photoelectrochemical water splitting devices — 215
J. Huran, P. Boháček, V. Sasinková, A. Kleinová, M. Mikolášek, A.P. Kobzev, L. Hrubčín, J. Arbet, and M. Sekáčová

Investigation of metal contacts on semi-insulating GaAs: physics, technology and applications   **219**
F. Dubecký, G. Vanko, D. Kindl, P. Hubík, E. Gombia, P. Boháček, M. Sekáčová, B. Zaťko

Nanostructuring by femtosecond laser ablation and RIE for MEMS and microfluidic systems fabrication   **223**
J. Zehetner, G. Vanko, J. Dzuba, T. Lalinský

Strain induced response of AlGaN/GaN high electron mobility transistor located on cantilever and membrane   **227**
J. Dzuba, G. Vanko, O. Babchenko, T. Lalinský, F. Horvát, M. Szarvas, T. Kováč, B. Hučko

Electrical Properties of Back-to-Back Metal/Insulator/Semiconductor Diodes Based on AlGaN/GaN heterostructure   **231**
J. Osvald

Evaluation of TIGBT field stop layer generated by helium implantation   **235**
R. Pechal, L. Dorňák, J. Vavro, A. Kozelský, A. Klimsza

Incremental Control Techniques for Layout Modification   **239**
P. Vacula, V. Kotě, A. Kubačák, M. Lžičař, R. Zelený, M. Husák, J. Jakovenko

The influence of ozone pre-treatment in HfO$_2$-based resistive switching memory structures   **243**
P. Benko, M. Mikolášek, L. Harmatha, K. Fröhlich

Author Index   **247**

PREFACE

The ASDAM 2016 Conference gets together researchers working in the field of semiconductor devices and microsystems again. It is already an established meeting of scientists prevalently from Europe but also from the USA and Asia. The new results of the research endeavour in the field of advanced semiconductor devices will be referred-to in the Conference sessions. The Conference has confirmed its reputation in the last few years as an already well-known meeting, not large in the number of participants but by the scientific level of their contributions and its nice surroundings providing very agreeable atmosphere for discussions and exchange of recently reached results of different laboratories.

ASDAM'16 is the 11th ASDAM Conference, being organized every second year since 1996 in Smolenice Castle, Slovakia. Invited lecturers open the Conference sessions describing the state of the art in the special topics of the Conference. This year they are concentrating on the technology and applications of III-Nitride nano-LED based single photon emitters, performance and variability of InGaAs and Si FinFETs in future digital applications, scaling limits of organic thin film transistors, nanocrystalline diamond films for electronic monitoring of gas and organic molecules, current status, challenges and opportunities of novel wide bandgap semiconductors for high power electronics as well as on the overview of GaN power device technology. The attention will also be paid to other topics as submicron and low-dimensional structures, photodetectors and radiation detectors, heterostructure device modelling and simulations, MEMS, etc.

The Programme Committee has received more than 70 abstracts out of which around 60 contributing papers divided into oral and poster presentations will be presented at the Conference. It is a success of the Conference that there will be 6 invited lectures given by the renowned scientists at the opening of single sessions. The organizers would like to thank all the authors and the Programme Committee members for making together the nice programme of the Conference.

The Conference has already had its successful history. We hope that the 11th ASDAM will prolong this series which has been far beyond the organizers expectations from the very beginning. We are looking forward to inspirational talks and discussions which will stimulate common research effort in our everyday work.

Welcome to ASDAM '16!

Bratislava, November 2016

Gabriel Vanko
General Chair

Štefan Haščík
Publication Chair

ASDAM 2016, The 11th International Conference on Advanced Semiconductor
Devices And Microsystems, November 13-16, 2016, Smolenice, Slovakia

Threshold voltage instabilities in AlGaN/GaN MOS-HEMTs with ALD-grown Al$_2$O$_3$ gate dielectrics: Relation to distribution of oxide/semiconductor interface state density

M. Ťapajna[1], L. Válik[1], D. Gregušová[1], K. Fröhlich[1], F. Gucmann[1], T. Hashizume[2], and J. Kuzmík[1]

[1]Institute of Electrical Engineering, Slovak Academy of Sciences, Dúbravská cesta 9, 841 04 Bratislava, Slovakia,
[2]Research Center for Integrated Quantum Electronics (RCIQE), Hokkaido University, 060-0814, Sapporo, and JST-CREST, 102-0075, Tokyo, Japan
e-mail: milan.tapajna@savba.sk

Metal-oxide-semiconductor (MOS) structure represents an important gate technology in GaN HEMTs. As oxide/semiconductor interface quality is remaining reliability concern, several techniques for determination of interface state density (D_{it}) has been proposed. In the literature, the hysteresis in C-V sweeps (or V_{th} shift, ΔV_{th}) is often interpreted as D_{it} in particular energy range in the semiconductor band-gap. In this work, we critically assessed a relevancy of relation between ΔV_{th} (measured at 25 and 125 °C) and experimentally determined D_{it} distribution, to point out possible pitfalls in the data interpretation. D_{it} distributions were measured by combination of complementary techniques and 1D simulations applied to state-of-the-art MOS-HEMT structures with Al$_2$O$_3$ films grown by ALD on AlGaN/GaN heterostructures. It is demonstrated that, apart from interface traps, also other parasitic effects related to border traps and oxide bulk traps can have dominant impact on ΔV_{th}. This means that ΔV_{th} could not be solely related to D_{it}, unless negligibility of other relevant effects is confirmed.

1. Introduction

Metal-oxide-semiconductor (MOS) gate structure is becoming a standard gate technology for AlGaN/GaN high electron mobility transistors (HEMTs) [1, 2]. However, MOS gate technology faces several challenges including relatively high density of oxide/semiconductor interface states (D_{it}), presence of border traps (BT) located in the oxide near its interface with semiconductor, and bulk traps distributed in the oxide. In particular, increased D_{it} can induce threshold voltage (V_{th}) instabilities in wide time range (from seconds to years) due to wide band gap nature of GaN based semiconductor. Therefore, several techniques for D_{it} determination in GaN MOS-HEMTs have been proposed in the literature [3, 4]. Commonly, simple capacitance-voltage (CV) hysteresis is used to assess the interface quality by relating extracted V_{th} shift (ΔV_{th}) to D_{it} in particular energy range. In this work, we investigate V_{th} instability in state-of-the-art MOS-HEMTs with Al$_2$O$_3$ gate dielectric grown by atomic layer deposition (ALD). In particular, relation of V_{th} instability to interface state density is deeply examined to point out possible pitfalls in the CV data interpretation. It was found that apart from D_{it}, also other effects related to BT emission and oxide bulk trapping can have a dominant impact on ΔV_{th}. Therefore, CV hysteresis alone may not always represent a reliable measure of the oxide/semiconductor interface quality in GaN MOS-HEMTs.

978-1-5090-3084-2/16 $31.00 © 2016 IEEE

2. Experimental details

Two large area MOS-HEMT structures with 20-nm thick Al_2O_3 films and different *net charge* at Al_2O_3/III-N interface (Q_{int}) were used in this study. In the first sample featuring Q_{int} close to zero (compensated interface charge), Al_2O_3 was grown by ALD using TMA and H_2O (referred to as H_2O-ALD) at 350 °C on $Al_{0.26}Ga_{0.74}N$/GaN heterostructure grown by MOCVD. To assure high oxide/semiconductor interface, protecting SiN layer was applied during Ohmic metallization step and N_2O-radical treatment (P_{RF}=200 W, 300 °C, 10 min) was applied prior to Al_2O_3 deposition [5]. The second sample showed high *negative* Q_{int} of density about 10^{13} cm^{-2}. Here, Al_2O_3 was grown by ALD using TMA and O_3 (sample O_3-ALD) at 100 °C on commercial GaN-cap/$Al_{0.29}Ga_{0.71}N$/GaN wafers. HCl pretreatment was applied prior to the oxide deposition.

The V_{th} shift (ΔV_{th}) was measured from a parallel shift of the CV curves in depletion part (see Fig. 1(a)) after double sweep CV measurements with sweep rate of 0.2 V/s at chuck temperature of 25 and 125 °C. ΔV_{th} as a function of maximum positive gate voltage ($V_{g,max}$) was extracted and compared to calculated shift in the CV curves, considering D_{it} determined experimentally. D_{it} was measured by combination of four complementary techniques: (i) V_{th}-transient as a function of temperature (25 to 200 °C, traps located ~0.5 to 1.1 eV below barrier conduction band (CB)), (ii) monochromatic light assisted V_{th}-transient (traps located 1.5 to 3.3 eV below CB)) [3], and (iii) CV frequency dispersion [4] or (iv) Terman method at positive V_g-s (for shallower traps). Using the experimental D_{it} distributions, CV hysteresis was simulated by using 1D Poisson solver that includes SRH statistics for interface traps capture/emission processes [6].

3. Results and discussion

3.1 Impact of border traps on ΔV_{th}

Fig. 1 shows CV measurements on H_2O-ALD structure as a function of $V_{g,max}$ (a) at 25 and 125 °C. An important feature of this sample is that it shows capacitance increase at positive V_g-s, which is indicative for spill-over of channel electrons into AlGaN (V_{spill}). Note

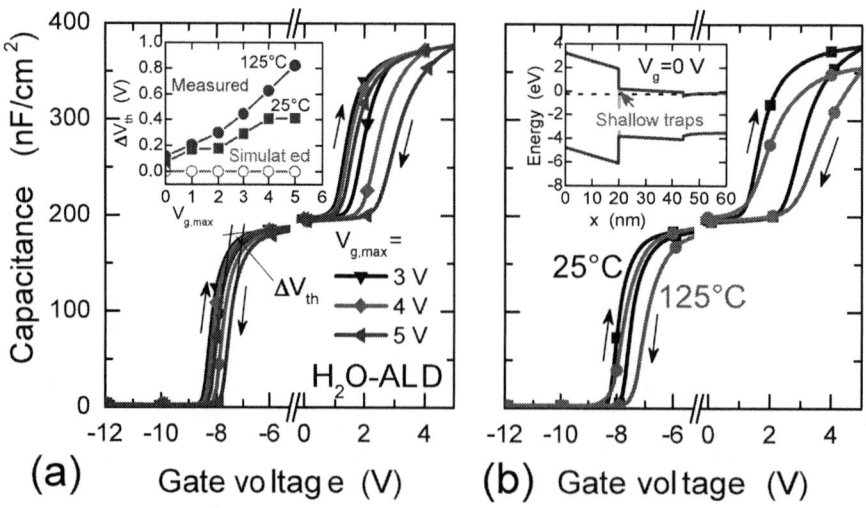

Fig. 1 (a) CV hysteresis measurements for sample H_2O-ALD as a function of maximum V_g. Inset shows the dependence of ΔV_{th} on $V_{g,max}$ as extracted from the experiment (full symbols) and simulations (open symbols) taking interface traps capture/emission into account. (b) Comparison of CV sweeps performed at 25 and 125 °C. Inset of (b) shows calculated band diagram of the gate structure in equilibrium.

that using extracted V_{th} (2DEG depletion) and V_{spill} (AlGaN polpulation), one can estimate charge distribution in the gate stack and Q_{int} based on the comparison between experimental and simulated CV curves (giving Q_{int} denisty of 6.5×10^{12} cm^{-2}, $i.e.$ close to full compensation of the interface charge). As can be inferred from the inset of Fig. 1(a), experimental ΔV_{th} monotonically increases with $V_{g,max}$ and with temperature up to 0.8 V, while essentially no ΔV_{th} was obtained from CV simulations with D_{it} profile determined for sample H$_2$O-ALD (shown in Fig. 3(a)).

The simulation results can be understood form the band diagram shown in the inset of Fig. 1(b). Here, E_F at oxide/barrier interface at V_g=0 V is located only 0.4 eV below AlGaN CB edge, being also consistent with relatively small V_{spill} (~1 V). This means that even though all interface states can capture electrons at V_g>0 V during positive sweep, relatively shallow interface states will have enough time to emit the carriers back to AlGaN CB during negative sweep (>50 s). Emission from deeper traps (>0.4 eV below CB) during negative sweep has eventually little effect on ΔV_{th} as these traps will be populated back at positive V_g-s. As a result, negligible effect of interface states on ΔV_{th} can be expected for this sample. Therefore, discrepancy between experimental and simulated ΔV_{th} indicates presence of other electron trapping centers. Negligible effect of trapping on the slope of the CV curves close to V_{th} and in spill-over points to border traps, i.e. charge levels in Al$_2$O$_3$ band gap, spatially located close to its interface with AlGaN barrier. Ionized BTs can capture electrons at high positive V_g-s, however, their emission is hampered by spatial separation.

3.2 Impact of oxide bulk traps on ΔV_{th}

Fig. 2(a) shows CV hysteresis measured on O$_3$-ALD structures at 25 and 125 °C as a function of $V_{g,max}$. Here, $V_{g,max}$ was limited by gate leakage onset at V_g=1.5 V. Interestingly, ΔV_{th} was found to be weakly dependent on $V_{g,max}$ and it increases at elevated temperature (inset of Fig. 2(a)). Moreover, CV curves measured at 125 °C shifted towards positive voltages compared to that measured at 25°C. Both effects are, however, inconsistent with impact of D_{it} on MOS-HEMTs capacitance behavior. First, ΔV_{th} is expected to increase with $V_{g,max}$ as a result of electron capture by deep traps above E_F at oxide/barrier interface. This is

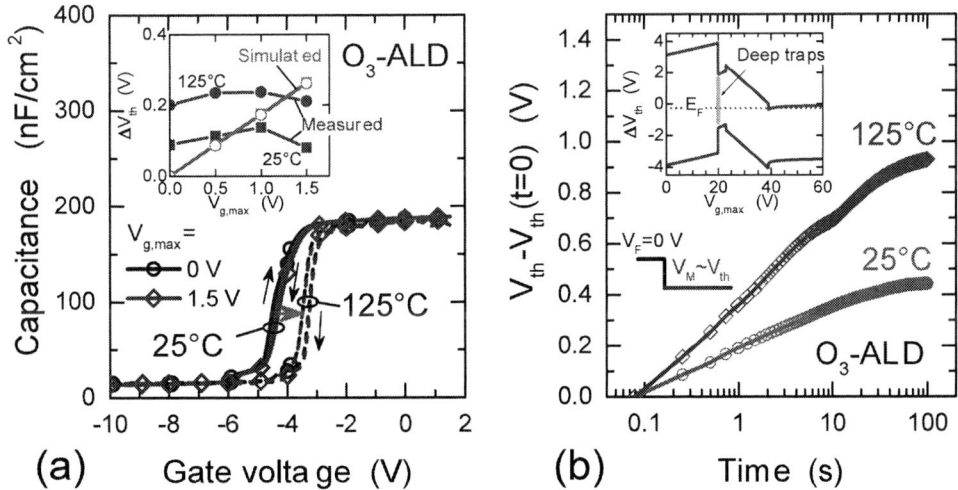

Fig. 2 (a) CV hysteresis of sample O$_3$-ALD as a function of $V_{g,max}$. Inset shows the dependence ΔV_{th} on $V_{g,max}$ extracted from the experiment (full symbols) and simulations (open symbols). (b) V_{th} transients measured after stepping $V_{g,F}$ from 0 to $V_{g,M}$~V_{th}. Inset of (b) shows calculated band diagram of the gate structure in equilibrium.

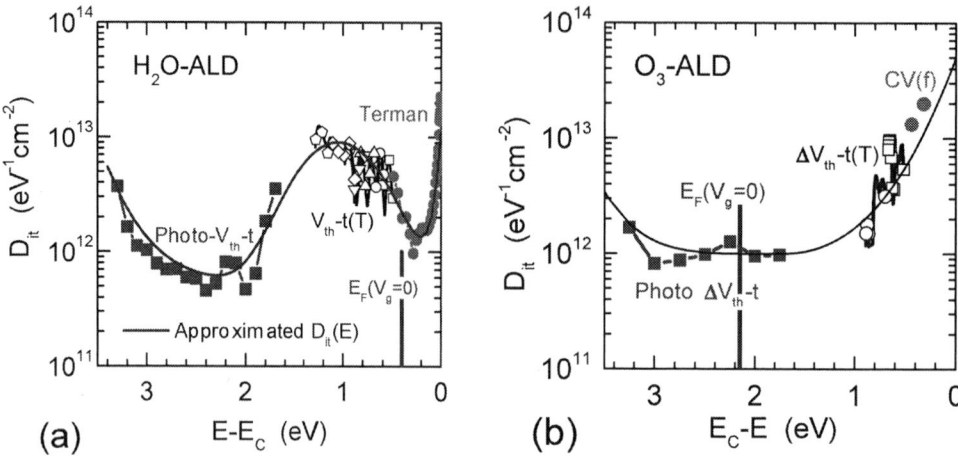

Fig. 3 D_{it} distributions of H_2O-ALD (a) and O_3-ALD (b) MOS-HEMT structures measured by four complementary techniques (lines with symbols) and their approximations used for CV simulations.

depicted in the inset of Fig. 2(b). Note that position of E_F results from relatively high Q_{int} at oxide/barrier interface mentioned above. Second, assuming interface states only, negligible or negative CV shift can be expected at elevated temperatures due to enhanced electron emission from interface states, in contrast to the experimental data. To understand this behavior, V_{th} transient was monitored after stepping V_g from 0 (equilibrium) to $V_g \sim V_{th}$ as shown in Fig. 2(b). Clearly, positive shift of V_{th} in time increasing with temperature was observed. This behavior is consistent with capture of electrons in the oxide bulk traps, resulting in negative charge built-up. Electrons are assumed to be injected from the metal gate into Al_2O_3 charge levels by thermionic field emission, as also proposed in [3]. Electron capture by oxide traps at negative V_g-s and partial re-emission upon positive V_g-s is in line with the observed ΔV_{th} that is mostly dependent on temperature rather than $V_{g,max}$.

4. Conclusions

It was demonstrated that among interface states, CV hysteresis of an MOS-HEMT structure can be also dominated by other capture/emission processes, such as those related to border traps and bulk oxide traps. Clearly, this needs to be taken into account for correct interpretation of the CV hysteresis, often used in the technology optimization effort.

Acknowledgement

This work was supported by V4-Japan joint call on advanced materials (project SAFEMOST) and national projects APVV 15-0031 and VEGA 2/0138/2014.

References

[1] K.-S. Im, J.-B. Ha, K.-W. Kim, *et al.*, *IEEE Electron Dev. Lett.* **31**, 192, 2010.
[2] D. Gregušová, M. Jurkovič, Š. Haščík, *et al.*, *Appl. Phys. Lett.* **104**, 013506, 2014.
[3] M. Ťapajna, M. Jurkovič, L. Válik, *et al.*, *Appl. Phys. Lett.* **102**, 243509, 2013.
[4] Y. Hori, Z. Yatabe, and T. Hashizume, *J. Appl. Phys.* **114**, 244503, 2013.
[5] Y. Hori, C. Mizue, and T. Hashizume, Japan. J. Appl. Phys. **49**, 080201, 2010.
[6] M. Miczek, C. Mizue, T. Hashizume, et al., *J. Appl. Phys.* **103**, 104510, 2008.

ASDAM 2016, The 11th International Conference on Advanced Semiconductor
Devices And Microsystems, November 13-16, 2016, Smolenice, Slovakia

Pseudo-vertical GaN-based trench gate metal oxide semiconductor field effect transistor

Rico Hentschel[1], Andre Wachowiak[1], Andreas Großer[1], Andreas Jahn[2], Ulrich Merkel[2], Ada Wille[3], Holger Kalisch[3], Andrei Vescan[3], Stefan Schmult[2] and Thomas Mikolajick[1,2]

[1] Namlab gGmbH, 01187 Dresden, Germany
[2] Institute of Semiconductors and Microsystems, TU Dresden, 01187 Dresden, Germany
[3] GaN Device Technology, RWTH Aachen University, 52074 Aachen, Germany
E-Mail: Rico.Hentschel@namlab.com

We report on the fabrication and characterisation of an enhancement mode n-channel pseudo-vertical GaN metal oxide semiconductor field effect transistor (MOSFET), which utilizes a high-k dielectric covered trench gate and a top side drain contact. The processing technology has been developed to be easily transferrable to a truly vertical MOSFET on GaN bulk material, as targeted for high-voltage power switching applications. Device functionality is demonstrated by linear transfer characteristic with a decent ON/OFF current ratio of 6 orders of magnitude and clear normally-off operation.

1.   Introduction

In the last years, tremendous progress has been made in the field of GaN based power devices. Although, mainly driven by the technology advancements of heterostructure based AlGaN/GaN high electron mobility transistors (HEMTs), this lateral device concept involves limitations in area related scaling of the breakdown voltage. From a general perspective, true vertical device topologies with backside drain contact have the advantage of area efficient scaling of the breakdown voltage by increasing the thickness of the drift region in vertical direction, an inherent physical separation of the low and high-voltage connections and more efficient thermal management. Motivated from these considerations, work on different vertical device concepts on GaN bulk material has been reported, with several contributions related to the vertical metal oxide semiconductor field effect transistor (MOSFET) [1, 2, 3]. The n-channel MOSFET concept with an inversion channel in p-GaN should ensure a high enough threshold voltage for high power applications. For our preliminary technology development, we used GaN template structures on sapphire substrates due to their better availability compared to GaN bulk material. This approach enables important understanding of basic properties of device processing, e.g. of the gate integration, but has limitations in addressing the high voltage behaviour due to a reduced drift layer thickness.

2.   Device Fabrication

The metal organic chemical vapour deposition (MOCVD) GaN template on sapphire contains, from the bottom upwards, an AlN-based nucleation layer, a 2.2 µm thick n^+-GaN layer (Si: $5 \cdot 10^{18}$ cm^{-3}), an 1.1 µm thick n^--GaN layer (Si: $2 \cdot 10^{16}$ cm^{-3}) and a 0.5 µm thick p-GaN layer (Mg: $3 \cdot 10^{18}$ cm^{-3}). A hydrogen out-diffusion anneal has been done at 750°C for 10 min in inert gas atmosphere after growth. The drift layer was kept thin in order to maintain the ability to etch the full depth down to the lower n^+-GaN, which acts as the pseudo-vertical drain layer. For a well-defined source sided p/n^+ junction, the MOCVD template was overgrown with 300 nm n^+-GaN (Si: $2 \cdot 10^{18}$ cm^{-3}) by plasma-assisted molecular beam epitaxy

978-1-5090-3084-2/16 $31.00 © 2016 IEEE

(PA-MBE). The layer stack has a dislocation density in the upper p- and n^+-GaN of about $1 \cdot 10^9$ cm^{-2}, as determined by atomic force microscopy (AFM).

After growth, device insulation and metal contacts to the different GaN layers were processed by alternating dry etching in Cl_2/BCl_3 chemistry and metal deposition via evaporation and lift-off technology. Ti/Al/Ni/Au and Ni/Au stacks were used as contact materials on n^+-GaN and p-GaN layers, respectively. Next, the trench gate module was fabricated with a PECVD Si_xN_y hard mask based dry etching step, followed by a wet chemical treatment with tetramethylammoniumhydroxide (TMAH). The process is similar as described in [1]. Subsequently, 30 nm of Al_2O_3 and a TiN gate electrode have been deposited by atomic layer deposition (ALD) on the obtained vertical gate trench. The Al_2O_3 does not only serve as gate dielectric, but also as a first passivation of the device. Fig. 1 shows the schematic cross section of the device and a scanning electron micrograph (SEM) of the gate trench after uniform Al_2O_3 and TiN deposition.

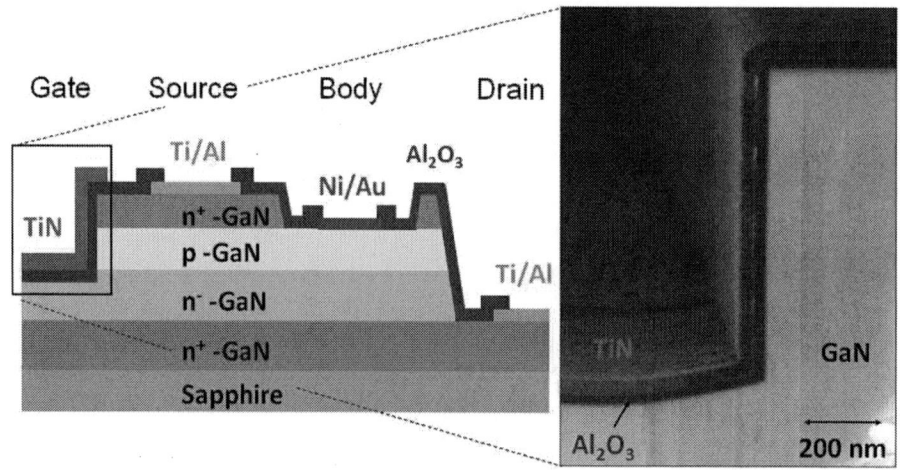

Fig. 1: Cross sectional schematic (not to scale) of the fabricated pseudo-vertical MOSFET and SEM cross section of the gate trench edge.

3. Results and Discussion

The transfer characteristics ($V_{DS} = 0.1$ V) in Fig. 2 and output characteristics in Fig. 3 demonstrate device functionality. We show normally-off operation with drain current ON/OFF ratio of up to 10^6. However, we do observe a significant shift of the threshold voltage towards higher voltages from the 1st measurement compared to the 2nd and subsequent measurements of the transfer curve, as shown in Fig. 2. The observed shift impedes the discussion of the expected threshold voltage. From a theoretical point of view, uncertainties arise from the unknown actual concentration of hydrogen-free Mg sites and the deep acceptor level of about 0.17 eV - 0.2 eV [4]. Since the metal contact to the etched p-GaN has a substantial non-linear behaviour, the desired information could not be extracted from $C(V)$ curves measured either on transistor or lateral MIS structures on etched p-GaN. In an alternative approach, we measured transfer characteristics with varying body bias, as shown in Fig. 4, for a different test device. From the shift of the transfer curves due to the applied body-source voltage, an estimation of the acceptor concentration of N_A^- of $1 \cdot 10^{18}$ cm^{-3} is calculated, when using an oxide capacitance of $C_{ox} = 2.6 \cdot 10^{-7}$ F/cm^2, which was extracted from $C(V)$ curves on planar MIS n-GaN structures processed in parallel on the same wafer and die [5]. The calculated value seems in a reasonable range compared to the chemical Mg concentration of the p-GaN layer of $\approx 3 \cdot 10^{18}$ cm^{-3}, when taking into account the partial passivation by residual H [6] and the additional activation of acceptors in the space charge region compared to the equilibrium case [7, 8]. Taking N_A^- of $1 \cdot 10^{18}$ cm^{-3} and applying the standard formula for the MOSFET threshold voltage [5] with the work function of TiN of $\approx$ 4.5 eV and neglecting any additional charge contributions, we obtain a threshold voltage of $V_{th} \approx 4$ V. Closer to the experimental value of the 1st curve is the threshold voltage $V_{th} \approx 7$ V obtained by taking the entire Mg concentration of $3 \cdot 10^{18}$ cm^{-3}. However, this comparison has to be interpreted with care, due to the mentioned difficulties and the uncertainty of the flat band voltage. The strong shift of the transfer curve towards higher gate-source voltages remains unclear. It would correspond to an effective negative sheet charge of $N > 5 \cdot 10^{12}$ cm^{-2} located at the high-k/GaN interface.

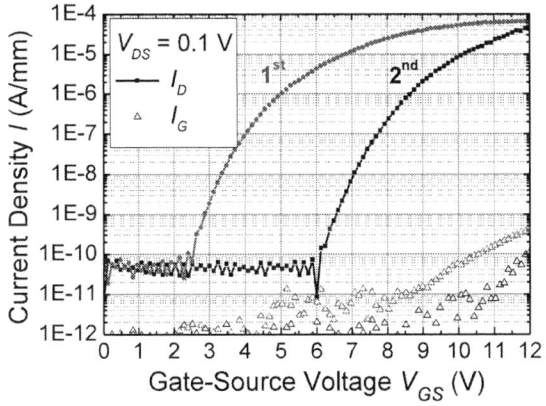 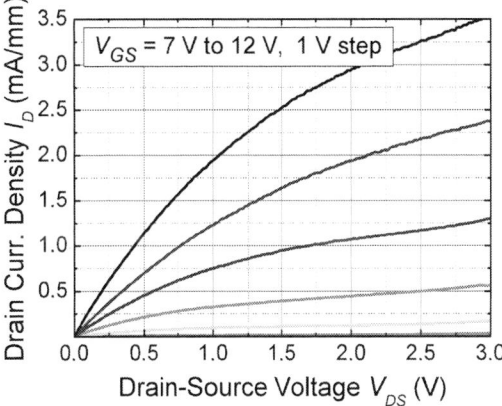

Fig. 2: Transfer characteristics $I_D(V_{GS})$, gate I_G and bulk leakage I_B of the device, each V_{GS} sweep for the respective curve starts at 0 V.

Fig. 3: Output characteristics $I_D(V_{DS})$ of the device measured after the initial threshold voltage shift measured from 0 V to 3 V.

Next, the OFF-current levels are analyzed in more detail. As seen from the transfer curves in Fig. 2, the drain OFF-current is not limited by the much smaller gate leakage and is independent of the gate-source voltage. In order to disentangle different contributions to the drain current, we measured all four terminal currents in the transistor OFF-state ($V_{GS} = 0$V) in dependence of the drain bias. In Fig. 5 these currents are shown and normalized to the active area of the bottom p/n$^-$-GaN junction of the test device. For a drain bias below 20V, the drain current equals the body current and moreover the current level is in agreement with the reverse junction leakage of the lower p/n$^-$ junction as measured on a separate diode test structure. For higher drain bias > 20V, the source current dominates the drain current, and the bulk current becomes negligible. It could not be verified, if this current stems from parasitic paths through the n$^+$/p/n$^-$ bulk structure, for example at locations of extended defects, or from the lateral device mesa edge. However, the later might be rather unlikely, since the source contact is guarded by the body contact towards the drain in the transistor design, (see Fig. 1) [9]. As can be seen in the output characteristics of Fig. 3, the ON-state current normalized to the channel width is relatively low with respect to the usage of a high-k gate dielectric, compared to a recently published work of a vertical MOSFET with low-k dielectric on bulk GaN [2]. Enhancement of the ON-state current is expected for further optimization of the final trench edge and the high-k/GaN interface, in order to improve the effective channel mobility. Series resistance contributions would be lowered in a truly vertical device with a backside drain contact.

In summary, the developed, pseudo-vertical MOSFET with trench gate complies with clear normally-off operation. While the ON-state needs further improvement with respect to the channel resistance and the observed threshold voltage shift, the OFF-state performance in the lower V_{DS} range is bound to the reverse leakage of the bottom p/n⁻ junction. The reduction of the active device area with respect to the upper and lower p/n junction and simultaneous use of GaN bulk material with a lower defect density should enable an OFF-state performance enhanced by a factor of around 100, due to decreased overall leakage. In addition, an enlargement of the drift layer compared to the presented pseudo-vertical approach is mandatory for achieving a breakdown voltage in the kV range [2, 3].

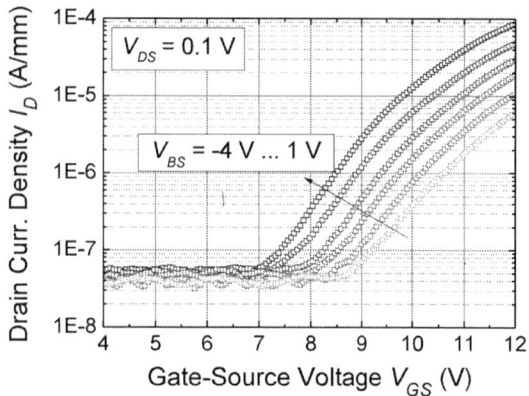

Fig. 4: Transfer characteristics $I_D(V_{GS})$ for different bulk voltages applied, each measured from 0V; subsequent measurements on a device with higher OFF-leakage.

Fig. 5: Output characteristics $I_D(V_{DS})$ in device OFF-state normalized to active area of p/n⁻ GaN junction area and diode reverse characteristics $I_{p/n^-}(V_{n^-})$.

Acknowledgement

This work was funded by the German Federal Ministry of Education and Research - BMBF (project: "ZweiGaN", no. 16ES0145K).

References

[1] Kodama, M., et al., *Appl. Phys. Expr.*, **1**(2), 021104, 2008.
[2] Oka, T., Ueno, Y., Ina, T. and Hasegawa, K., *Appl. Phys. Expr.*, **7**(2), 021002, 2014.
[3] Oka, T., Ina, T., Ueno, Y., & Nishii, J., *Appl. Phys. Expr.*, **8**(5), 054101, 2015.
[4] Brochen, S., Brault, J., Chenot, S., Dussaigne, A., Leroux, M., and Damilano, B., *Appl. Phys. Lett.*, **103**(3), 032102, 2013.
[5] Sze S. M., *Physics of Semiconductor Devices*, John Wiley & Sons, New York, 1981.
[6] Pearton, S. J., Zolper, J. C., Shul, R. J., and Ren, F., *Journal of Appl. Phys.*, **86**(1), 1-78, 1999.
[7] Myers, S. M. and Wright, A. F., *Journal of Appl. Phys.*, **90**(11), 5612-5622, 2001.
[8] Weigel, M., Stenzel, R., Klix, W., Hentschel, R., Wachowiak, A., Mikolajick, T., *in Proc. of the 40th WOCSDICE Conference*, Aveiro, Portugal, 2016, pp. W27-W28.
[9] Qi, M., et al., *Appl. Phys. Lett.*, **107**(23), 232101, 2015.

ASDAM 2016, The 11th International Conference on Advanced Semiconductor
Devices And Microsystems, November 13-16, 2016, Smolenice, Slovakia

DC and pulsed IV characterisation of AlGaN/GaN MOS-HEMT structures with Al_2O_3 gate dielectric prepared by various techniques

F. Gucmann[1], D. Gregušová[1], L. Válik[1], M. Ťapajna[1], Š. Haščík[1], K. Hušeková[1],
K. Fröhlich[1], O. Pohorelec[1,2], and J. Kuzmík[1]

[1]Institute of Electrical Engineering, SAS, Dúbravská cesta 9, 841 04 Bratislava, Slovakia
[2] Faculty of Electrical Engineering and Information Technology, Institute of Electronics
and Photonics, Slovak University of Technology, Ilkovičova 3, 812 19 Bratislava, Slovakia
e-mail: filip.gucmann@savba.sk

AlGaN/GaN metal-oxide-semiconductor high electron mobility transistors (MOS-HEMTs) represent an important technology for future high-efficient power and RF electronics. Due to oxide/semiconductor interface issues, fabrication of reliable MOS gate stack is still challenging, however. In this work we investigated the influence of gate oxide preparation technique to static and pulsed-mode operation of Al_2O_3/GaN/AlGaN/GaN MOS-HEMTs. Devices with gate oxide prepared by high-temperature MOCVD and low-temperature ALD using water vapour or ozone as oxidants are compared in terms of dynamic on-state resistance (R_{DSon}), threshold voltage shift (ΔV_{th}), and Al_2O_3/GaN interface charge density (N_{it}).

1. Introduction

Nowadays, a significant attention is paid to AlGaN/GaN metal-oxide-semiconductor high electron mobility transistors (MOS-HEMTs) in the field of material and device research. Offering exceptionally high breakdown voltages and low on-state resistance, these devices provide great potential for high-power switching applications, e.g. in highly efficient DC-DC current converters. Providing high values of electron mobility (μ_n) and improved electromagnetic radiation hardness, devices for high-power and high-frequency devices (for air- and spaceborne applications) can be built. An important technology aspect of AlGaN/GaN MOS-HEMT fabrication is gate stack preparation, which still encompasses serious challenges. As widely discussed in literature [1-2], number of trapping centres are generated after gate oxide deposition. These include oxide/semiconductor interface states, border traps located close to this interface and bulk traps spread across the oxide layer. Capture/emission processes related to these trap states are responsible for device threshold voltage (V_{th}) instabilities and decreased performance, especially in desired pulsed-mode operation [3]. In this work we investigate the influence of various Al_2O_3 gate oxide deposition techniques on DC and pulsed-mode operation of AlGaN/GaN MOS-HEMTs by means of gate-lag current-voltage (*IV*) measurements. Al_2O_3 films grown by two modes of low-temperature (100 °C) atomic layer deposition (ALD) using different oxidant (water vapour vs. ozone) and high-temperature (HT) (600 °C) metal-organic chemical-vapour deposition (MOCVD) are studied and compared.

2. Experimental details

GaN/$Al_{0.24}Ga_{0.76}$N/GaN MOS-HEMT structures with 10 nm Al_2O_3 gate oxide were prepared upon MOCVD-grown heterostructure shown in Fig. 1. Substrate drastic chemical cleaning (1HF : 2H_2O, 120 s; 1HCl : 2H_2O, 120 s; 2NH_4OH : 1H_2O_2, 600 s) preceded mesa structures isolation by reactive ion etching in $SiCl_4$. Alloyed ohmic contacts comprised standard e-gun evaporated 30 nm Ti/180 nm Al/40nm Ni/50 nm Au metal stack annealed

978-1-5090-3084-2/16 $31.00 © 2016 IEEE

at 850 °C for 60 s with fast temperature ramp in nitrogen ambient. Subsequently, three different deposition techniques combined with pre-deposition treatment in diluted hydrochloride acid (HCl-PDT) (1HCl : 2H₂O, 30 s) for native oxide removal, followed by surface ex-situ O_2 plasma treatment (O_2-PDT) (P_{RF}=50 W, 90 s) were used for gate oxide preparation: (i) HCl-PDT, O_2-PDT, low-temperature (LT) (100 °C) ALD utilising ozone as oxidant (O_3-ALD) (ii) HCl-PDT, O_2-PDT, LT (100 °C) ALD utilising water vapour as oxidant (H_2O-ALD), and (iii) *no* HCL-PDT, O_2-PDT, HT (600 °C) liquid injection MOCVD. Gate metallisation stack (40 nm Ti/50 nm Au) was evaporated by e-gun and patterned by lift-off. Finally, Al_2O_3 was removed from source and drain contacts and pad metallisation (10 nm Ti/300 nm Au) was deposited.

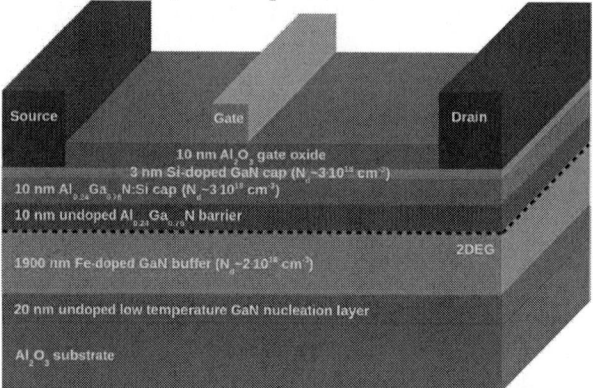

Fig. 1. Schematic cross-section of fabricated MOS-HEMT device showing epitaxial structure.

Room temperature static and pulsed (100 ns pulses on the gate, duty cycle 1%) *IV* curves were acquired using Keithley 4200 measuring unit. Quiescent point (V_Q) for pulsed analysis was set to -4 V. The difference in on-state resistance between static and pulsed mode (ΔR_{DSon}) was extracted from the linear part of output characteristics. Hysteresis (ΔV_{GS}) in the static mode was defined at the fixed value of drain current (I_D=0.1 A/mm), while the threshold voltage shifts ΔV_{th} were taken as the difference between: (i) V_{th} of the forward- and backward-swept static transfer characteristics, denounced as ΔV_{th}(DC^{fw}-DC^{bw}) and (ii) V_{th} of the forward-swept static and pulsed transfer characteristics ΔV_{th}(DC^{fw}-pulsed) . Assuming only oxide/semiconductor interface states, one can define

$$N_{it}^{slow\ states} = 1/q \cdot C_{ox} \cdot (V_{th}^{DC\ forward} - V_{th}^{DC\ backward}) \tag{1}$$
$$N_{it}^{fast\ states} = 1/q \cdot C_{ox} \cdot (V_{th}^{DC\ forward} - V_{th}^{pulsed}) \tag{2}$$

Where C_{ox} represents the gate oxide capacitance. Corresponding Al_2O_3/GaN interface states sheet charge densities (N_{it}) were extracted assuming (1) and (2) for slow ($N_{it}^{slow\ states}$) and fast states ($N_{it}^{fast\ states}$), respectively.

3. Results and discussion

Fig. 2, 3, and 4 show measured data for O_3-ALD, H_2O-ALD, and MOCVD MOS-HEMT devices, respectively. Static and pulsed output characteristics, transfer characteristics, and transconductances are compared consecutively in panels (a), (b), and (c), respectively. O_3-ALD and MOCVD samples exhibit negligible ΔV_{th} (DC^{fw}-pulsed) compared to H_2O-ALD (0.07 and -0.05 V compared to -0.67 V, respectively) and, consequently, significantly reduced $N_{it}^{fast\ states}$ (3.48·10¹¹ and 3.32·10¹¹ cm⁻² compared to 6.54·10¹² cm⁻²). Positive ΔV_{th} (DC^{fw}-pulsed) of O_3-ALD sample may relate to enhanced injection of electrons from the gate electrode to the oxide [4]. On the contrary O_3-ALD showed slightly higher $N_{it}^{slow\ states}$ as compared to H_2O-ALD (3.13·10¹² cm⁻² compared to

$2.79 \cdot 10^{12}$ cm^{-2}). Expectedly, pulsed output characteristics show suppressed self-heating effect and in case of O$_3$-ALD and MOCVD samples no notable gate-lag (current collapse). On the contrary, H$_2$O-ALD showed significant reduction of drain current (I_D) for $V_{GS}>$-1 V, indicating influence of increased $N_{it}^{fast\ states}$. In contrast to low ΔV_{th} (DC^{fw}-pulsed), O$_3$-ALD exhibited significantly higher ΔR_{DSon}, than the other samples however, which may relate to trapping in the gate-to-drain region. MOCVD sample showed best performance in all investigated aspects. Tab. 1 summarises the extracted data.

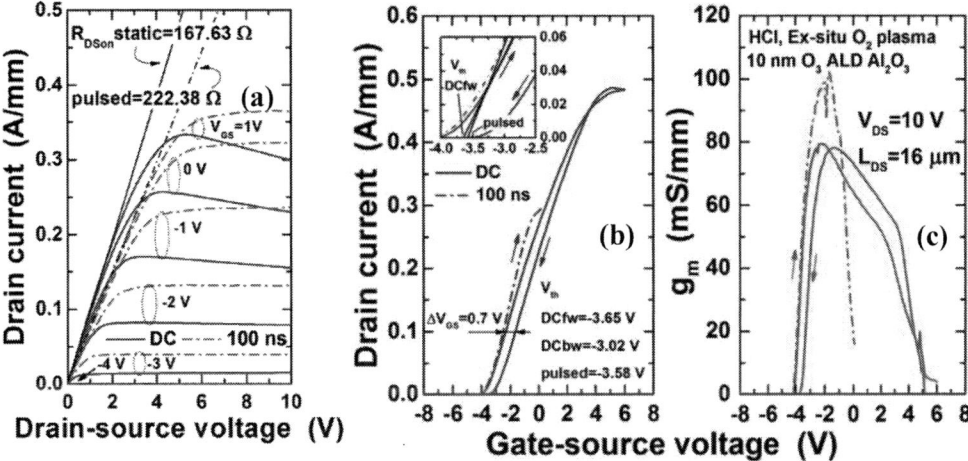

Fig. 2. O$_3$-ALD MOS-HFET. Static (solid lines) and pulsed (dash-dotted lines) *IV* curves. (a) Output characteristics. V_{GS}=1÷-4 V, ΔV_{GS}=-1 V. (b) Transfer characteristics. Inset shows detail of pinch-off area with corresponding V_{th} determined from linear fit. (c) Transconductance.

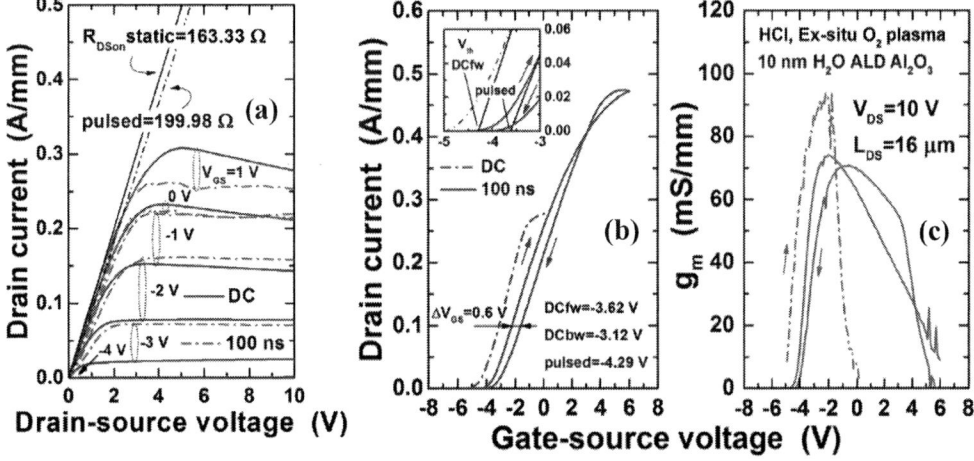

Fig. 3. H$_2$O-ALD MOS-HFET. Static (solid lines) and pulsed (dash-dotted lines) *IV* curves. (a) Output characteristics. V_{GS}=1÷-4 V, ΔV_{GS}=-1 V. (b) Transfer characteristics. Inset shows detail of pinch-off area with corresponding V_{th} determined from linear fit. (c) Transconductance.

Tab. 1. Summary of extracted data.

Sample	ΔR_{DSon} dynamic-static (Ω)	ΔV_{GS} (DC^{fw}-DC^{bw}) (V)	ΔV_{th} (DC^{fw}-DC^{bw}) (V)	ΔV_{th} (DC^{fw}-pulsed) (V)	N_{int} slow states (cm^{-2})	N_{int} fast states (cm^{-2})
O$_3$-ALD 100 °C	54.75	0.7	0.63	0.07	$3.13 \cdot 10^{12}$	$3.48 \cdot 10^{11}$
H$_2$O-ALD 100 °C	36.68	0.6	0.5	-0.67	$2.79 \cdot 10^{12}$	$6.54 \cdot 10^{12}$
MOCVD 600 °C	4.93	0.2	0.29	-0.05	$1.92 \cdot 10^{12}$	$3.32 \cdot 10^{11}$

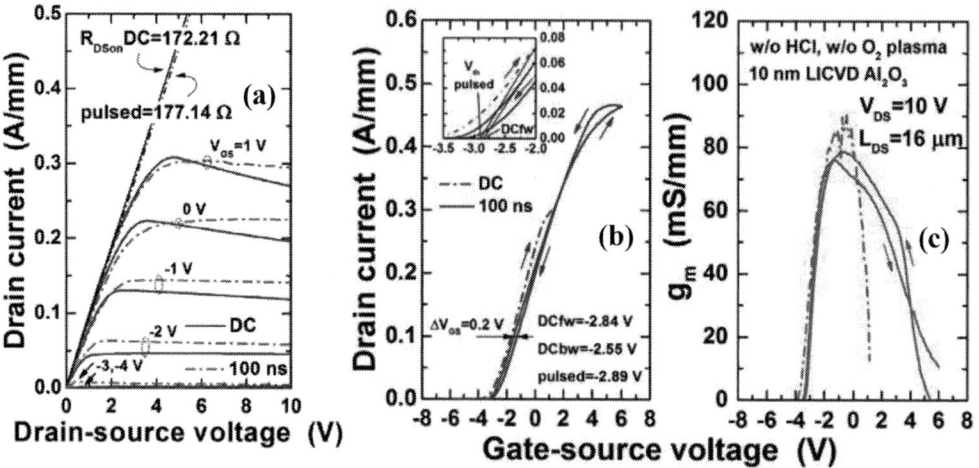

Fig. 4. MOCVD MOS-HFET. Static (solid lines) and pulsed (dash-dotted lines) *IV* curves. (a) Output characteristics. V_{GS}=1÷-4 V, ΔV_{GS}=-1 V. (b) Transfer characteristics. Inset shows detail of pinch-off area with corresponding V_{th} determined from linear fit. (c) Transconductance.

4. Conclusions

Al$_2$O$_3$/GaN/AlGaN/GaN MOS-HEMT devices were processed an analysed in static and pulsed-mode operation. V_{th} shifts were used for extraction of slow and fast Al$_2$O$_3$/GaN interface state densities attributed to observed V_{th} shifts. Devices with MOCVD-grown gate oxide showed negligible V_{th}, indicating high quality of the oxide/semiconductor interface and the oxide layer itself. Comparison of devices with gate oxide prepared by the two ALD modes is not conclusive. Nevertheless, O$_3$-ALD-prepared MOS-HEMTs showed better performance in pulsed mode and lowered density of fast states (and opposite sign of ΔV_{th} (DC^{fw}-pulsed)) than those prepared by H$_2$O-ALD. Dynamic on-state resistance, however, was significantly higher in case of O$_3$-ALD. This suggests co-existence of two competing capture/emission processes taking part in various measures. At negative gate bias, electrons from gate metal are injected into oxide and trapped inside bulk trapping centres, while interface states electron capture/emission occurs at more positive gate bias [4]. While latter effect might have similar impact in both intrinsic and extrinsic area, former has stronger influence under the gate electrode.

Acknowledgement

This work was supported by V4-Japan joint call on advanced materials (project SAFEMOST) and national projects APVV-15-0243 and APVV-15-0673.

References

[1] Y. Yue, Y Hao, J. Zhang, *et al., IEEE Electron Device Lett.* **29**, 838, 2008.

[2] X. Sun, O. I. Saadat, K. S. Chang-Liao, *et al., Appl. Phys. Lett.* **102**, 103504, 2013.

[3] G. Meneghesso, F. Rampazzo, P. Kordoš, *et al., IEEE Trans Electron Devices* **53**, 2932, 2006.

[4] M. Ťapajna, M. Jurkovič, L. Válik, *et al., Appl. Phys. Lett.* **102**, 243509, 2013.

ASDAM 2016, The 11th International Conference on Advanced Semiconductor
Devices And Microsystems, November 13-16, 2016, Smolenice, Slovakia

Analysis of Hybrid Pair Si IGBT-SiC diode for Very High Power Applications

Muhammad Nawaz, [1]Filippo Chimento

ABB Corporate Research, Forskargränd 7, SE – 721 78 Västerås, Sweden

[1]ABB Italy, Via San Giorgio 642, 52028, Terranuova Bracciolini, (AR), Italy

e-mail: muhammad.nawaz@se.abb.com   filippo.chimento@it.abb.com

This paper deals with static and dynamic measurements performed for silicon based hybrid IGBT power modules. Hybrid power modules were obtained from Mitsubishi with voltage rating of 1200 V and current rating of 800 A in half bridge configuration mode where silicon carbide based Schottky diodes are used as freewheeling diodes across silicon based IGBTs. The power modules are fabricated in flexible configuration so as to allow either full power module with 1200 V and 800 A as a single power switch or placed in two parallel legs with 1200 V and 400 A switches. First results from engineering samples show overall good confidence level that resulted in an on-resistance (R_{ON}) of 8.0 – 10.0 mΩ, blocking voltage of 1700 V and a threshold voltage around 5.0 – 5.5 V at 300 K as promised by the manufacturer from most of the transistor samples.

1.   Introduction

In order to prevent from global warming, reduction in greenhouse gases is pre-requisite conditional requirement. In order to achieve this objective, power electronics circuitry needs to be energy-efficient. Here the improved efficiency of power inverters with more system compactness, relaxed cooling requirement and lower environmental impact are critical parameters that can only be achieved through the use of innovative power device concept, circuits, controllers and other components used in the power system [1 – 6]. Since power semiconductor devices are the main building blocks used in inverter circuitry, there is increasing demand for low loss power devices that realize higher efficiency [7]. Note that the silicon (Si) based insulated gate bipolar transistor (IGBT) and silicon based freewheeling diodes (FWD) chips are nowadays common in diverse power applications [6]. However, Si based IGBTs and diodes are approaching towards their theoretical performance limits where future breakthroughs that can be achieved using silicon material from the point of view of lower loss cannot be expected. For this reason, wide bandgap (WBG) semiconductor such as silicon carbide (SiC) [1 - 5, 7] that exhibit superior material physical properties (e.g., bandgap, thermal conductivity, carrier velocity etc.,) compared to Si counterpart offer potential replacement to basic building block of inverter system which is based on Si-IGBT and Si-diodes across today.

More interesting, is that in other high power applications such as motor drives, welding equipment, traction, HVDC (High voltage DC) and FACTS (Flexible AC transmission systems), insulated gate bipolar transistors (IGBTs) are considered key components as well. High voltage state of the art thick epitaxy based SiC devices such as 21 kV BJTs [12], 20 kV GTOs [13], 27 kV n-IGBTs [14] and 27 kV PIN diodes [15] have been presented recently. A carrier life time of 27 µs (as grown samples have a life time of 3 µs) was obtained using special carrier life time enhancement process for PIN diodes [15]. While a number of SiC

978-1-5090-3084-2/16 $31.00 © 2016 IEEE   13

discrete devices (i.e., BJTs, JFETs, MOSFETs) and full SiC MOSFET power modules [4, 11] with varying current and voltage ratings in the range of 1.2 − 1.7 kV are now commercially available in the market, cost and reliability concern are the major issues that still prohibit them to be used in the field. With this objective in mind, hybrid combination from the point of view of device technological maturity, cost, reliability and lower losses, hybrid pair based on Si-IGBT and SiC based diodes [8 - 10] offer promising alternatives for variety of applications discussed earlier. These hybrid pairs are now available in the commercial market from various vendors covering wide spectrum of applications.

In this work, hybrid power modules (CMH1200DC-34S) based on Si-IGBTs and SiC diodes from Mitsubishi, have been evaluated to get an understanding of their potential benefits compared to full Si-IGBT modules. The hybrid power module has a current and voltage rating of 1200 A and 1700 V respectively. Static and dynamic measurements with in-house Tek 371A curve tracer and double pulse tester have been performed at various temperatures.

2. Module description

Figure 1 illustrates the physical footprint of the Si-IGBT hybrid power module and electrical configuration is shown in figure 2. The hybrid power module [8],[9] is composed of two independent power switches that can be connected in half bridge mode (e.g., connecting either terminal 2 to terminal 4 or terminal 3 to terminal 1 as shown in Fig. 1 (b)). The hybrid power module package uses AlSiC base plate configured in two separate power switches with a maximum power dissipation of 8300 W. The power module has blocking voltage of 1700 V (with shorted gate-emitter terminals) and draws a collector-emitter current of 1200 A (under continuous DC operation) and 2400 A (under pulsed operation). The hybrid module has short circuit survivability time of 10 μs and maximum junction temperature of 175 °C as reported in the datasheet. Hybrid power module uses AlSiC as a base plate on which Si based IGBTs and SiC based Schottky diodes are placed. A maximum thermal resistance from junction to case (R_{thj-c}) of 18 and 36 K/kW is reported for Si-IGBT and SiC-diode respectively, in the datasheet. Note that an overall parasitic stray inductance of the power module is 30 nH. Furthermore, a total gate charge of 12 μC is reported in the datasheet at V_{CC} of 850 V, I_{CE} of 1200 A and V_{GE} of 15.0 V.

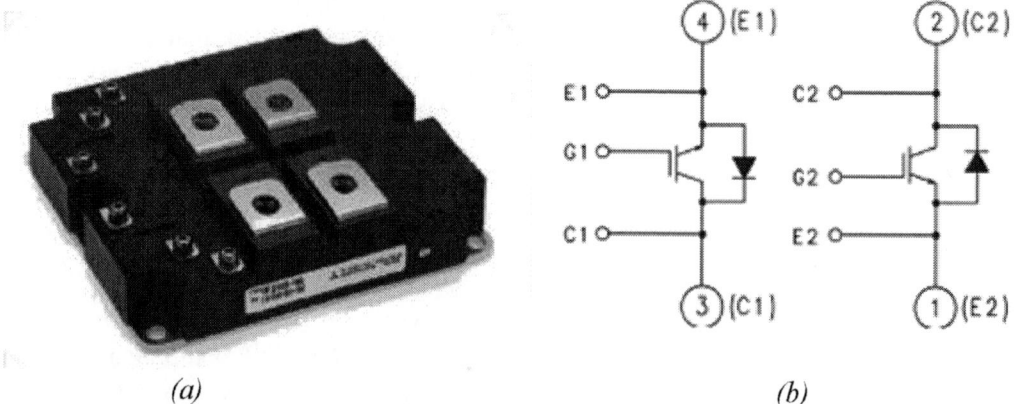

(a)   (b)

Fig. 1. Top view of Mitsubishi hybrid power module (a) CMH1200DC-34S (i.e., Si based IGBT + SiC based Schottky diode). Electrical circuit configuration (b) of Mitsubishi hybrid power module CMH1200DC-34S (i.e., Si based IGBT + SiC based diode).

3. Results and discussion

Static characteristics of hybrid power modules have been performed using Tektronix curve tracer with current measurement capability of up to 300 A. Current Vs voltage transistor characteristics at various gate biases (up to gate bias of only 10 V due to limited current capability of curve tracer) are illustrated in Fig. 2 at various temperatures and at different values of gate bias voltage. All tested power modules show normal transistor behaviour. At lower gate biases (say 8 and 10 V) close to threshold voltage, the collector current increases with temperature in the saturation region (due to increase in the channel mobility) while drop in the collector current is witnessed at higher gate biases (due to overall bulk mobility in the drift part of the device) as expected. Blocking voltage measurements, leakage current and ON-resistance extraction is illustrated in figure 3. Increase in the on-resistance (extracted in the linear region of I_{CE} versus V_{CE} characteristics at 8 V and 10 V gate bias) with temperature is obtained due to drop in the bulk carrier mobility and drop in the blocking voltage as a result of increased leakage current with temperature is observed. For a given collector bias, total leakage current of both parts of IGBT (i.e., full module under gate-emitter shorted) increases significantly with temperature as a result of increased in the intrinsic carrier concentration. An increase in the leakage over 100 time is observed with temperature variation form 300 to 400 K at 1.1 kV. This significant increase in the leakage current with temperature limits the Si based IGBT devices to be used below the specified temperature of 400 K as observed in our measurements. An increase in the On-resistance from 6.0 to 15 Ω is observed with temperature variation from 300 to 400 K at V_{GE} of 10.0 V. The observed hysteresis in the blocking voltage characteristics (Fig. 3a) is just due to the usage of Tek 371a curve tracer for these measurements. Note that static measurements have only been performed up to a gate bias of 10.0 V contrary to a nominal gate bias of 15 V for Si based IGBTs, which means that the on-resistance value will slightly be better than the one shown in this paper.

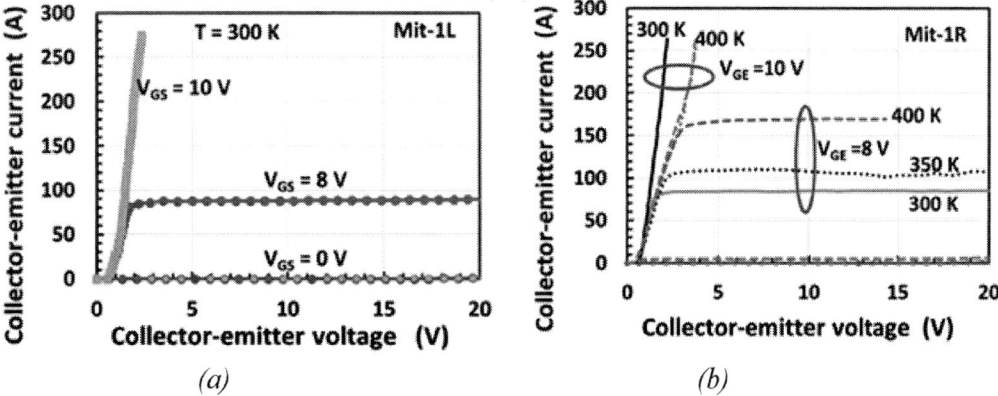

(a)   *(b)*

Fig. 2. Static characterization of Si based hybrid IGBT power module. Note that 1L (a) is right position, 1R (b) is left position on the power module at various temperatures.

Dynamic tests are performed using CT concept commercial gate driver originally designed for 3.3 kV Si-IGBT [6]. Figure 4 illustrate the full turn-on and turn-off transient for 1000 V supply voltage at 300 K using 200 μF DC capacitor and 100 μH load inductance. While keeping the same pulse sequence, load inductance has been decreased to get higher load current in several test runs. A turn off voltage overshoot of 300 V is noticed for 1000 V supply voltage. Interestingly, reverse recovery behavior of the top diode was negligible. A voltage slope dV/dt of approximately of 5.5 kV/μs and -3.3 kV/μs is obtained during turn off and turn on respectively, at 1000 V and 25 ℃. Similarly, a current slope di/dt of approximately -3.2 kA/μs and 0.92 kA/μs is witnessed during turn off and turn on phase respectively, at 1000 V and 25 ℃. Similarly and more specifically, a current rise time of 160 ns and fall time of 170 ns is achieved during turn on and turn off transient respectively.

Energy losses are also plotted as a function of load current for various supply voltages at 300 K as shown in figure 4d. As a result of the performed characterization a quasi linear increase in the total losses is obtained with the increase of load current as expected. Total energy losses were found out to be 337 mJ and 467 mJ at 800 V (600 A) and 1000 V (755 A) respectively, at 300 K. Extracted energy losses is also reported in figure 5 for various temperatures. For fixed load inductance (i.e., load current) and supply voltage, energy losses increases linearly with temperature as expected. Note that under similar test conditions, full SiC MOSFET modules resulted in superior switching loss performance with turn on loss of 31 mJ and turn off loss of 101 mJ at 1000 V and 645 A for room temperature [7]. Moreover, switching losses remain overall insensitive with variation of temperature using full SiC-MOSFET modules contrary [7] to Si based IGBT modules with Si and/or SiC external Schottky diode.

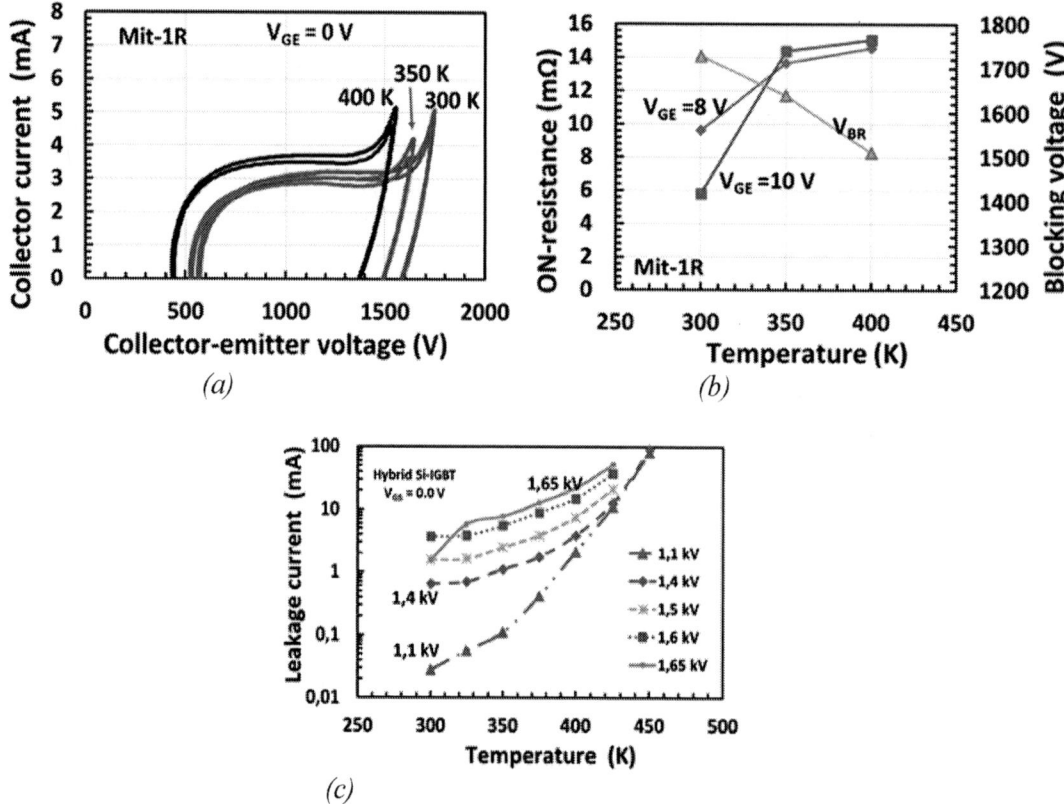

Fig. 3. Blocking voltage (a), ON-resistance (b) and leakage current (c) of Si based hybrid IGBT power modules at various temperatures.

Device measurement statistics show that almost all purchased modules have been working according to the datasheet claims with almost uniform static and dynamic characteristics and hence module manufacturing process shows fairly good confidence level with quite sense of process maturity and yield. On the other side, module design and external PIN connections have been found out user-friendly indeed and suitable for the use in very high power applications such as FACTS and HVDC.

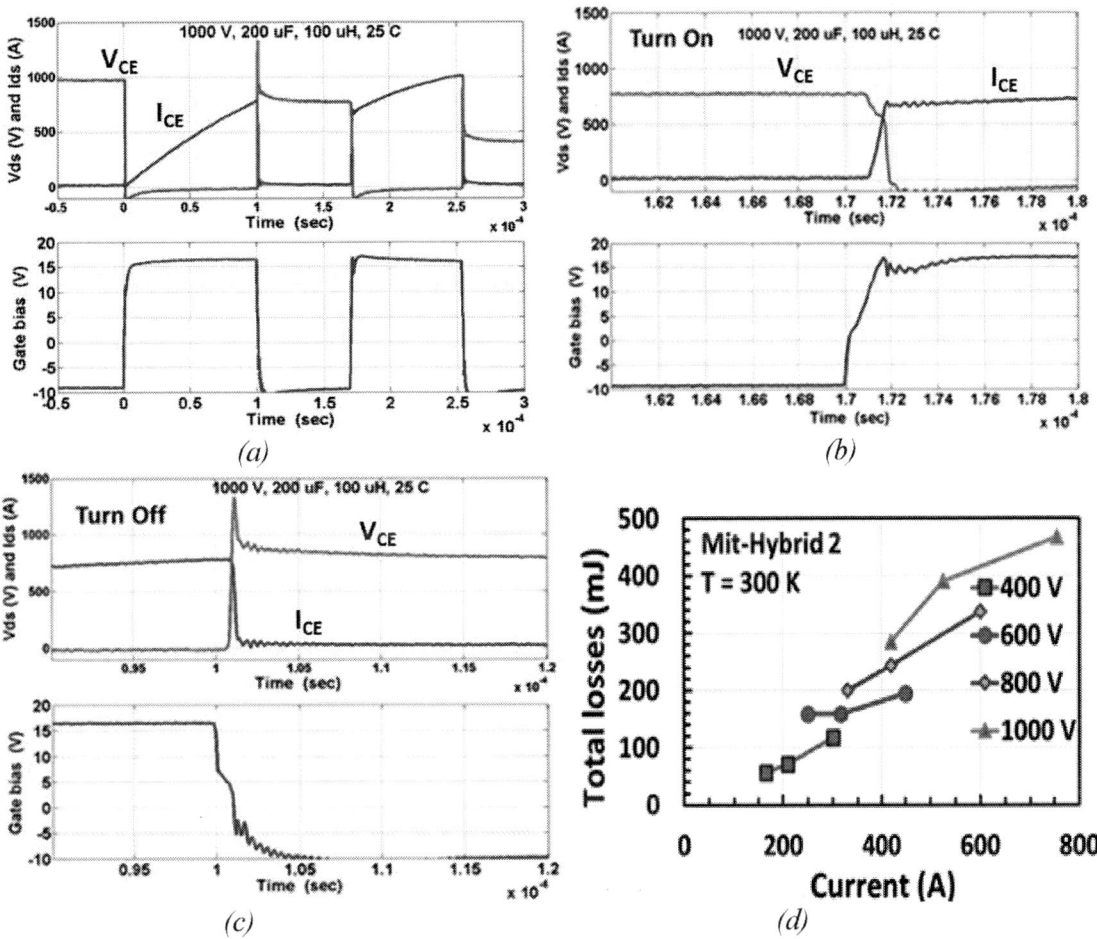

Fig. 4. Full double pulse transient sequence (a) at 1000 V using 200 µF DC capacitor and 100 µH load inductance. Also shown is the zoomed in view of turn on (b), turn off (c) loss at 1000 V supply voltage and total energy losses at 300 K for various supply voltages.

Parameters Mit-Hyb 2	400V/300A 100uH, 25°C 200 µF	600V/450A 100uH, 25°C 200 µF	800 V/600A, 100uH, 25°C 200 µF	1000 V/755A 100uH, 25°C 200 µF
E_{ON} (mJ)	64	74	131	165
E_{OFF} (mJ)	53	120	206	302
E_{total} (mJ)	117	194	337	467

Parameters Mit-Hyb 2	1000V/755A 100uH, 25 °C 200 µF	1000V/755A 100uH, 75 °C 200 µF	1000 V/755A, 100µH, 150 °C 200 µF
E_{ON} (mJ)	165	166	167
E_{OFF} (mJ)	302	313	381
E_{total} (mJ)	467	479	548

Fig. 5: Energy loss extraction of hybrid power module at different temperatures.

4. Conclusions

Hybrid power modules from Mitsubishi (*CMH1200DC-34S*) have been evaluated in this work. These modules has a current rating of 1200 A and voltage rating of 1700 V. ON-resistance of these modules was $8 - 10$ mΩ at 300 K at 10 V gate bias. Total energy losses were found out to be 337 mJ and 467 mJ at 800 V (600A) and 1000 V (755A) respectively, at 300 K. Overall, the module performance is according to the manufacturer claim as witnessed from our static and dynamic measurements. This data will provide a useful guideline for the power converter design.

References

[1] V. Pala, E. V. Brunt, L. Cheng, M. O'Loughlin, J. Richmond, A. Burk, S. T. Allen, D. Grider, J. W. Palmour, and C.J. Scozzie, "10 kV and 15 kV Silicon carbide power MOSFETs for next-generation energy conversion and transmission systems," IEEE Energy Conversion Congress and Exposition (ECCE), 2014, pp. 449-454.

[2] T. Funaki et al., "Power conversion with SiC devices at extremely high ambient temperatures," IEEE Transactions on Power Electronics, Vol. 22, no. 4, 2007, pp. 1321 – 1329.

[3] O. Harmon, T. Basler and F. Björk, "Advantages of the 1200 V SiC Schottky Diode with MPS Design", In Bodo power systems, pp. 34 – 37, Dec. 2015.

[4] M. Nawaz, "Predicting potential of 4H-SiC power devices over 10 kV", In Proc. IEEE Power Electronics and Drives (PEDS-2013), 2013, pp. 1291 – 1296.

[5] M. Nawaz, "On the evaluation of gate dielectrics for 4H-SiC based power MOSFETs," Active and Passive Electronic Components, Vol. 2015, Article ID 651527, 2015, 12 pages.

[6] F. Chimento, M. Nawaz, N. Chen, L. Wang, "Dynamic characterization of parallel-connected high-power IGBT modules", Proceedings of the 2013 IEEE Energy Conversion Congress and Exposition . Denver (CO) USA, September 2013, pp. 4263-4269.

[7] M. Nawaz, F. Chimento and K. Ilves, "Static and dynamic performance assessment of commercial SiC MOSFET power modules, In proceeding. ECCE'15, 2015, pp. 4899 – 4906.

[8] Mitsubishi Si IGBT hybrid power module datasheet "CMH1200DC-34S", 18th Oct 2011.

[9] M. Besacier, M. Coyaud, J.L. Shanen, J. Roudet, "Hybrid Si-SiC fast switching cell modelling and characterisation including parasitic environment description by PEEC method", In proceeding. of PESC 2002.

[10] Alvarez, R. Filsecker, F. Bernet, " Characterization of a new 4.5 kVpress pack SPT+ IGBT for medium voltage converters", Proc. of the 1st IEEE Energy Conversion Congress and Exposition, 2009. ECCE 2009, San Jose (CA), 20-24 Sept. 2009, pp. 3954 - 3962.

[11] J. L. Hostetler, P. Alexandrov, X. Li, L. Fursin and A. Bhalla, " 6.5 kV SiC normally-off JFETs – TechnoloJ. gy status," IEEE Workshop on Wide Bandgap Power Devices and Applications (WiPDA), 2014, pp. 143-146.

[12] H. Miyake , H. N.T. Okuda , T. Kimoto and J. Suda "21-kV SiC BJTs with space-modulated junction termination extension", IEEE Electron Device Lett., Vol. 33, 2012, pp. 1598 -1600.

[13] L. Cheng et al., "20 kV, 2 cm^2, 4H-SiC gate turn-off thyristors for advanced pulsed power applications", IEEE 19th Pulsed Power Conference (PPC), 2013, pp. 1 – 4.

[14] E. Van Brunt, L. Cheng, M. O'Loughlin, J. Richmond, V. Pala, J. W. Palmour, C. W. Tipton and C. Scozzie, "27 kV, 20 A 4H-SiC n-IGBTs" Materials Science Forum, Vol. 821-823, 2015, pp 847-850.

[15] Naoki Kaji et al., "Ultrahigh-voltage SiC p-i-n diodes with improved forward characteristics", IEEE, Trans On Electron Devices, Vol. 62, No. 2, pp. 374 – 381, Feb 2015.

ASDAM 2016, The 11th International Conference on Advanced Semiconductor Devices And Microsystems, November 13-16, 2016, Smolenice, Slovakia

SiO2/4H-SiC Interface Traps Effects on the Input Capacitance of DMOSFET

Gian-Domenico Licciardo and Luigi Di Benedetto

Department of Industrial Engineering, University of Salerno
Via Giovanni Paolo II, 132 84084 Fisciano (SA), Italy
e-mail: gdlicciardo@unisa.it and ldibenedetto@unisa.it

In this paper, an analytical model is presented to describe the input capacitance of Power-MOSFETs in 4H polytype of Silicon Carbide (4H-SiC). In order to provide an instrument for accurate interpretations of C-V measurements and for a deeper understanding of the device operations, the model describes the charge variations induced by the presence of the oxide-semiconductor interface trapped charge. Their energy dependence has been accounted to describe the charge dynamics into the channel and the accumulation layer and proved by comparisons with numerical simulations.

1. Introduction

The 4H polytype of silicon carbide (4H-SiC) is currently one of the most promising wide band-gap semiconductor that, thanks to good electrical and physical parameters, is well suited for the implementation of power devices [1]. Power JFETs [2],[3] and Vertical DMOSFETs (VDMOS) [4],[5] are two of the most interesting devices, although VDMOS better combines the advantages of a voltage-controlled device to good figure-of-merits, enabled by the JFET-like structure and the possibility to manage high blocking-voltages [6], as well as to avoid problems related to the doping diffusion and bulk lifetime that affect bipolar devices [7]-[12]. However, the operation of the VDMOS is significantly dependent on the quality of the interface between gate oxide and semiconductor, which presents a relevant density of defects [4],[13] so that the principal electrical parameters of the device are altered, like the threshold voltage, V_{TH}, and the mobility in the channel and accumulation layer [4],[14],[15]. Therefore, both static and switching operations of the VDMOS, are strongly influenced by the charge dynamics in the channel and the accumulation region, which have repercussions also on C-V measurements [16]. Following up the needs of accurate quantitative descriptions, in this work an analytical model of the input capacitance of VDMOS in 4H-SiC, C_{ISS}, is presented, which turns useful for accurate interpretations of C-V measurements, as well as for the correct modelling of the switching behaviour of the device. The accuracy of the model is verified by comparisons with numerical simulations.

2. Analytical model

2.1 Interface trapped charge

The inset of Fig. 1 shows the 2D structure of the half-device used for model and simulations, completed with the equivalent capacitances involved in the following analysis. The trap distribution induced at the insulator-semiconductor interface, D_{it}, has been described in terms of the energy trap, E_t, by the superposition of four functions, symmetrically placed with respect to the mid-gap: $D_{it}(E_t) = D_{it,TA}(E_t) + D_{it,TD}(E_t) + D_{it,MA} + D_{it,MD}$. As shown in Fig. 1, the

978-1-5090-3084-2/16 $31.00 © 2016 IEEE

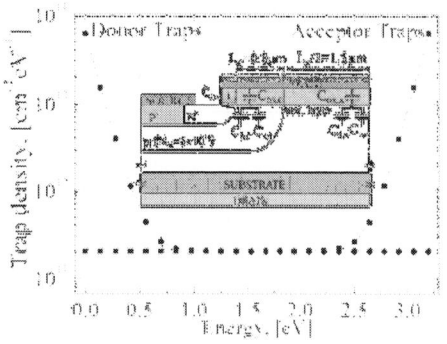

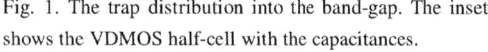

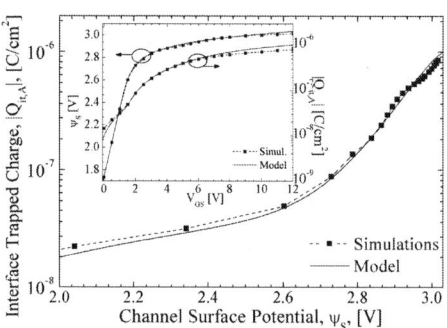

Fig. 1. The trap distribution into the band-gap. The inset shows the VDMOS half-cell with the capacitances.

Fig. 2. Comparisons between analytical and simulated $Q_{it,A}$-ψ_S curves. The inset compares ψ_S-V_{GS} and $Q_{it,A}$-V_{GS}

term $D_{it,TA}(E_t) = D_{it,T0}e^{\frac{Et-E_C}{WTA}}$ ($D_{it,TD}(E_t) = D_{it,T0}e^{-\frac{Et-E_V}{WTD}}$) is the tails of acceptors (donors) distribution in the upper (lower) half-gap, while $D_{it,MA(D)}$ is the deep level distribution, assumed constant near the mid-gap [4,14]. Since the Fermi level moves in the upper half of the band-gap when the device is forward biased, the total trap distribution can be approximated to $D_{it}(E_t) \approx D_{it,TA}(E_t) + D_{it,MA}$. The related trapped charge density, $Q_{a,it}$, can be calculated as a function of the surface potential in the channel, ψ_S, by means of the Fermi-Dirac distribution of probability as:

$$Q_{a,it}(\psi_S,T) = -q\int_{E_i}^{E_C} D_{it}(E_t)\left(1 + \frac{n_i}{n_S(\psi_S,T)}e^{\frac{Et-Ei}{k_BT}}\right)^{-1}dE_t. \tag{1}$$

where $n_S(\psi_S) = n_i^2 N_W^{-1}e^{\frac{\psi_S}{V_t}}$ is the electron density at the oxide interface, E_i is the intrinsic Fermi level, assumed as the neutral energy value for the imposed symmetry of trap distribution and k_B is the Boltzmann constant. By substituting E_i with the lower limit of the conduction band, E_C, the trapped charge density can be rewritten as:

$$\begin{cases} Q_{a,it}(\psi_S,T) = Q_{it,TA}(\psi_S,T) + Q_{it,MA}(\psi_S,T) \\[2mm] Q_{it,TA}(\psi_S,T) = -qD_{it,T0}WTA\left[2F1\left(1,\frac{k_BT}{WTA};\frac{k_BT+WTA}{WTA};-\frac{N_CN_W}{n_i^2}e^{\frac{\psi_S}{V_t}}\right) + \right. \\[3mm] \qquad \left. -e^{\frac{E_V-E_C}{WTA}}2F1\left(1,\frac{k_BT}{WTA};\frac{k_BT+WTA}{WTA};-\frac{N_CN_W}{n_i^2}e^{-\frac{0.5\times V_G+\psi_S}{V_t}}\right)\right] \\[3mm] Q_{it,MA}(\psi_S,T) = D_{it,MA}\left[\frac{V_G}{2} - V_t\ln\left(1+\frac{N_CN_W}{n_i^2}e^{\frac{\psi_S}{V_t}}\right) + V_t\ln\left(1+\frac{N_CN_W}{n_i^2}e^{-\frac{0.5\times V_G+\psi_S}{V_t}}\right)\right]. \end{cases} \tag{2}$$

where N_W is the channel doping, coincident with that of the p-type well, N_C is the effective density of states, $V_G = V_{G0} - 3.3\times 10^{-4}(T-300)$, with V_{G0}=3.2V, is the temperature dependent band-gap, $V_t = k_BTq^{-1}$ the thermal voltage and $2F1(a,b;c;z) = \sum_{i=0}^{\infty}\frac{(a)_i (b)_i z^i}{(c)_i i!}$, being $(n)_i$ the Pochhammer symbol. Fig. 2 shows the accuracy of (2) by comparing analytical and simulated curves of $Q_{a,it}$ when $D_{it,T0} = 5\times 10^{13}cm^{-2}eV^{-1}$, $WTA = 0.09eV$ and $D_{it,MA(D)} = 2\times 10^{11}cm^{-2}eV^{-1}$

978-1-5090-3084-2/16 $31.00 © 2016 IEEE

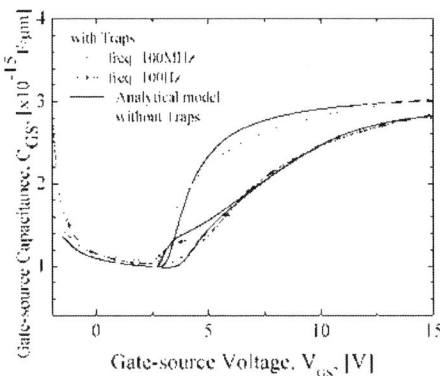 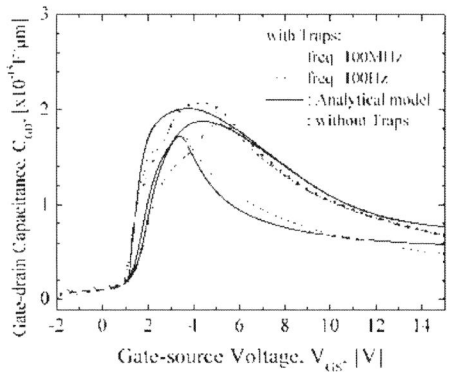

Fig. 3. Model-simulation comparisons of C_{GS}-V_{GS}, with/without interface traps, at 100Hz and 100MHz.

Fig. 4. Model-simulation comparisons of C_{GD}-V_{GS}, with/without interface traps, at 100Hz and 100MHz.

[4]. Considering that, for the presence of the energy dependent traps, an explicit expression of $\psi_S(V_{GS})$ cannot be found, (2) has been expressed as a function of the gate voltage by pre-calculating the pairs (V_{GS}, ψ_S) in the channel from the following equation [12],

$$\begin{cases} V_{GS} = V_{FB,C}(T) - V_{traps}(\psi_{S,C}, T) + \psi_{S,C} + \dfrac{\sqrt{2\varepsilon q N_W V_t}}{C_{OX}} F\left(\dfrac{\psi_{S,C}}{V_t}, \dfrac{n_i}{N_W}\right) \\ F\left(\dfrac{\psi_{S,C}}{V_t}, \dfrac{n_i}{N_W}\right) = \left[\dfrac{\psi_{S,C}}{V_t} + \left(e^{-\frac{\psi_{S,C}}{V_t}} - 1\right) + \left(\dfrac{n_i}{N_W}\right)^2 \left(e^{\frac{\psi_{S,C}}{V_t}} - 1\right)\right]^{\frac{1}{2}}. \end{cases} \tag{3}$$

where $V_{FB,C} = \Phi_m - \chi - V_t \ln\left(N_W n_i^{-1}\right) - 0.5 V_G$ is the flat-band voltage, $\psi_{S,C}$ is ψ_S in the channel, Φ_m is the gate-metal work-function, χ, the electron affinity, $C_{OX} = \varepsilon_{OX} t_{OX}^{-1}$ and $V_{traps}(\psi_{S,C}, T) = Q_{a,it}(\psi_{S,C}, T) C_{OX}^{-1}$. The surface potential in the accumulation region, $\psi_{S,A}$, can also be calculated by using (3) and modifying the flat-band voltage expression as $V_{FB,A} = \Phi_m - \chi + V_t \ln\left(N_D n_i^{-1}\right) - 0.5 V_G$. The accuracy of (3) is shown in the inset of Fig. 2, where the analytical curves of ψ_S and $Q_{a,it}$ as functions of V_{GS} are compared with numerical simulations. Eq. (2) and (3) allows calculating V_{TH} with much more accuracy, as $V_{TH} = V_{GS}(\psi_{S0}, T) = V_{FB,C}(T) - V_{traps}(V_{GS}, T) + \psi_{S0}(T) + C_{OX}^{-1}\sqrt{2\varepsilon q N_W \psi_{S0}(T)}$ where $\psi_{S0} = 2V_t \log\left(n_i N_W^{-1}\right)$ is defined as the potential value at which the strong inversion occurs and the other symbols have their usual meanings.

2.2 Input capacitance

In order to derive the expression of C_{ISS}, it has been expressed as the superposition of the gate-source, C_{GS}, and the gate-drain, C_{GD}, capacitances, which in turn have been calculated by the superposition of four contributions, shown in the inset of Fig. 1, and derived as follows. The charge into the channel, at each biasing conditions, can be expressed as $\left|Q_{S,C}\right| = \left|2V_T \varepsilon_S L_{Deb,C}^{-1} F(\psi_{S,C})\right|$, where $L_{Deb,C} = \sqrt{V_t \varepsilon N_W^{-1} q^{-1}}$ is the Debye length, and from this the lumped capacitance, $C_{S,C} = \left|dQ_{S,C}/d\psi_{S,C}\right| L_C Z$, where L_C and Z are the length and width of the channel, respectively. Eq. 2 allows calculating the lumped capacitance related to the interface trapped charge, $C_{it,C} = \left| dQ_{it,a} / d\psi_{S,C} \right| L_C Z$ that is in parallel to $C_{S,C}$. The oxide capacitance, $C_{OX,C} = \varepsilon_{ox} t_{ox}^{-1} L_C Z$, where t_{ox} is the oxide thickness, is placed in series to the previous parallel capacitances. Finally, the overlapping capacitance between gate and source, $C_{OV} = \varepsilon_{ox} t_{ox}^{-1} L_{OV} Z$, sums to the previous contributions. C_{GD} can be calculated in a similar way, considering that when the channel moves from the accumulation to strong inversion, the accumulation layer moves from weak depletion to the strong accumulation. Therefore, similar quantities can be calculated with reference to the accumulation layer. The resulting C_{GS}, C_{GD} and C_{ISS} capacitances are calculated as:

$$C_{GS} = C_{OV} + \frac{C_{OX,C}\left(C_{it,C} + C_{S,C}\right)}{C_{OX,C} + C_{it,C} + C_{S,C}}, \quad C_{GD} = \frac{C_{OX,A}\left(C_{it,A} + C_{S,A}\right)}{C_{OX,A} + C_{it,A} + C_{S,A}}, \quad C_{ISS} = C_{GS} + C_{GD}. \quad (4)$$

In Fig. 3 and 4, comparisons of the model with numerical simulations, using physical parameters taken from [2]-[4], prove their accuracy. $C_{S,C(A)}$ dominates from accumulation to weak inversion, the oxide capacitance become prevalent at $V_{GS} > V_{TH}$ and the trap capacitance, $C_{it,C(A)}$, determines the knee around the V_{TH} value that tends to disappear when the measurement frequency increases because of the reduced portion of the band-gap, interested by the AC stimulus. At strong inversion, after the knee, $C_{GS} \approx C_{OV} + C_{OX}$ begins to increase with V_{GS}, while C_{GD} tends to vanishes, because of the annulment of the potential barrier between the channel and the accumulation region close to the p-well.

3. Conclusions

An analytical model of the input capacitance of VDMOS in 4H-SiC is presented that accurately describes the oxide/semiconductor interface trapped charge and its dependence from the surface potential and gate voltage.

References

[1] K. Hamada *et al.*, *IEEE Trans. on Electron Devices*, **62**, 2, 278-285, 2015.
[2] S. Bellone *et al.*, *Solid State Electron.*, **109**, 17–24, 2015.
[3] S. Bellone *et al.*, *Solid State Electron.*, **120**, 6–12, 2016.
[4] G.D. Licciardo *et al.*, *IEEE Trans on Power Elect.*, **30**, 10, 5800-5809, 2015.
[5] A. Saha *et al.*, *IEEE Trans. on Electron Dev.*, **54**, 10, 2786–2791, 2007.
[6] L. Di Benedetto *et al.*, *IEEE Trans. on Electron. Devices*, **63**, 9, 3795-3799, 2016.
[7] S. Bellone *et al.*, *IEEE Trans. on Electron Dev.*, **56**, 12, 2902-2911, 2009.
[8] S. Bellone *et al.*, *Proceedings of the International Semiconductor Conference (CAS)*, **2**, 405-408, 2010.
[9] S. Bellone *et al.*, *IEEE Trans. on Instrum. and Meas.*, **57**, 6, 1112-1117, 2008.
[10] S. Bellone *et al.*, *IEEE Trans. on Electron Dev.*, **54**, 11, 2998-3006, 2007.
[11] L. Di Benedetto *et al.*, *IEEE Electron. Dev. Lett.*, **35**, 2, 244-246, 2014.
[12] S. Bellone *et al.*, *IEEE Trans. on Electron Dev.*, **59**, 9, 2546-2549, 2012.
[13] A. Castellazzi *et al.*, *Microelectronics Reliability*, **52**, 9, 2414–2419, 2012.
[14] G.D. Licciardo *et al.*, *IEEE Trans. on Electron. Devices*, **63**, 4, 1783-1787, 2016.
[15] S. Potbhare *et al.*, *IEEE Trans. Electron Devices*, **55**, 8, 2029-2040, 2008.
[16] L. Di Benedetto *et al.*, *IEEE Electron. Dev. Lett.*, **37**, 2016. *Doi: 10.1109/LED.2016.2613821*

ASDAM 2016, The 11th International Conference on Advanced Semiconductor Devices And Microsystems, November 13-16, 2016, Smolenice, Slovakia

Optimization of semi-insulating GaAs detector for thermal neutron detection

A. Šagátová[1,2], B. Zaťko[3], K. Sedlačková[1], V. Nečas[1], P. Boháček[3]

[1]Slovak University of Technology in Bratislava, Faculty of Electrical Engineering and Information Technology, Ilkovičova 3, SK-812 19 Bratislava, Slovak Republic
[2]University Centre of Electron Accelerators, Slovak Medical University,
Ku kyselke 497, SK-911 06 Trenčín, Slovak Republic
[3]Institute of Electrical Engineering, Slovak Academy of Sciences,
Dúbravská cesta 9, SK-841 04 Bratislava, Slovakia
e-mail: andrea.sagatova@stuba.sk

We have studied the semi-insulating (SI) GaAs Schottky diode with ^6LiF conversion layer as a detector of thermal neutrons. Two parameters of detector were optimized: the active detector volume modified by the reverse voltage applied on barrier and the thickness of conversion layer. Increasing the reverse voltage, the active volume of detector spreads into depth of GaAs substrate. Optimal reverse voltage was determined to be 50 V, as the responsible active detector thickness effectively absorbs the neutron products but the accompanying gamma rays from neutron source are absorbed with very low probability. The optimal thickness of ^6LiF conversion layer was set to 26 μm according to our experiments, when the layer reached the maximum conversion efficiency of neutrons.

1. Introduction

Gallium Arsenide is one of the ideal semiconductor materials for detectors of ionizing radiation. Its wide bandgap energy (1.42 eV) enables detector operation at RT (room temperature). Relatively high atomic numbers of Ga (31) and As (33) make detector relatively effective for X-ray and gamma ray detection. High mobility of charged carriers ($\mu_{electrons} >$ 8000 $cm^2V^{-1}s^{-1}$ and $\mu_{holes} = 400$ $cm^2V^{-1}s^{-1}$ at RT) leads to high reaction rate of detector. Moreover, the operating expenses are low thanks the low price of base material and its high radiation hardness [1-3]. The applications of GaAs detector are wide. It can be used for detection of photons, charged particles and with appropriate conversion layer also as a detector of thermal or fast neutrons. Small dimensional (< 100 μm) semiconductor detectors can be arranged into two-dimensional fields and thus create a position sensitive sensor for digital radiographic imaging. Recently, the classical films from X-ray radiography are being replaced by such digital sensors. However, the X-ray radiography utilizes the high attenuation coefficient of X-rays in high-density materials but is insufficient in imaging of low-density materials. On the other hand, neutrons are very effective in imaging low-density materials even those packed inside of high-density ones. This way neutronography represents a complementary method for X-ray radiography. Semiconductor based neutron position sensitive sensors show very good properties for neutron imaging. They are investigated at only a few institutions and are usually based on silicon detectors [4, 5]. In this paper we are studying the capability of GaAs Schottky barrier detector with ^6LiF conversion layer to detect thermal neutrons and we are optimizing the thickness of the conversion layer and the voltage applied on the barrier to improve the signal from neutrons.

2.   Detector preparation

We have prepared investigated detector structures from bulk VGF (Vertical Gradient Freeze) SI (semi-insulating) GaAs wafer (producer: Wafertech UK) at the Institute of Electrical Engineering SAS in Piešťany. SI GaAs is overcompensated material and its typical resistivity is of 1×10^7 Ωcm - 1×10^8 Ωcm. The galvanomagnetic measurements showed the resistivity of 9.1×10^7 Ωcm and the electron Hall mobility of 6285 cm^2/Vs at RT. The wafer was polished from both sides to the thickness of 350 µm. On the top side of GaAs wafer the circle blocking Ti/Pt/Au (10 nm/30 nm/90 nm) Schottky contact, of 0.6 cm diameter, was evaporated. A whole area quasi ohmic contact from Ni/AuGe/Au (30 nm/50 nm/90 nm) multilayer was formed on the back-side of the wafer. GaAs detector is not sensitive to neutrons by itself. A neutron conversion layer has to be applied to transform neutrons to other particles, detectable by the structure. The best choice to convert thermal neutrons seems to be the layer of ^{6}LiF [4, 6]. We have sprayed the ^{6}LiF layer (enriched by ^{6}Li isotope to 95%) of a thickness in the range from 5 to 61 µm on the top Schottky contact of our detector. Here the thermal neutrons interact with ^{6}Li producing 2.05 MeV alpha particles and 2.73 MeV tritons which are particles detectable by SI GaAs detector, if they reach the active detector volume:

$$^6Li + n \rightarrow {}^4He(2.05 MeV) + {}^3H(2.73 MeV) \qquad (1).$$

The response of detector to neutrons was measured using a source of fast neutrons, the ^{239}PuBe with fluency of 1.7×10^7 neutrons per second into solid angle 4π. The fast neutrons were moderated by pyrolytic graphite to thermal ones. The distance between the detector and the source centre in graphite container was about 6 cm. However, not only neutrons were impinging the detector, as each source of neutrons produces also gamma rays by activation of the source shielding.

3.   Optimization of reverse voltage

Only the part of the detector structure created by depletion region is sensitive to ionizing radiation. In the case of SI GaAs the depletion region is penetrating into the depth of substrate with increasing applied reverse voltage linearly for voltage higher than 20 V [7]. The products of neutron interaction with ^{6}LiF conversion layer (Eq. 1) have rather short range in detector volume. The 2.05 MeV alphas have the range of 5.4 µm and the 2.73 MeV tritons of 29 µm in GaAs. It means that even thin active volume of detector (at 50 V) will sufficiently register neutron products. On the other hand, the gamma rays from neutron source background are registered with higher probability by the detector with larger active thickness. These facts have manifested in measured spectra. We have used reverse voltages from 50 V to 250 V with 50 V step in our experiments. In Fig. 1 the spectra measured with conversion layer are compared to spectra obtained using bare detector (gamma background) at two different reverse voltages 50 V and 250 V. At 250 V (Fig. 1a) the signal from the neutron products, the alphas (scatter), is of the same height as the gamma background (black line). On the contrary, the neutron response was very well separated from gamma background at 50 V (Fig. 1b). Optimal reverse voltage for registering thermal neutrons was determined to 50 V, effectively detecting neutron products and suppressing the gamma rays. The originality of two peaks in spectra measured by GaAs detector with conversion layer was verified by measuring the alpha particles of known energy, the 5.48 MeV alphas from ^{241}Am. In Fig. 2 a very good linearity of calibration curve obtained from 2.05 MeV alpha peak, 2.73 MeV triton peak and 5.48 MeV ^{241}Am peak can be observed.

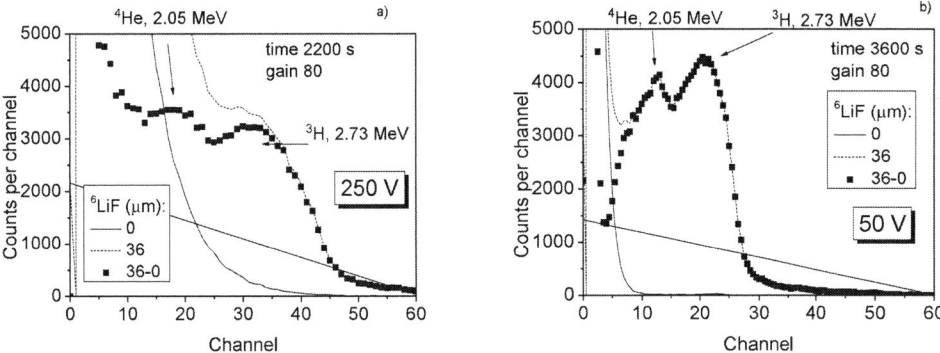

Figure 1. The spectra of thermal neutron response (scatter) obtained by SI GaAs detector with 36 μm thick ^{6}LiF layer as measured spectrum (line) minus gamma background (black line): a) at 250V reverse bias voltage on detector and b) 50 V reverse bias voltage.

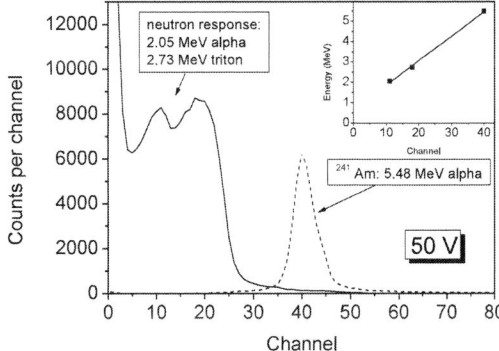

Figure 2. Energy calibration of GaAs neutron detector at 50 V reverse voltage. Calibration curve obtained from two peaks of neutron response (2.05 MeV alpha & 2.73 triton) and the 5.48 MeV alpha particles from ^{241}Am.

4. Optimization of conversion layer thickness

The conversion layer is essential for registering neutrons with GaAs detector. Depositing thicker conversion layer on detector surface the probability of neutron interaction with its atoms increases. The problem is the range of products of these interactions in the layer itself. The alphas from Eq. 1 have the range of 6.05 μm and tritons of 33 μm in ^{6}LiF layer. Too thick conversion layer will absorb them. According to MCNPX simulations, the optimal thickness of ^{6}LiF layer is 25 μm [7], which is proved also by following experiment. In Fig. 3 we can see the spectra measured by SI GaAs detector with ^{6}LiF conversion layer at the ^{239}PuBe neutron source. The bare detector response is compared to spectra measured by detector with conversion layer of thicknesses in the range from 5 to 61 μm. One can observe, that the height of peaks steeply increases with increasing thickness up to 26 μm followed by their decrease. In Fig 4 the integral of counts in alpha and triton peaks is depicted as a function of ^{6}LiF layer thickness proving this observation. It can be seen that the conversion layer with the thickness of 26 μm converted maximum number of neutrons to secondary particles.

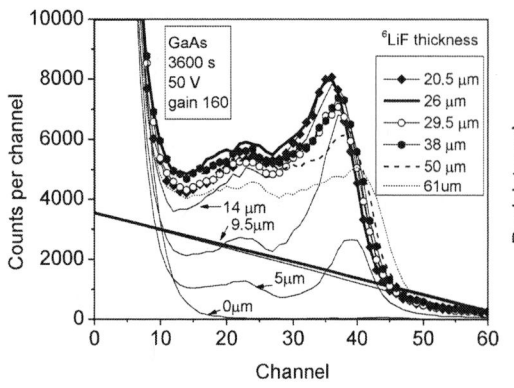

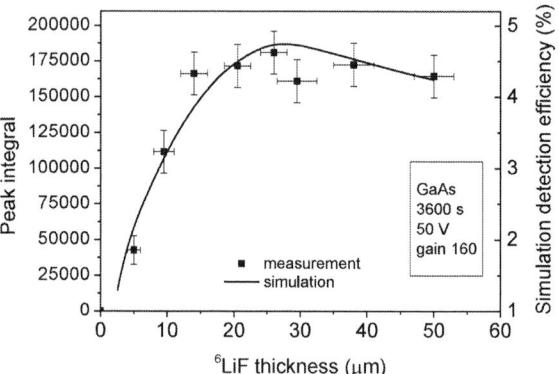

Figure 3. The response of moderated [239]PuBe neutrons measured by SI GaAs detector with [6]LiF conversion layer of various thicknesses in the range from 5 to 61 μm compared to bare detector signal.

Figure 4. The number of tritons and alpha particles registered as a response of detector to thermal neutrons compared to simulation in MCNPX.

4.   Conclusions

The SI GaAs Schottky barrier structure with [6]LiF conversion layer was tested as a detector of moderated neutrons from [239]PuBe source in graphite container. First the optimal reverse voltage was chosen to be 50 V, enabling effective registration of products of neutrons from conversion layer contrary to gamma ray background of the source. Then the conversion layers of various thicknesses were deposited on the top Schottky contact of the structure optimizing its ideal thickness. The highest response from neutrons was obtained by 26 μm thick layer applied which is in good agreement with simulation in MCNPX code assuming optimal thickness of 25 μm.

Acknowledgement

This work was partially supported by the Slovak Grant Agency for Science through the grant 2/0152/16, by the Slovak Research and Development Agency under contract No. APVV-0321-11 and by the Project Research and Development Centre for Advanced X-ray Technologies (ITMS code 26220220170) of the Research & Development Operational Program funded by the European Regional Development Fund (0.7).

References

[1] V. Linhart *et al, Nucl. Instrum. & Meth. in Phys. Res.* A**563** (2006) 66.

[2] K. Afanaciev *et al, JINST* 7 (2012) P11022.

[3] A. Sagatova *et al, Applied Surface Science* XXX (2016) XXX, DOI
      10.1016/j.apsusc.2016.08.167, in press.

[4] J. Jakubek *et al, Nucl. Instrum. & Meth. in Phys. Res.* A**560** (2006) 143.

[5] J. Uher *et al, Nucl. Instrum. & Meth. in Phys. Res.* A**576** (2007) 32.

[6] D.S. McGreor *et al, Nucl. Science Symposium Conference Record,* San Diego CA, IEEE
      Vol.4 (2001) 2454. DOI 10.1109/NSSMIC.2001.1009315.

[7] K. Sedlackova *et al, Proc. of the 22th International Workshop on Applied Physics of
      Condensed Matter (APCOM 2016),* Štrbské Pleso, Slovak Republic, 22.-24.6.2016.,
      eds. J. Vajda and I. Jamnický, Bratislava: FEI STU (2016) 126. ISBN 978-80-227-
      4572-7, internet: *http://kf.elf.stuba.sk/~apcom/proceedings/pdf/126_sedlackova.pdf*

ASDAM 2016, The 11th International Conference on Advanced Semiconductor
Devices And Microsystems, November 13-16, 2016, Smolenice, Slovakia

Towards III-nitride nano-LED based single photon emitters: technology and applications

H. Hardtdegen[a,b], and M. Mikulics[a,b]

[a] *Peter Grünberg Institute (PGI-9), Forschungszentrum Jülich, 52425 Jülich, Germany*
[b] *JARA – Fundamentals of Future Information Technology*

Three alternative device concepts for single photon emitters based on III-nitride nano-LEDs are introduced, their technology reported and the applications they are suitable for presented. The first concept is a vertical device concept and is based on mesoscopic sized (InGa)N nano-pyramids prepared by bottom-up selective area metalorganic vapor phase epitaxy. The emission of the emitters is controlled by the composition of the nano-pyramids and their size and can be tuned to the telecommunication wavelength range useable for highly secure data communication. Furthermore, a hybrid device platform was devised which consisted of a top-down etched nano-LED and a mesoscopic sized nanocrystal. The primary emission of the LED is used to induce emission from the crystal. The emission is tunable by the crystal`s band gap together with its diameter for crystal sizes at which quantum confinement effects are to be expected. Beside the high device efficiency, the broad range of emission wavelengths achievable characterizes this approach. The third approach employs the top-down formed nano-LED photon emitters for lithography. Here, the emission energy of the emitter is utilized to induce the chemical reaction in the photo resist chosen. Ultimately, only one photon is needed to change one chemical bond. This would then allow a scaling of lithography down to the molecular size. All three photon emitters were integrated into high frequency layouts suitable for DC and HF characterization/operation.

1. Introduction

During the last decades, the demand for fast and highly secure data communication has increased immensely and is one of the main drivers for technological developments in information technology (IT). At the same time power consumption allotted to IT has increased from a few percent to about 20 % of the total power consumption in industrialized countries. In addition, the awareness for the global limitation of resources has grown. The consumption of resources therefore needs to be born in mind when alternative device concepts are developed for a so-called "Green" Emerging Technologies (GET).

Key elements for a future low energy consumption optoelectronics as well as for highly secure, fast and efficient optical data communication [1-4] are single photon sources (SPS). Preferably, the SPS should operate at room temperature. Their emission wavelength should be tunable for the application envisaged, they should be highly efficient and they should be integrated into semiconductor technology. The most suitable choice of materials for such emitters are group III nitrides and for device structures nano-light-emitting diodes (nano-LEDs). Their emission can be tuned by the composition of the group III-N alloys from the infrared, through the telecommunication down to the ultra-violet wavelength range. Due to the large quantum confinement in their nanostructures, single photon emission is observable even at room temperature. In addition, narrow emission is to be expected and high-speed process possible. LED structures are a standard technology and highly efficient.

978-1-5090-3084-2/16 $31.00 © 2016 IEEE

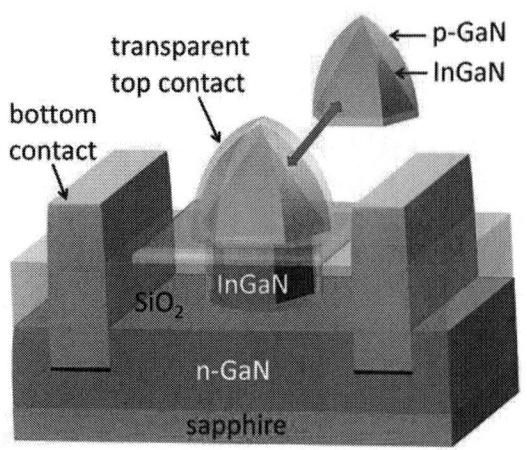

Figure 1. Principle schematics of the vertical integration technology based on InGaN pyramids of mesoscopic size (grown by SA-MOVPE).

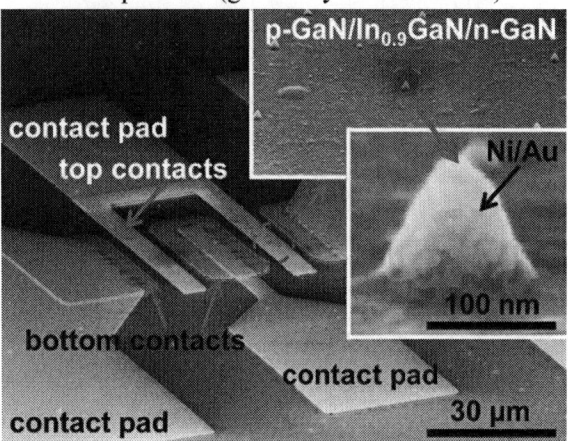

Figure 2. Scanning electron micrograph of integrated p-GaN/$In_{0.9}Ga_{0.1}$N/n-GaN p-i-n nanostructures. The inset shows the hexagonally arranged structures and a single nanopyramid with Ni/Au top contact.

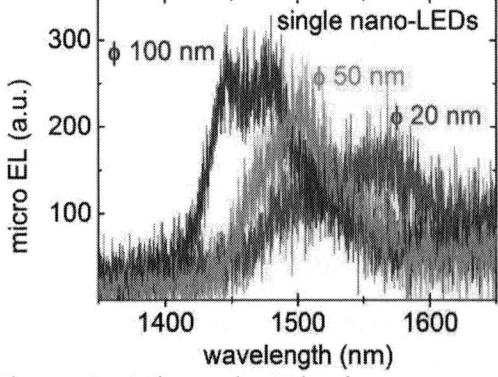

Figure 3. Micro electroluminescence measurements for single 20 nm, 50 nm and 100 nm (diameter) vertically integrated p-GaN/$In_{0.9}Ga_{0.1}$N/n-GaN LED structures.

For this contribution, we chose three alternative SPS device concepts we developed for so-called "Green" Emerging Technologies (GET). The first approach is a bottom-up approach based on undoped InGaN nanopyramids encompassed in n-doped and p-doped GaN. The emitters were tuned to emit in the telecommunication wavelength range and are therefore to be applied to optical data communication and can be used for optical interconnects. The second approach consists of a hybrid device platform. Nano-LEDs are prepared in a top down approach and used as the primary excitation source to electro-optically pump nanocrystals as the secondary excitation sources, which serve as SPS. This approach is innovative due to its simplicity and suitability for the mass-production of SPS of selectable wavelengths. For the third approach, the unusual application of the SPS is in the center of focus. Nano-LEDs were developed as sources for a mask-less lithography.

2. Single photon emitters based on InGaN nanopyramids

Figure 1 presents the novel vertical device concept [5,6]. For this approach, we started with an n-doped GaN layer deposited by metalorganic vapor phase epitaxy (MOVPE) on a (0001) sapphire substrate. The GaN template was covered by a SiO_2 layer, in which hexagonally arranged holes with diameters between 20 nm and 100 nm and with a pitch of 3 μm were prepared by means of e-beam lithography followed by developing and reactive ion etching. The etching process was optimized carefully with respect to the etching depth and the retention of an as perfect and clean as possible GaN surface for epitaxy. (InGa)N nanopyramids (red) were then deposited by selective area (SA) MOVPE [7–9] on such templates. Subsequently, p-doped GaN (green) was grown to cap the pyramids. The recessed bottom contacts

978-1-5090-3084-2/16 $31.00 © 2016 IEEE   28

were prepared by different lithography, dry etching and annealing steps. Also the required (semi-)transparent top contact is challenge. Here, a thin semi-transparent Ni/Au film was employed for the top contact. The nano-pyramids were integrated into a device layout suitable for DC testing and future high-frequency operation [10]. Figure 2 presents a scanning electron microscopy (SEM) image of the vertically integrated nano-LEDs.

A series of (InGa)N nano-pyramids with differing In content were grown using a variation of the In precursor to total group III precursor molar flow ratio between 0 and 100 %. Photoluminescence studies disclosed that a composition of 90 % leads to emission in the desirable telecommunication wavelength range [6]. This composition / molar flow ratio was employed for our further studies. Next, electroluminescence (EL) studies were carried out. The arrays of nano-LEDs based on $In_{0.9}Ga_{0.1}N$ nano-pyramids formed in three different sized holes in the mask (20 nm, 50 nm and 100 nm in diameter) were found to exhibit EL in the same intensity range for each respective nanopyramid size. The EL of single nano-LEDs were compared next and is presented in Figure 3. Obviously, the emission wavelength increases systematically as the structure size becomes smaller. The size of the nano-LEDs can be used additionally to tune the emission wavelength. The long term stability of the devices was tested by performing reliability studies. It was found, that the EL intensity decreases only moderately even for the smallest nano-pyramid size without any indication of degradation for at least 1000 hours [6].

With this novel new device concept, a promising technological route to future InGaN based low energy consumption optoelectronics is demonstrated suitable for operation in the telecommunication wavelength range.

3.   Hybrid optoelectronic platform for single photon emitters

Lately, we devised a novel technology suitable for the mass-production of SPS [11,12]. A hybrid device platform was introduced (Figure 4). The hybrid consists of InGaN nano-LEDs and nanocrystals (CdSe). The latter are to serve as the single photon emitters and are electro-optically driven by the former. The approach is characterized by the large variety of mesoscopic structured materials and nanocrystals which can be chosen as the SPS. In addition, it is simple to get access to the emission of such small structures, since no technological difficulties need to be overcome related to contacting them.

At first, InGaN/GaN multi quantum well (MQW) structures encompassed in n- and p-doped GaN layers were deposited by MOVPE [7, 13–16]. In the next step, they were structured to nanowires using lithography and dry chemical etching techniques. The procedure is reported in Mikulics and Hardtdegen [17]. The optimization of the dry etching step is especially crucial for their efficient and controlled emission intensity and wavelength. Smooth nanowire side walls and a minimization of etching induced defects are to be attained so that non-radiative recombination are avoided [18,19]. The nano-LEDs are isolated using a spin–on glass (hydrogen-silsesquioxane), which develops into SiO_x upon annealing. Here, a smooth surface is important, pinholes should be avoided and low leakage currents need to be obtained. The control of glass thickness and etching mask diameter and thickness are a prerequisite for the formation of suitable aperture sizes. Then the bottom contacts were formed by lithography and dry etching steps followed by metallization with a 35 nm Ti/200 nm A l/ 40 nm Ni/100 nm Au layer stack. At last the Ni etching cap was removed leaving behind the p-GaN surface ready for the formation of the nano-LED`s top contact and apertures for the nano-crystals. Ni/Au top contacts were formed to the nano-LEDs. Hereafter, CdSe nano-crystals, which were colloidally dispersed in toluene, were injected and distributed across the nano-LED area. In a nutshell, the particles were integrated into the hybrid device platform exhibiting a layout suitable for future DC and HF testing. An SEM micrograph of the whole device platform is presented in Figure 5.

978-1-5090-3084-2/16 $31.00 © 2016 IEEE

Figure 6 (left) presents the micro-EL spectrum recorded for a nano-LED with a diameter of 100 nm. The emission wavelength of 402 nm with a FWHM of 60 meV is suitable for the electro-optical pumping of the 4 nm sized CdSe nanocrystal sitting at its top. The initiated secondary emission recorded using micro PL is shown on the right hand of Figure 6. The spectrally sharp nano-crystal emission at 540 nm with a FWHM of 30 meV indicates that the Bohr-radius of the excitons exceeds the nanocrystal size for this small object. The results demonstrate that the nano-LEDs/nano-crystal hybrid device platform can be used to produce SPS [12]. Their integration into a high frequency device layout degrades neither their optical nor their electrical properties.

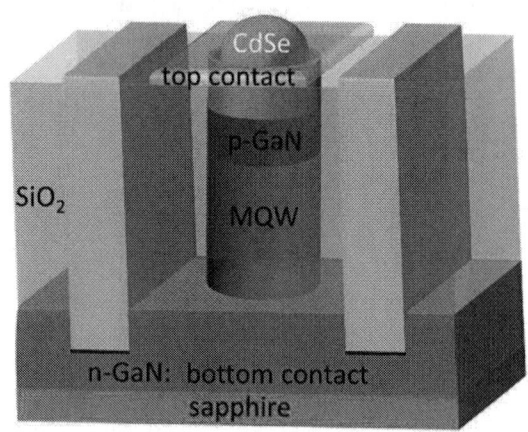

Figure 4. Principal schematics of the hybrid/III-nitride based nano-LED structure.

4. Mask-less lithography based on SPS

At last, we propose a completely novel SPS application for a future mask-less lithography [17,20] The concept is based on SPS as the illumination sources to initiate a chemical reaction in a photo resist. The sources are to be in close contact mode to the photo resist. They are positioned into arrays and are to be individually addressable. The patterns are produced by driving the respective LEDs in the array, i.e. the electrically driven array forms the illumination source and the mask simultaneously. Figure 7 presents the schematics of the application and the patterned resist after lithography and successive developing.

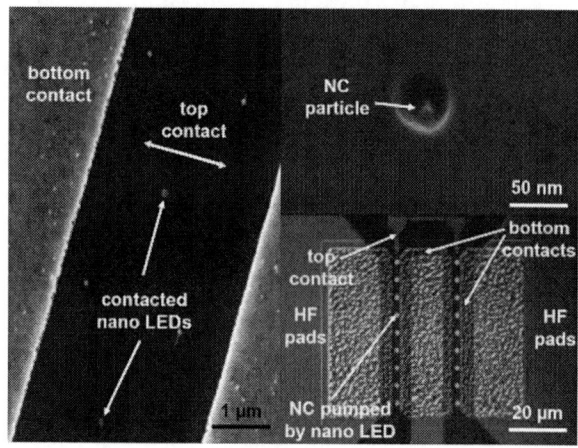

Figure 5. SEM micrographs: (a) nano-LEDs with NCs (b) nanocrystal (NC) positioned in SiO$_2$ hole structure. (c) fully integrated NC/nano-LED structures in the device layout suitable for DC and HF characterization.

An array of nano-LEDs with a diameter of 100 nm was prepared starting with a p-GaN / InGaN/GaN MQW /n-GaN structure deposited by MOVPE. The technological procedure for the nano-LEDs technology is similar to the one described above. In addition to the optimization of the dry etching process, special care needed to be taken in the control of the spin-on glass thickness. The Ni cap must protrude the

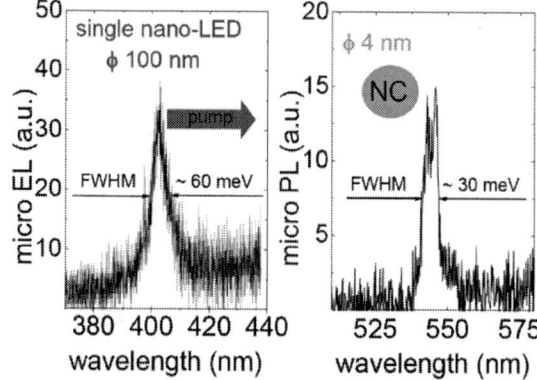

Figure 6. Micro-EL measurement of a single nano-LED with a diameter of 100 nm (left) used to pump a CdSe nanocrystal, which then emits at 540nm.

978-1-5090-3084-2/16 $31.00 © 2016 IEEE

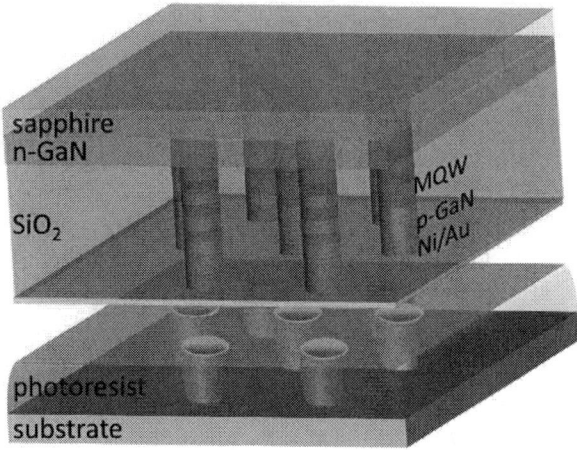

Figure 7. Principal schematics of the nano-LED assisted lithography.

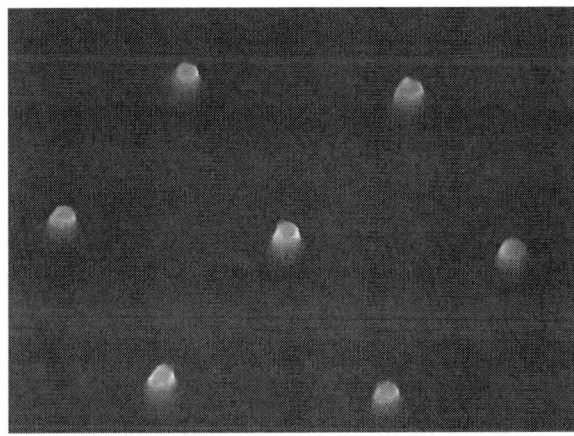

Figure 8. Scanning electron micrograph (SEM) image of the III-nitride based single photon emitters/nano-LEDs exhibiting a pitch of 3µm.

.

Figure 9. SEM image of the patterned photo resist-hole structures- after exposure and subsequent developing.

surface of the annealed SiO_x completely and the oxide should be flush with the top of the p-GaN layer. Bottom and top contacts are formed as described above. An SEM micrograph of the nano-LEDs integrated into the device layout is presented in Figure 8. The LEDs were driven simultaneously in the array and characterized by micro EL before their application in lithography. The intensity of the nano-LEDs emitting at 400 nm is quite homogeneously distributed across the entire array. Successively, the nano-LEDs were operated to perform lithography. A conventional resist (AZ 5206) was chosen to prove the principle of the concept. The wavelength of the nano-LEDs is sufficient to induce the photochemical reaction in the resist chosen. Figure 9 depicts the outcome after illumination and successive developing. The structured photo resist proves the principle of this lithographical approach. The diameter of the apexes and their volume are controllable by tuning photon counts from a single photon source.

In future, novel photo resists must be developed, which can ensure a single reaction between a single molecule and a single photon. The nano-LEDs should be further developed so that they can function in the "single photon regime" at room temperature. In addition, the SPS should be driven by means of a cross-bar geometry as the driving scheme. Since the size of the pattern can be tuned by the photon counts (i.e. the illumination time) emitted from the SPS, an instantaneous and flexible pattern modification in the lithographical process can achieved by a suitable driving algorithm. Lastly, reliability studies need to be carried out before a real lithography tool can become commercially available.

5. Conclusions

These key "Green" Emerging Technologies (GET) form the foundation for a large range of low energy consuming (opto)electronic applications which range from photon flux arrays for metrology, through quantum cryptography, molecular electronics and ultimately-targeted at a photon-based processor and memory operation. Our current achievements demonstrate a proof of principle and are primarily focused on reaching important milestones- to stimulate and or to initialize essential impulses in the development of further unconventional technological innovations.

Aknowledgements

The authors greatly acknowledge the German Ministry of Education and Research (BMBF) for financial support within the projects EPHQUAM (16BL0904) and QPENS (13N9898).

References

[1]   B. Lounis and M. Orrit, Rep. Prog. Phys. **68**, 1129 (2005).

[2]   N. Gisin, G. Ribordy, W. Tittel, and H. Zbinden, Rev. Mod. Phys. **74**, 145 (2002).

[3]   G. S. Buller and R. J. Collins, Meas. Sci. Technol. **21**, 012002 (2010).

[4]   S. Kück, A. L. Migdall, I. Pietro Degiovanni, and J. Y. Cheung, J. Mod. Opt. **59**, 1455 (2012).

[5]   A. Winden, M. Mikulics, D. Grützmacher, and H. Hardtdegen, Nanotechnology **24,** 405302 (2013).

[6]   M. Mikulics, A. Winden, M. Marso, A. Moonshiram, H. Lüth, D. Grützmacher, and H. Hardtdegen, Appl. Phys. Lett. **109**, 041103 (2016).

[7]   H. Hardtdegen, M. Hollfelder, R. Meyer, R. Carius, H. Münder, S. Frohnhoff, D. Szynka, and H. Lüth, J. Cryst. Growth **124**, 420 (1992).

[8]   Y. S. Cho, H. Hardtdegen, N. Kaluza, N. Thillosen, R. Steins, Z. Sofer, and H. Lüth, Phys. status solidi **3**, 1408 (2006).

[9]   A. Winden, M. Mikulics, T. Stoica, M. von der Ahe, G. Mussler, A. Haab, D. Grützmacher, and H. Hardtdegen, J. Cryst. Growth **370**, 336 (2013).

[10]   A. Winden, M. Mikulics, A. Haab, D. Grützmacher, and H. Hardtdegen, Jpn. J. Appl. Phys. **52**, 08JF05 (2013).

[11]   M. Mikulics and H. Hardtdegen, Patent specification, DE102012025088 (A1, 2012), WO2014094705 A1 (2013).

[12]   M. Mikulics, Y. C. Arango, A. Winden, R. Adam, A. Hardtdegen, D. Grützmacher, E. Plinski, D. Gregušová, J. Novák, P. Kordoš, A. Moonshiram, M. Marso, Z. Sofer, H. Lüth, and H. Hardtdegen, Appl. Phys. Lett. **108**, 061107 (2016).

[13]   H. Hardtdegen, M. Pristovsek, H. Menhal, J.-T. Zettler, W. Richter, and D. Schmitz, J. Cryst. Growth, **195**, 211 (1998).

[14]   H. Hardtdegen, N. Kaluza, R. Schmidt, R. Steins, E. V. Yakovlev, R. A. Talalaev, Y. N. Makarov, and J.-T. Zettler, Phys. Status Solidi **201**, 312 (2004).

[15]   H. Hardtdegen, N. Kaluza, R. Steins, R. Schmidt, K. Wirtz, E. V. Yakovlev, R. A. Talalaev, and Y. N. Makarov, J. Cryst. Growth **272**, 407 (2004).

[16]   H. Hardtdegen, N. Kaluza, R. Steins, P. Javorka, K. Wirtz, A. Alam, T. Schmitt, and R. Beccard, Phys. status solidi **202**, 744 (2005).

[17]   M. Mikulics and H. Hardtdegen, Nanotechnology **26**, 185302 (2015).

[18]   M. Mikulics, H. Hardtdegen, D. Gregušová, Z. Sofer, P. Šimek, S. Trellenkamp, D. Grützmacher, H. Lüth, P. Kordoš, and M. Marso, Semicond. Sci. Technol. **27**, 105008 (2012).

[19]   M. Mikulics, H. Hardtdegen, A. Winden, A. Fox, M. Marso, Z. Sofer, H. Lüth, D. Grützmacher, and P. Kordoš, Phys. status solidi **9**, 911 (2012).

[20]   M. Mikulics and H. Hardtdegen, Germany, patent specification, DE2012101617820120816 (2014).

ASDAM 2016, The 11th International Conference on Advanced Semiconductor
Devices And Microsystems, November 13-16, 2016, Smolenice, Slovakia

GaAs-based photodetector with applied PDMS membrane with photonic crystal in the surface

D. Pudiš[1], M. Tłaczała[2], L. Suslik[1], W. Dawidowski[2], B. Ściana[2], J. Kovac[3],
J. Kovac jr.[3], M. Goraus[1], P. Gašo[1], J. Ďurišová[1], I. Zborowska-Lindert[2]

1. Dept. of Physics, University of Žilina, Žilina, Slovakia
2. Faculty of Microsystem Electronics and Photonics, Wrocław University of Science and
Technology, Janiszewskiego 11/17, 50-372 Wrocław, Poland
3. Inst. of Electronics and Photonics, Slovak University of Technology, Bratislava, Slovakia
e-mail: pudis@fyzika.uniza.sk

*In this paper we present GaAs-based photodiode with implemented
polydimethylsiloxane (PDMS) membranes with patterned photonic crystal (PhC)
structures on surface. Two-dimensional (2D) PhC surface relief structures of
square symmetry were analyzed by atomic force microscope and optical effect of
2D PhC structure was investigated from goniophotometer photoresponse
measurements. Spatial modulation of light coupling with square symmetry was
documented for irradiation the GaAs-based photodiode with red light.*

1. Introduction

Unique features of photonic crystals (PhCs) started their employment for light
manipulation on the chip. Using the PhCs in conventional optic and optoelectronic devices
leads to interesting improvements of their optical properties. The PhC were successfully
implemented in different devices as light emitting diodes (LED) [1], lasers and optical
waveguides with photonic structure [2] and photodetectors with PhC [3]. Especially, effects
on the emission properties of LEDs are intensively studied from the point of view of an
improvement of light extraction efficiency and modification of radiation pattern. Wavelength
and directional selective optical properties may be also interesting in solar cells and
photodetectors [3].

Most of applications use the surface patterning of the LED chip [1]. Another treatment
uses the patterned polymer membranes from polydimethylsiloxane (PDMS), which can be
directly applied on the chip. The PDMS offers interesting elastic properties and is good
shaping by imprinting techniques [4]. Optically the PDMS shows high transparency in visible
range of spectrum. By its patterning and simple positioning on the device surface one can
achieve original optical properties of optoelectronic devices. It was shown, that
two-dimensional (2D) PhC PDMS structures applied in the LED surface can improve the light
extraction from the LED and modify a far-field pattern. We used the same method for
fabrication and application of PhC PDMS membranes on the photodiode chip.

In this paper, we propose concept of application and fabrication of PhC patterned on surface
of PDMS membranes for using in GaAs-based photodiode. By PDMS patterning and its
application on the photodiode we expect the modification of spatial light coupling resulting in
spatial modulation of photoresponse characteristics of GaAs-based photodiode.

For PhC PDMS membrane fabrication, we used interference laser lithography for thin
photoresist layer patterning in combination with PDMS embossing. 2D PhC surface relief

978-1-5090-3084-2/16 $31.00 © 2016 IEEE   33

structures of square symmetry were patterned in thin PDMS membranes and applied on the photodiode chip.

2. Experimental

For the implementation of PDMS membranes with surface PhC structure, the GaAs-based photodiode with InGaAsN active region was prepared. The epitaxial structure of GaAs-based photodiode was grown by atmospheric pressure metal organic vapor phase epitaxy (AP-MOVPE) using AIX 200R&D horizontal reactor on silicon doped GaAs substrate. The growth temperature was 585°C and 670°C for InGaAsN and GaAs epilayers, respectively. Detailed information about the growth process conditions and precursors were published in [5-7]. The active region of p-i-n GaAs-based device consist of undoped InGaAsN active layer sandwiched between silicon doped n-type GaAs buffer and Zn doped p-type InGaAsN layer, capped by heavily Zn doped p-type GaAs contact layer, as is shown in scheme in Fig. 1a. The MESA shape of the device was defined by optical lithography and wet chemical etching. Metallic p-type (Pt/Ti/Pt/Au) and n-type (AuGe/Ni/Au) contacts were deposited on the top of the structure around the MESA.

The patterned PDMS membranes were prepared by combination of interference laser lithography of thin photoresist layer and embossing process of liquid PDMS. The thin photoresist layer of AZ 5214E of thickness about 2 μm was spin coated on Si substrate using SPIN 150 coater with post-baking at 65 °C for 2 minutes and at 100 °C for 3 minutes to remove the solvent. For patterning, we exposed photoresist layer by one-dimensional optical field created by interference lithography in Mach-Zehnder configuration using Toptica Blue Mode laser emitted at wavelength of 403 nm. 2D PhC structure of square symmetry and period of 1600 nm was prepared by double exposure process at defined beam angle $2\theta = 14.5°$ with sample rotation $\alpha = 90°$ [8]. After exposure, the exposed photoresist sample was developed in AZ 400K developer for 10 s and rinsed in deionized water.

p+ GaAs:Zn	~ 40 nm
p InGaAsN:Zn	~ 30 nm
i InGaAsN	~ 130 nm
n GaAs:Si	~ 200 nm
substrate GaAs:Si	~350 μm

a)

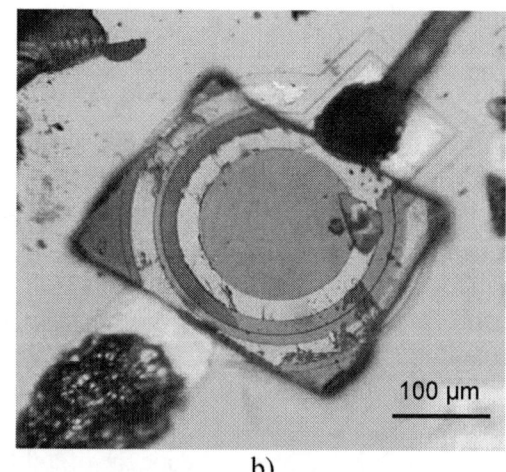

b)

Fig. 1. a) Scheme of the epitaxial structure of InGaAsN/GaAs photodiode and b) optical microscope image of PDMS membrane applied on the photodiode chip.

In the next step, liquid PDMS was prepared from components of Sylgard 184 elastomer and curing agent at ratio 10:1. For uniform surface, the 120 μm thin PDMS layer was spin coated at 2000 rpm on patterned photoresist sample. Subsequently, the sample with PDMS layer was cured for 45 min at 75°C. Finally, the cured PDMS membrane was mechanically separated from photoresist layer. In the next step, the 2D PhC PDMS membrane was directly applied on the surface of the photodiode chip with upward oriented patterned part as is shown in Fig. 1b. PDMS membranes adheres self on the photodiode chip. In the optical characterization we focused on far-field measurements to show the spatial modulation of light coupling into the photodiode chip.

3. Results

Embossed PDMS membranes were analyzed by atomic force microscope (AFM). Fig. 2 shows homogeneous planar PhC structure of embossed PDMS membrane with square symmetry. PDMS surface well reproduces the original photoresist master with period of 1610 nm and maximal depth of 420 nm as was confirmed from AFM. Thickness of PDMS membranes is app. 120 μm.

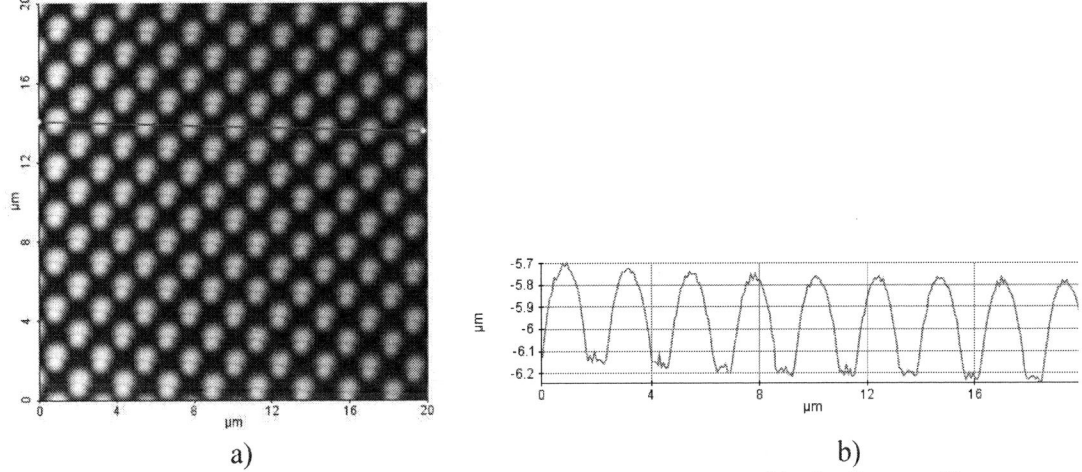

a)   b)

Fig. 2. a) AFM image of 2D PhC structure of square symmetry and b) its line profile.

PDMS membranes with membrane with PhC of period 1600 nm and square symmetry were applied on the photodiode surface. Effect of 2D PhC PDMS membranes on light coupling in photodiode was investigated. Photoresposne spatial diagrams of photodiode with PhC PDMS membrane was measured by goniophotometer using the irradiating by optical fiber from 2 cm distance and different spectral illuminations (LEDs emitting at central emission wavelength at 625 nm and 780 nm). This system enables measurement in spherical coordinates by moving the fiber in azimuthal and elevation angles. Measurement was performed in complete 3D space around the photodiode chip with 5 degree resolution. In measured 3D spatial photoresponse diagrams the photodiode with PhC structure shows spatial maxima given by light diffraction on PhC structure (Fig. 3a). It was documented also from cross sections where two different radiation wavelengths are compared. Shorter wavelength (625 nm) shows higher diffraction creating local maxima at higher angles (Fig. 3b). These characteristics document effect of patterned PhC PDMS membrane on spatial photoresponse of used photodiode.

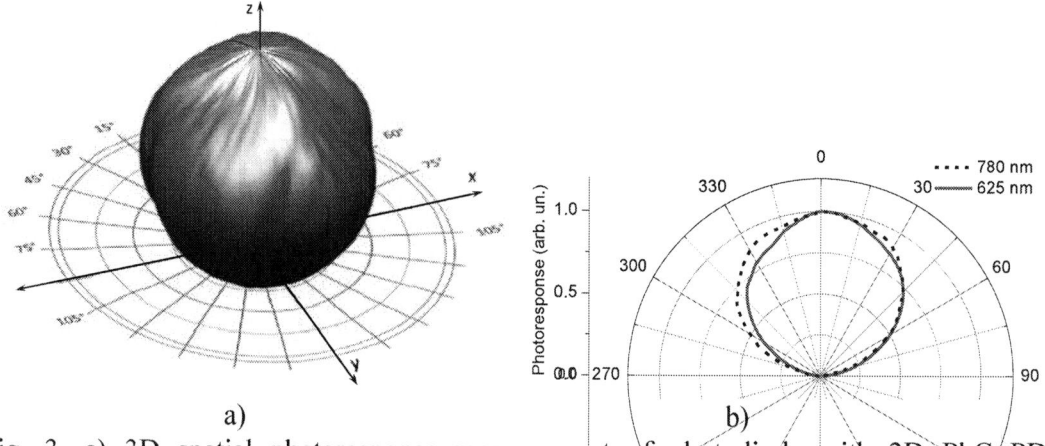

Fig. 3. a) 3D spatial photoresponse measurement of photodiode with 2D PhC PDMS membrane irradiated by LED source with central emission wavelength of 625 nm and b) cross-sections of radiation patterns for different wavelengths (625 and 780 nm).

4. Conclusions

In this paper we presented lithography process combined with embossing technique for fabrication of different PhC structures on surface of thin PDMS membranes. 2D PhC surface relief structures of period 1600 nm were patterned in thin PDMS membranes with depth up to 420 nm. From spatial photoresponse measurements the diffraction effect of PhC structure was clearly identified.

Results presented in this paper document the possibility of 2D PhC structures patterned on PDMS membranes, what may be highly attractive for improvement of optical properties of photodiodes and solar cells.

Acknowledgement

This work was supported by the Slovak National Grant Agency under the projects No. VEGA 1/0491/14 and 1/0278/15 and the Slovak Research and Development Agency under the project No. APVV 0395-12. This work was co-financed by Wrocław University of Science and Technology statutory grants.

References

[1]   E. Matioli, E. Rangel, M. Iza et al., *Appl Phys Lett* **96**, 031108, 2010.
[2]   A. Mekis, J. C. Chen, I. Kurland et al., *Phys Rev Lett* **77**, 3787, 1996.
[3]   J. K. Yang, M. K. Seo, I. K. Hwang et al., *Appl Phys Lett* **93**, 211103-1, 2008.
[4]   Nusil Silicone Technologies. Product Profile LS-6943.
       [Online], http://nusil.com/library /products/LS-6943P.pdf
[5]   W. Dawidowski et al., *Int J Electron Telecommun* **60**, 151, 2014.
[6]   B. Ściana et al., *Proc. SPIE* **8902**, 89020J, 2013.
[7]   W. Dawidowski et al., *Solid-State Electronics* **120**, 13, 2016.
[8]   N. D. Lai, W. P. Liang, J. H. Lin et al., *Opt Express* **13**, 9605, 2005.

ASDAM 2016, The 11th International Conference on Advanced Semiconductor Devices And Microsystems, November 13-16, 2016, Smolenice, Slovakia

Improving Light Extraction Efficiency of PhC LED

P. Hronec, J. Kováč, J. Škriniarová, J. Kováč jr., F. Uherek

Institute of Electronics and Photonics, Faculty of Electrical Engineering and Information Technology, Slovak University of Technology in Bratislava, Ilkovičova 3, 812 19 Bratislava, Slovak Republic,
e-mail: frantisek.uherek@stuba.sk

In this work we focus on improving light extraction efficiency of PhC LED. Studied LED structure was based on GaAs/AlGaAs material system. 1D PhC was patterned on the top of the LED structure with 3 different periods (500 nm, 600 nm, 800 nm). Several studies confirm improving light extraction efficiency of studied LED structure using PhC patterning. In this work, additional thin Au or ZnO layer was deposited on the top of the patterned PhC. Additional layers improved light extraction efficiency (LEE) for PhC LED. The best results were achieved for Au deposited PhC LED sample with LEE enhancement of 59%.

1. Introduction

In conventional semiconductor light emitting diodes (LEDs), the majority of generated light is trapped in high-refractive index confinement layers due to the total internal reflection at the semiconductor/air interface. It was confirmed that photonic crystal (PhC) structures are very useful in the surface of different semiconductors [1] for fabrication of semiconductor based PhC devices as LEDs with light emission enhancement [2] and photodetectors [3].

Many groups reported the usage of PhC patterning for light extraction efficiency (LEE) improvement of LEDs. For maximum extraction efficiency, there must be a strong coupling between the trapped waveguide modes and PhC structure [2, 4]. The period of patterned PhC should be related to the photonic band gap. Generally, the PhC period used in semiconductor based LEDs are about several hundreds of nanometers. Firstly, the depth of etching plays very important role in cross-coupling of trapped modes with the leaky Bloch modes associated with the PhC. Lewins et al. investigated in detail the effect of PhC depth on LED extraction efficiency [5]. Likewise the size of PhC holes, mostly known as fill factor, is related to the photonic band gap of PhC and determines the extraction efficiency. From the point of PhC shape and geometry, the wide variety of PhC and photonic quasi-crystals were investigated. Moreover, the choice of PhC geometry plays more important role in determining of far-field radiation pattern diagram of the LED. Photonic quasi-crystals with high level of symmetry produce less directional emission [6].

Such a type of PhCs needs to be defined by lithography process, achieving hundreds of nanometers resolution. PhC patterned LEDs in this paper were prepared by Electron Beam Direct Write Lithography (EBDWL). This technique allows precise shape and geometry patterning. Improvement of light extraction efficiency of LED with PhC prepared by this technique was studied in earlier papers [7]. In this paper, an additional improvement of light extraction efficiency was studied.

978-1-5090-3084-2/16 $31.00 © 2016 IEEE

2. Experimental

Studied patterned LED is $Al_{0.295}Ga_{0.705}As/GaAs$ based structure with $Al_{0.295}Ga_{0.705}As$ active region consisting of three GaAs quantum wells. The structure arrangement with layers thicknesses is shown in Fig. 1a. The steps for the LED production: upper ring metallization: AuBe (1% Be) alloy for p-type ohmic contact (upper contact), bottom metallization: AuGeNi (12% Ge, 1% Ni) alloy for n-type ohmic contact (bottom contact), contact annealing (at $420^{\circ}C$ for 2 min in forming gas) and MESA etching was performed by wet chemical etching (Fig 1b).

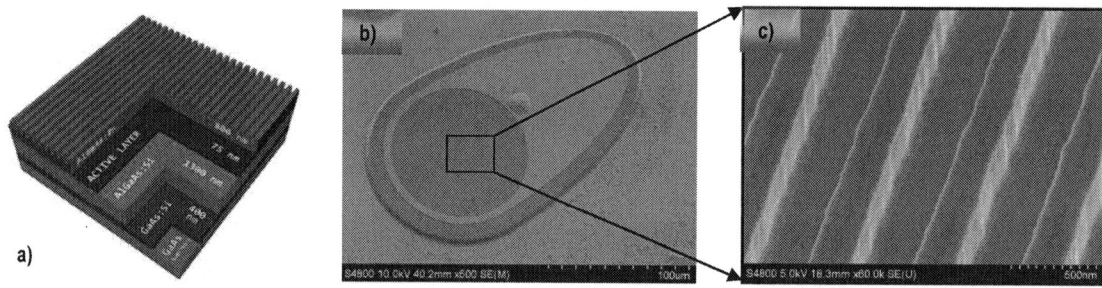

Fig. 1 The LED structure with 1D PhC structure in the LED surface a) schematic illustration with layer arrangement b) top view imaged by SEM microscope c) detail view of 1D PhC structure

We used optical and ion beam lithography techniques for patterning of the surface of GaAs/AlGaAs-based LEDs. The structure was patterned in the upper confinement AlGaAs layer. Surface of the 1D PhC structure was analyzed by employment of scanning electron microscope microscope (Fig 1c). The patterns in the XR 1541 negative e-beam resist have been exposed by the ZBA 21 variable shaped e-beam pattern generator using 40 kV accelerating voltage in 50 nm resolution mode. The resist was exposed with incrementing doses ranging from 260–500 μCcm^{-2} and its initial thickness was 400 nm.

3. Results

Three different sets of samples PhC LED based on AlGaAs/GaAs MQW LED structure were processed. 1D PhC prepared by EBDWL was patterned on the top of LED samples in each set. Additionally, in one set a thin Au layer was deposited on the top of patterned PhC and in the last set a thin ZnO layer was deposited instead of Au layer. Each set of samples composes of 1D PhC with 3 different periods (500 nm, 600 nm, 800 nm) with 4 samples from each of them.

To confirm LEE of studied PhC LED samples the light-current (L-I) characteristics of studied PhC LED samples were measured. To neglect heating of the sample L-I characteristics were measured in the current range of 1-10 mA. The central emitting wavelength of this LED is about 845 nm. The measured L-I characteristics for all sets of samples are shown in Fig. 2. For PhC LED with ~ 100nm thick ZnO layer the best LEE enhancement was achieved for PhC period of 500 nm. Further, the best results at all were achieved for PhC LED with Au layer with maximum LEE of 59%. The improvement of LEE enhancement of this sample over bare PhC LED can be assigned to decrease ohmic contact resistance as well as improved LED lateral current distribution. Fig. 3 shows average LEE enhancement for investigated structures for given PhC periods. Evaluated results were interpolated by second order polynomial equation. For bare PhC LED as well as PhC LED with thin (>5nm) Au layer the best LEE enhancement was achieved for PhC period of 600 nm. Interpolated dependency of LEE on PhC period for PhC LED samples with ZnO layer doesn't show any local maximum but there is a local minimum for LEE for the PhC period of 730 nm.

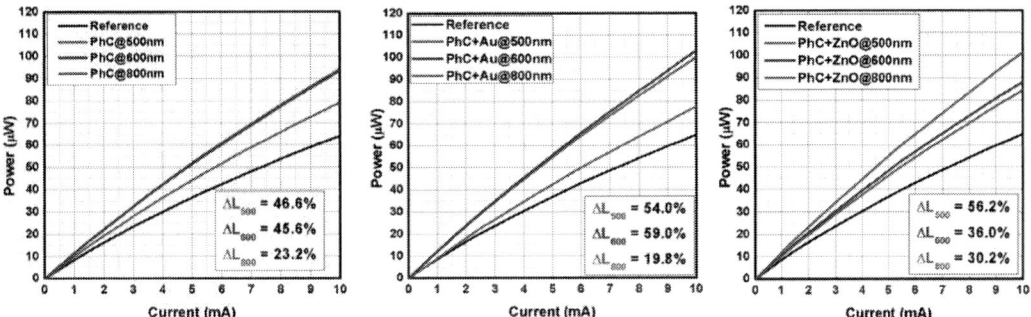

Fig. 2   Light-current characteristics of LEDs with various PhC period prepared by EBDWL without conductive layer (left), covered by thin Au layer (middle), and covered by thin ZnO layer (right).

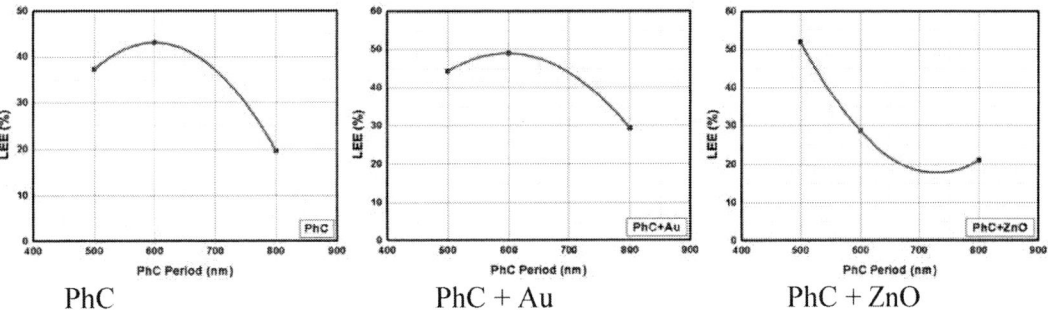

PhC   PhC + Au   PhC + ZnO

Fig. 3   Dependency of light extraction enhancement on PhC period for LED samples with different type of top layer. Bare PhC (left), PhC with Au layer (middle), and PhC with ZnO layer (right).

4.   Conclusions

EBDWL was used to prepare 1D PhC LED samples with bare PhC. Additionally, in one set of samples a thin Au and ZnO layer was deposited on the top of patterned PhC. L-I characteristics showed highest LEE enhancement for 600 nm PhC period in case of bare and Au deposited PhC and for 500 nm PhC period in case of ZnO deposited PhC. The best results were achieved for Au deposited PhC LED sample with LEE enhancement of 59%. This can be referred to decreased ohmic contact resistance as well as improved LED lateral current distribution.

978-1-5090-3084-2/16 $31.00 © 2016 IEEE   39

Acknowledgement

This work was supported by the APVV-0395-12, APVV 14-0297 and VEGA-1/0739/16 projects. We thank our colleagues from Institute of Informatics, Slovak Academy of Sciences, Dúbravská cesta 9, 845 07 Bratislava, Slovak Republic who provided Electron Beam Direct Write (EBDW) lithography.

References

[1] H. Benisty, C. Weisbuch, D. Labilloy, M Rattier, *Photonic crystals in two-dimensions based on semiconductors: fabrication, physics and technology*, Appl. Surf. Sci., **164** (2000) 205–218.

[2] E. Matioli , E. Rangel, M. Iza, B. Fleury, N. Pfaff, J. Speck, E. Hu, C. Weisbuch, *High extraction efficiency light-emitting diodes based on embedded air-gap photonic-crystals*, Appl. Phys. Lett. **96** (2010) 031108.

[3] M. Malekmohammad, M. Soltanolkotabi, R. Asadi, M.H. Naderi, A. Erfanian, M. Zahedinejad, S. Bagheri, M. Khaje, *Hybrid structure for efficiency enhancement of photodetectors*, Appl. Surf. Sci., **264** (2013) 1-6.

[4] M. D. B. Charlton, P. A. Shields, D. W. E. Allsop, W. N. Wang, *High-efficiency photonic quasi-crystal light emitting diodes incorporating buried photonic crystal structures*, Proc. SPIE 7784 (2010) 778407.

[5] C. J. Lewins, E. D. Le Boulbar, S. M. Lis, P. R. Edwards, R. W. Martin, P. A. Shields, D. W. E. Allsopp, *Strong Photonic Crystal Behavior in Regular Arrays of Core-Shell and Quantum Disc InGaN/GaN Nanorod LEDs*, J. Appl. Phys. **116** (2014) 044305.

[6] M. D. B. Charlton, P. A. Shields, D. W. E. Allsop, W. N Wang, *High-efficiency photonic quasi-crystal light emitting diodes incorporating burier photonic crystal*, Proc. of SPIE7784 (2010) 778407.

[7] P. Hronec, A. Kuzma, J. Škriniarová, J. Kováč, A. Benčurová, Š. Haščík, P. Nemec, *Optical Properties of LEDs with Patterned 1D Photonic Crystal*, Proc. SPIE 9556, Nanoengineering: Fabrication, Properties, Optics, and Devices XII, 2015

ASDAM 2016, The 11th International Conference on Advanced Semiconductor Devices And Microsystems, November 13-16, 2016, Smolenice, Slovakia

Photoluminescence of InGaN/GaN MQW structures - technological aspects

A. Hospodková, J. Pangrác, J. Oswald, K. Kuldová, M. Zíková, and E. Hulicius

Institute of Physics, CAS, v.v.i.,
162 00, Cukrovarnická 10, Prague 6, Czech Republic
e-mail: hospodko@fzu.cz

In this work results obtained on several types of InGaN/GaN multiple quantum well (MQW) scintillator structures are presented. Luminescence properties of scintillator structures with different number of QWs and different growth rate of QWs were measured. We show that the growth rate and QW number are very important parameters to increase the QW excitonic luminescence. Photoluminescence and cathodoluminescence are compared and discussed.

1. Introduction

Large band gap semiconductors such as GaN or ZnO are suitable for scintillator and detector structures for ionizing radiation detection. While ZnO is used in scintillators for several decades, the promising application of GaN epitaxial layers in scintillator structures has attracted scientific attention in last ten years [1]. GaN can be prepared with higher crystallographic quality and homogeneous epitaxial layers, which brings an advantage of better emission homogeneity over large area, high signal to noise ratio and narrower spectral range in comparison to ZnO. GaN has also an advantage of high radiation resistance. So the detectors of ionizing radiation and scintillators seem to be a new and perspective application of nitride semiconductors [2].

Similarly to LED, scintillator efficiency can be significantly improved by incorporation of InGaN/GaN multiple quantum well (MQW) structures into the active region [3]. However, the design of scintillator structures obeys different design requirements, like higher number of QWs, only one type of doping, considerably thicker active part of the structure or proper placement of active region with respect to the penetration depth of detected ionizing radiation. Some aspects of scintillator design can take advantage of well-developed LED technology, such as improvements of QW design for enhancement of electron-hole wave function overlap and optimization of buffer/nucleation/coalescence layer growth.

In this work two types of scintillator structures with 10 and 30 QWs and different QW growth rate are studied by photoluminescence (PL) and cathodoluminescence (CL). We try to improve the scintillator properties by changing of technological parameters during the structure growth.

2. Experimental

All structures were prepared on Aixtron 3x2 CCS MOVPE system equipped by Laytec EpiCurveTT apparatus for in situ measurement of reflectivity and true wafer temperature. Trimethylgallium (TMGa) and ammonia (NH_3) were used as precursors with a hydrogen carrier gas for the growth of buffer layers, triethylgallium (TEGa), trimethylindium (TMIn) and NH_3 with nitrogen carrier gas were used for the growth of MQW region including barriers. Sapphire substrate with c-plane orientation was baked out at 1045 °C for 300 s and afterwards nitrified by the NH_3 injection at 528 °C, which was also the temperature used for

978-1-5090-3084-2/16 $31.00 © 2016 IEEE

the nucleation layer growth. After the coalescence, the growth was finished by 2.4 μm of GaN high temperature buffer layer grown under the same growth conditions, last 300 nm of the buffer layer was n-type doped using SiH4. The growth parameters, which were kept fixed for the growth of the buffer layer, are growth temperature 1036 °C, reactor pressure 250 mbar, total flow through the reactor 8 slpm, TMGa flow $1.1 \cdot 10^{-4}$ mol/min and V/III ratio 1286. Several parameters of the MQW growth were changed to investigate the influence of growth parameters on the PL properties of InGaN/GaN MQW. The changed parameters were a growth rate and number of QW layers. Growth parameters of GaN barriers were kept constant with the growth temperature 800 °C, reactor pressure 400 mbar, growth rate 0.1 nm/s and TEGa flow rate $8.38 \cdot 10^{-6}$ mol/min.

The PL spectra were measured at RT using SDL1 monochromator by semiconductor laser LD375 $\lambda = 375$ nm. Due to the different decay times of excitonic and defect luminescence bands and the dependence of peak intensity of particular luminescence peak on intensity of excitation [3] the excitation intensity was kept fixed for all measurements, so that all PL results are comparable.

3. Results and discussion

In this work we have concentrated our effort on increase of the intensity of fast excitonic QW emission and decrease of the luminescence of QW defect band, which has slower luminescence response and is not desired for fast scintillator applications. Two possible approaches are shown in this work: increasing the number of InGaN QWs and decreasing the QW growth rate. Both approaches have surprisingly mutual relation.

3.1 Number of QW layers

We have observed that increasing number of QW significantly increases the intensity of excitonic PL maximum, while the intensity of defect band luminescence remains almost unaffected (Fig.1) or in some cases even decreased with increased number of QWs.

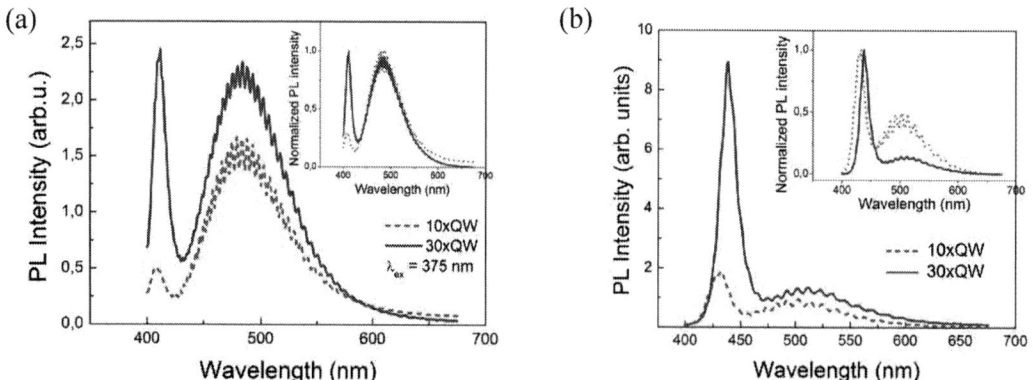

Fig. 1. Comparison of PL spectra of samples with 10 and 30 QWs prepared with (a) faster and (b) slower growth rate of InGaN QW.

The possible explanation could be that the defect luminescence originates dominantly from the deepest or from the top QWs. In this case only the excitonic peak would be enhanced with the number of QW layers.

To elucidate the space origin of the defect band luminescence, we have measured the samples with PL results presented in Fig. 1(a) also by CL. An advantage of CL is partial

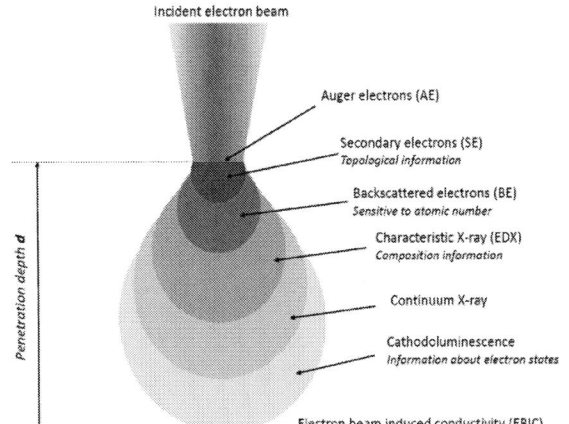

Fig. 2 Scheme of different type of radiation response to primary electron beam penetrating to different depth.

spatial resolution of luminescence, see Fig. 2. The penetration depth can be controlled by incident electron energy according to the formula $d = 10.46\ U^{1.68}$ (where U is acceleration voltage in kV and d is penetration depth in nm) [4].

CL spectra were measured for different penetration depth a) 156 nm (electron energy 5 keV, penetration under the 10 QW active region and into the middle of the 30 QW active region), see Fig. 3(a) and b) penetration depth 420 nm (electron energy 9 keV, penetration below both types of active region), see Fig. 3(b).

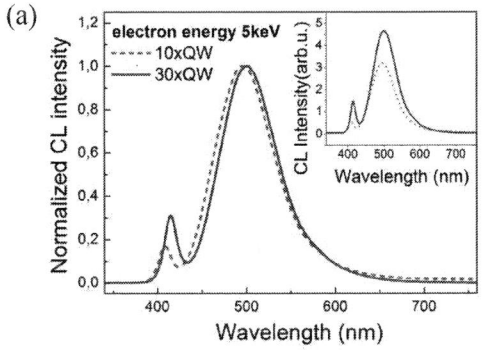
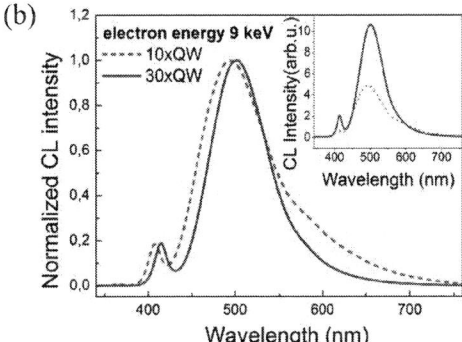

Fig. 3. Comparison of CL spectra of samples with 30 QW for incident electron energy (a) 5 keV and (b) 9 keV.

It can be seen in inset of Fig. 3(a) that for the penetration depth around 150 nm the excitonic maximum is three times more intensive for sample with 30 QW active region than for 10 QWs, which is much lower than in the case of PL. The intensity of defect band is lower for 30 QWs. The enhancement of both maxima with number of QW is the same when excitation of luminescence takes place in GaN bellow the MQW active region, see Fig. 3(b). Since PL excitation is more effective in the upper part of the active region, from the comparison of PL and CL with different penetration depths we can conclude that deeper part of MQW active region has higher contribution to defect band luminescence.

3.2 Growth rate of QW layers

We have found that decreased InGaN QW growth rate is very efficient way to increase the excitonic PL maximum efficiency. However, excitonic luminescence enhancement was more significant for samples with higher number of QWs in active region, see Fig. 4. This phenomenon may be attributed to improved crystallographic quality of upper QW layers in the case of higher QW number and lower growth rate of InGaN QWs. The defect band luminescence is suppressed with decreasing the growth rate.

978-1-5090-3084-2/16 $31.00 © 2016 IEEE

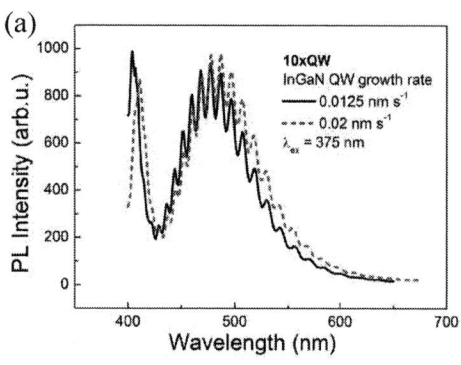 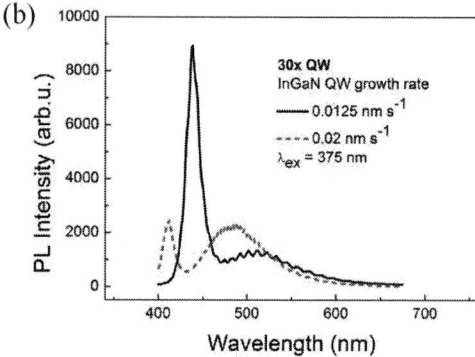

Fig. 4. Comparison of PL spectra of samples with different growth rates for samples with (a) 10 and (b) 30 InGaN QWs in active region.

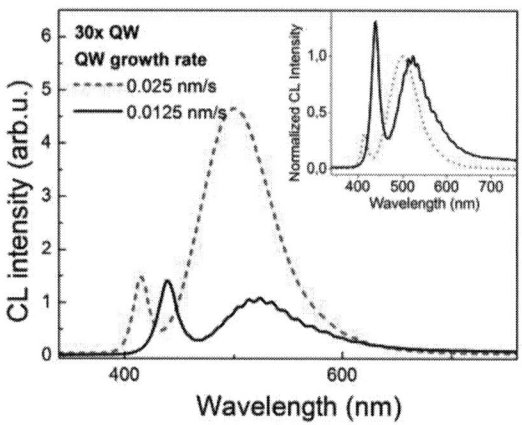

Fig. 5. CL spectra for 5 kev electrons of two samples with 30 QWs prepared with different growth rates of InGaN QW.

While for the PL of samples with 30 QWs the excitonic luminescence was enhanced, in the case of CL with penetration depth around 150 nm the excitonic luminescence remained unchanged, but the defect band was significantly suppressed with decreased growth rate, see Fig. 5. Since this type of excitation is most efficient for QWs in the central part of the active region, we can suppose that the suppression of the defect band by the slower growth rate is very efficient for these types of QWs.

4. Conclusions

We have shown that increasing number of QWs can significantly improve the luminescence efficiency of scintillator not only because of increasing the thickness of the active region, but also by improving the quality and luminescence efficiency of upper QWs. Further improvement can be achieved by decreasing the InGaN QW growth rate, which is even more efficient for structures with higher number of QWs in active region.

Acknowledgement

The authors acknowledge support from NPU LO1603 – ASTRANIT and by the GACR project no. 16-15569S.

References

[1]   P. Pittet et al., *Opt. Mater.* **31**, 1421, 2009.
[2]   G. Balakrishnan, *Nanotechnology* **26**, (viewpoint) 09050, 2015.
[3]   A. Hospodková et al., *Nanotechnology* **25**, 455501, 2014.
[4]   O. Kurniawan, and V.L.S. Ong, *Scanning* **29**, 280, 2007.

ASDAM 2016, The 11th International Conference on Advanced Semiconductor
Devices And Microsystems, November 13-16, 2016, Smolenice, Slovakia

ELECTRICAL CHARACTERISATION OF MIS PHOTOANODES ANNEALED UNDER DIFFERENT CONDITIONS FOR SOLAR FUEL GENERATION

M. Mikolášek[1*], J. Racko[1], V. Řeháček[1], L. Harmatha[1],
M. Ťapajna[2], and K. Fröhlich[2]

[1]*Institute of Electronics and Photonics, Slovak University of Technology,
Ilkovičova 3, 812 19 Bratislava, Slovakia*
[2]*Institute of Electrical Engineering, Slovak Academy of Sciences,
Dúbravská cesta 9, 841 04 Bratislava, Slovakia*
*e-mail: miroslav.mikolasek@stuba.sk

The impact of post-deposition annealing of silicon based metal-insulator-semiconductor structures with SiO_2/TiO_2 dielectric layers is inspected by means of electrical characterisation. It is shown that annealing at 400 °C in the forming gas results into the photovoltaic response of MIS structure. Such behaviour is facilitated by the tunnelling of holes through SiO_2. Annealing at higher temperature or in the air ambient exhibit negligible photovoltaic response due to grown of additional SiO_2 at the interface and/or low interface quality.

1. Introduction

The photo-electrochemical (PEC) conversion of solar energy into the chemical energy of the solar fuels such as hydrogen or hydrocarbon address several energy and sustainability challenges by providing a way of storing solar energy when it is available for use when it is needed and by providing source of energy which can replace fossil fuels. The PEC hydrogen generation is based on water splitting reactions and require structures with high photo-generation of carriers and high stability in the water electrolyte [1]. Metal insulator semiconductor (MIS) structure consisting of silicon, which provide high photocurrent and metal oxide, which provide protection against the harsh environment is an attractive concept with perspective to attain high performance and stability [2-4]. This paper is focused on electrical characterization of $Ni/TiO_2/SiO_2/Si(n)$ MIS photoanodes prepared by atomic layer deposition of TiO_2 and provide insight on the impact of different annealing conditions on the carrier transport through the TiO_2/SiO_2 layers with the perspective for PEC applications.

2. Experimental

The MIS structures were prepared by atomic layer deposition (ALD) of 10.3 nm thin TiO_2 layer on n-type silicon substrate (n-Si) with native SiO_2 layer. The silicon wafer with orientation (100), thickness of 625 μm and resistance of 5-8 Ωcm was used as a substrate. The TiO_2 film was grown by thermal ALD at 150 °C using titanium isopropoxide and water as precursor and reactant, respectively. Three annealing conditions were used for 60 minutes long post-deposition processing of $TiO_2/SiO_2/n$-Si MIS structures i) forming gas annealing at 400 °C (FG400), ii) forming gas annealing at 600 °C (FG600) and iii) air gas annealing at 400 °C (ATM400). A mixture of 10% H_2 + 90% N_2 was used as a forming gas. After annealing, full area Al bottom contact and Ni top contacts with area 0.25 mm^2 were prepared by evaporation. The electrical behavior of MIS structures were studied by capacitance-voltage (C-V) and impedance spectroscopy (IS) by using LCR AGILENT 4284A. Current-voltage (I-V) measurements were carried out under dark and light of halogen lamp by Keithley 2612.

978-1-5090-3084-2/16 $31.00 © 2016 IEEE   45

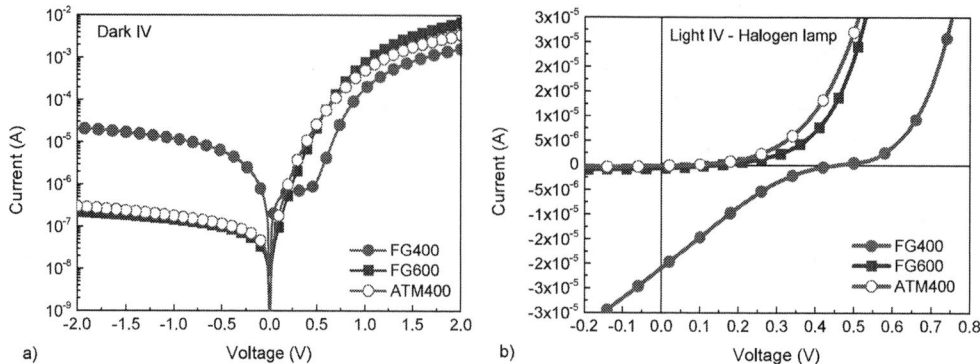

Fig. 1 a) Dark *I-V* and b) light *I-V* characteristics of Ni/TiO$_2$/SiO$_2$/Si(n) MIS structures annealed at 400 and 600 °C in forming gas (FGA) and at 400 °C in the air.

3. Results and discussion

Figs. 1a and b show dark and light *I-V* curves of FG400, FG600, ATM400 samples, respectively. Comparing such samples, only FG400 exhibits photovoltaic response with open circuit voltage, $V_{OC} \sim 0.45$ V and short circuit current of $J_{SC} \sim 20$ µA. It is necessary to stress that due to not proper solar grid, these values are used only for comparison reason. Two conditions are required for manifestation of the photovoltaic behaviour: a) low recombination of photo-generated carriers (in our case holes), which means low density of defect states at the interface and b) effective transport of these carriers through dielectric layers of the MIS sample. According to our observation, only structure FG400 exhibits conditions that meet such requirements. The difference in the current transport behaviour is further studied through the dark *I-V* measurements (Fig. 1a). While all samples exhibit rectification behaviour, only sample FG400 shows two regions in the forward biased semi-logarithmic *I-V* representation. Such regions indicate the presence of two transport mechanisms in the structure. High ideality factor of $n = 26$ in the linear low voltage region of the sample FG400 suggests presence of tunnelling. While the TiO$_2$ has much higher thickness of 10.3 nm as well as higher conductance due to the oxygen vacancies presented in the film compared to SiO$_2$, we can assume that such tunnelling facilitate transport of carriers through a thin SiO$_2$ layer at the interface. Other transport mechanism for holes is assumed to be assisted by the defects in the TiO$_2$ layer. In case of samples FG600 and ATM400, negligible photovoltaic response indicates insufficient transport of photo-generated holes and/or high recombination at the interface.

Capacitance-voltage (*C-V*) and impedance characterisation (IS) were carried out to further elucidate the transport properties in the samples. Figs. 2a-f show *C-V* curves measured as a function of frequency and Nyquist plots measured as a function of temperature. Comparing *C-V* curves, the increase of capacitance upon the decrease of frequency indicate high conductance of the dielectric layers for FG400 and ATM400. On the other hand, sample FG600 exhibits saturation of capacitance and indicate dielectric behaviour of the SiO$_2$/TiO$_2$ layers. This is reflected also in the Nyquist plot in Fig.2f showing high resistance behaviour. For sample FG400, we can observe increase of the capacitance upon the decrease of the frequency in the inversion region. Such behaviour is linked with the minority carrier and indicate better interface properties of FG400 compared to FG600 and ATM400.

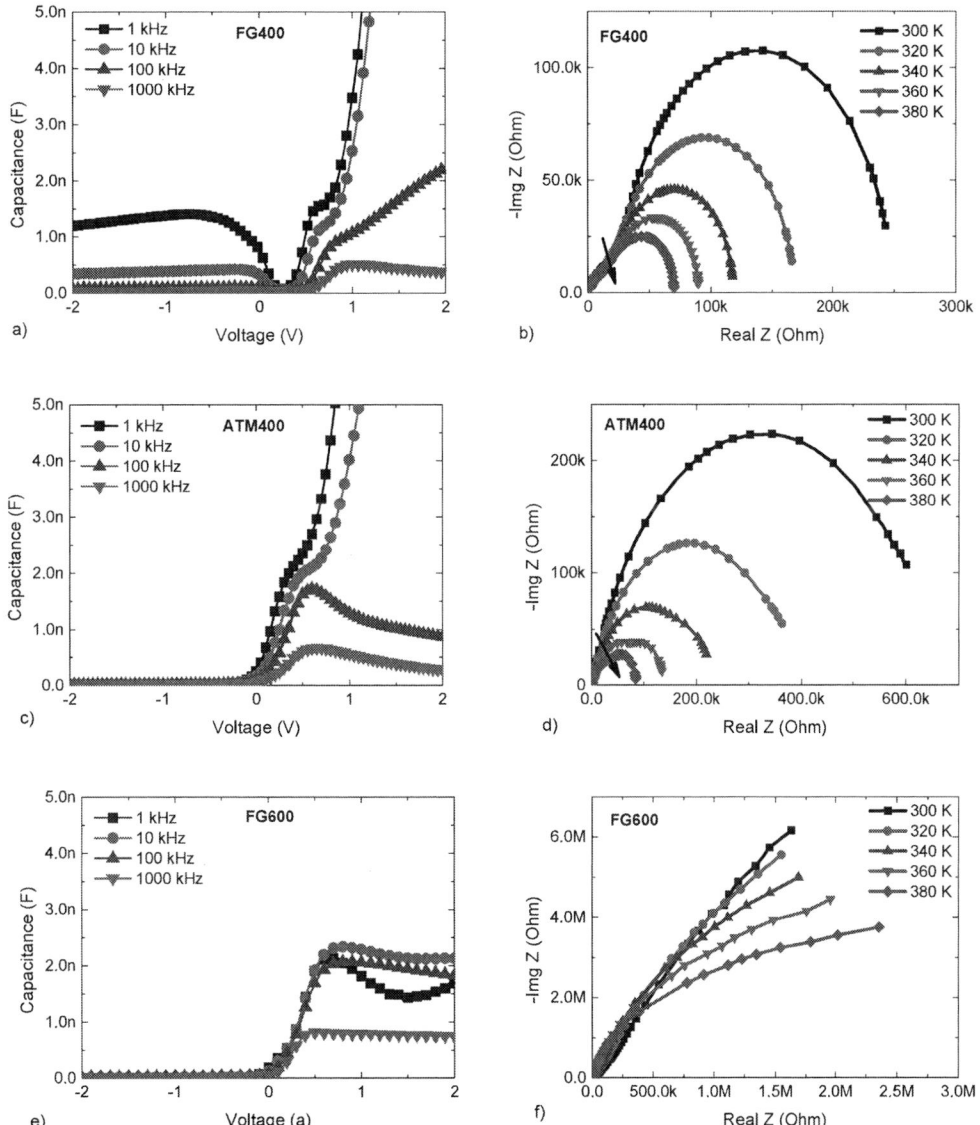

Fig. 2 *C-V* characteristics measured at varied frequencies (a, c, e) and Nyquist plots measured at varied temperatures (b, d, f) of samples FG400, FG600 and ATM400.

It can be concluded that annealing at 400 °C in the gas ambient containing hydrogen allows to maintain low defect states at the interface. On the other hand, increase of the annealing temperature or change of the annealing gas do not provide interface with low defect state density.

Nyquist data presented in Figs. 2b and d can be fitted by equivalent circuit with two parallel RC elements and one CPE-R element connected in series showing three relaxation processes. CPE represents constant phase element and, in our case, shows capacitance like behaviour. Detailed study of impedance characterisation will be presented elsewhere. The main difference between the Nyquist plot of FG400 and ATM400 stems from low temperature dependence of the high frequency data for FG400. This is in accordance with the assumed tunnelling transport, which facilitate transport of carriers in sample FG400. In addition, IS analysis allowed to estimate the thickness of SiO_2 layers, showing increase of SiO_2 layer thickness from ~2 nm for sample FG400 to ~4 nm for sample ATM400. We can suppose that annealing of samples in the air ambient allows diffusion of oxygen to the interface and grown of additional SiO_2. Similar behaviour was observed in [5], where annealing of 100-nm thick TiO_2 layer in the air ambient led to the grown of SiO_2 interlayer. Thicker SiO_2 layer suppress tunnelling of carriers and together with defect states at the interface can be responsible for negligible photovoltaic response of sample ATM400.

4. Conclusion

This paper shows electrical characterisation of MIS photoanodes with SiO_2 and TiO_2 dielectric layers annealed at different ambient and temperature conditions. Annealing in the forming gas at 400 °C revealed the highest photovoltaic response, which was attributed to both low defect states at the interface and tunnelling of photo-generated holes through the dielectric layer. Annealing in the forming gas at 600 °C or in the air at 400 °C revealed low interface quality resulting into negligible photovoltaic response. Moreover, impedance characterisation shows increase of the SiO_2 interlayer upon the annealing in the air ambient, which suppressed tunnelling through SiO_2. Based on this study, the post-deposition annealing in the FG at 400 °C should result in the best photo-electrochemical response of MIS base PEC photoanodes.

Acknowledgement

This work was supported by the Scientific Grant Agency of the Ministry of Education of the Slovak Republic and of the Slovak Academy of Sciences VEGA 1/0651/16 and 2/0138/14. We thank Denis Hruban and Pavol Dohanyos-Zelenyanszki for help with electrical characterisation of samples and Martin Weis for contacts preparation.

References

[1] Y. W. Chen, J. D. Prange, S. Dühnen, Y. Park, M. Gunji, Ch. ED Chidsey, and P. C. McIntyre, *Nature materials,* **10**(7), 539-544, 2011.

[2] H. Shu, N. S. Lewis, J. W. Ager, J. Yang, J. R. McKone, and N.C. Strandwitz, *The Journal of Physical Chemistry C,* **119**(43), 24201-24228, 2015.

[3] H. Shu, M. H. Richter, M. F. Lichterman, J. Beardslee, T. Mayer, B.S. Brunschwig, and N. S. Lewis, *The Journal of Physical Chemistry C,* **120**(6), 3117-3129, 2016.

[4] P.F. Satterthwaite, A.G. Scheuermann, P.K. Hurley, C.E. Chidsey, and P.C. McIntyre, *ACS applied materials & interfaces*, **8**(20), 13140-13149, 2016

[5] M. McDowell, M. F. Lichterman, A. I. Carim, R. Liu, S. Hu, B. S. Brunschwig, and N. S. Lewis, *ACS applied materials & interfaces* **7**(28), 15189-15199, 2015.

ASDAM 2016, The 11th International Conference on Advanced Semiconductor Devices And Microsystems, November 13-16, 2016, Smolenice, Slovakia

Electro-Physical 2-D/3-D TCAD Analysis of reduce the temperature-related negative effects on Silicon Heterojunction Solar Cell for Concentrated Photovoltaic

P. Príbitný, M. Mikolášek, J. Marek, A. Chvála and D. Donoval

Institute of Electronics and Photonics, Slovak University of Technology, Bratislava, Slovakia,

Ilkovičova 3, 812 19 Bratislava 1

e-mail: patrik.pribytny@stuba.sk

This study attempts to analyze a small-area silicon solar cell, which is designed for operation under medium concentration. By using the TCAD [1] tool and by adopting calibrated physical models, the front contact grid of the cell has been optimized to operate in the range from 100–200 suns. Furthermore, a comparison between different geometric grids has been discussed. The main figures of merit of the solar cell have been calculated in the 1–300 suns concentration range. This study attempts to analyze the temperature and self-heating effect by using 3-D simulation methodology and introduces a method to reduce the temperature-related negative effects on solar cells. Moreover, a new concentrator system for indoor cell characterization has been outlined. The analysis of optical and electrical behavior can prove to be helpful during the design and optimization of parameters and geometry from semiconductor layers, metallization, package, and up to textured surface.

1. Introduction

Over the past few years, the photovoltaic industry has been able to significantly reduce production costs, thereby making grid parity closer and closer. At the same time, the market continues to push even further towards photovoltaic products with higher (alternative: increased) efficiency and lower production costs. There are several main factors behind this growing interest. Firstly: the economic advantage of reducing the required amount of cell material per watt has been achieved; and secondly: the fact that the exploitation of high-efficiency and custom-designed cells under concentrating light, which are still too expensive to be used in user applications. The latter group includes so-called high-efficiency solar cells, which are the focus of this study. Amorphous silicon is still an interesting material for developing Concentration Photovoltaic cells working at low and medium concentration range. In this work we describe the modeling, design and functional characterization of a small-area silicon solar cell, which is suitable for Concentration Photovoltaic cell applications up to 200 suns. Three-dimensional (3-D) numerical simulations with calibrated physical models have been performed for both the cell design and a deep understanding of its performance. We used a full 3-D approach, where a global model includes a part of a solar cell with a textured surface (Fig. 1). Due to device complexity, many of them cannot be simulated in the full 3-D setup within a reasonable amount of time. Therefore, derived solutions are proposed, which are based on the combined-mode setup coupling the 3-D model of the solar cell to a "standard" TCAD model of the active device. This approach involves the coupling of optical and electro-physical simulations in the package.

978-1-5090-3084-2/16 $31.00 © 2016 IEEE

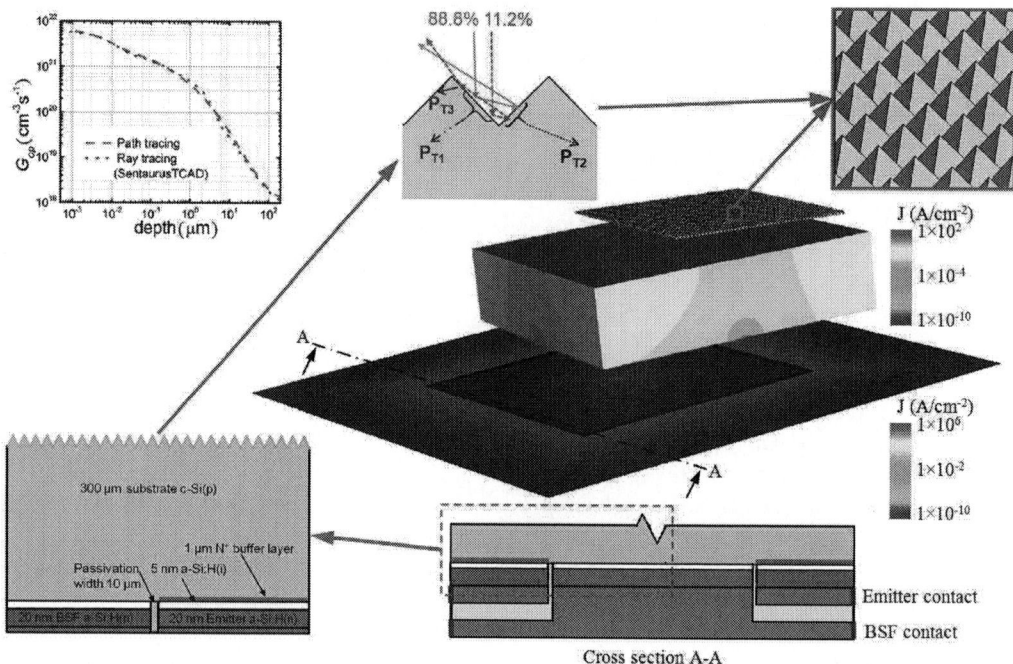

Fig. 1 Full 3-D approach, where global model includes a part of solar cell with textured surface. Cross-section of the simulated structure of BC-SHJ with implanted buffer layer. The rear a-Si:H(p) p-contact is a point contact surrounded by an a-Si:H(n) emitter [2].

3. Simulation methodologies

In this paper, a 3-D simulation analysis of the solar cell includes a complete description of the semiconductor structure in terms of optical profile, and an electro-physical modeling of the device using drift-diffusion transport. We present a 3-D simulation study of BC-SJH on p-type substrate with an incorporated implanted N+ buffer layer in the emitter structure. The a-Si:H(p) p-contact is formed as a point contact and the a-Si:H(n) emitter represents the area around this point contact. The a-Si:H(n)/c-Si(p) emitter interface is passivated by a 5 nm thick layer of a-Si:H(i). A 1 µm thick implanted N^+ buffer layer is formed in the silicon substrate of the emitter interface. Thus the layer sequence of the emitter structure is as follows: a-Si:H(n)/a-Si:H(i)/c-Si(N^+)/c-Si(p). In the sequel this emitter structure is labeled as a-Si:H(n)/c-Si(p) for simplicity reasons. The full 3-D modeling provides accurate results (parameters of solar cell) and a complex interaction of current distribution on the device electrical behavior. However, such an approach can be applied to the part of cell alone, but it is difficult to apply to the full system. Building a TCAD model from a complex device design (thousands of cells or stripes) would easily lead to several millions of mesh nodes without even getting a sufficiently accurate grid. To bring local generated current information on the cell into the electro-physical simulation of the device, a possible solution is the splitting of the single node into several ones, each of them being connected to a single 2-D or 3-D solar cell. The partitioning of the original problem into a multiple mixed-mode approach seems to be the most attractive one, since it couples the accuracy of TCAD simulation of the device embedded into a solar cell and a short simulation time, while preserving information on the non-uniformity of generated current with self-consistent coupling to TCAD elementary devices. Specifically, the simulations have allowed the development and optimization of a front contact grid scheme and the design of the cell operating under medium concentration. [3]

In our new setup, the single node is discretized into three nodes: one is placed on the surface, one fills the light area, and one is located at the interface between contact and silicon. Therefore, the simulation requires solving the drift-diffusion and the Poisson equations in three solar cells instead of one, and carrier continuity equation is solved in the solar cell as previously.

4.   Solar cell structure and problem description

Under concentrated light, the cell has been characterized within the range 1–300 suns, by using the experimental setup and conditions and by considering the AM1.5D spectrum. During the measurements, the back-side, which holds the test cell, is kept at a constant temperature of 298 K. However, due to the cell self-heating the effective cell temperature is higher. The numerical simulations take into account the self-heating effect. Fig. 2 shows the calculated V_{OC} versus sun concentration (SC) in the case of both self-heating and constant temperature analyses, compared with experimental data. Assuming a constant temperature, V_{OC} is expected to increase almost logarithmically with irradiance. On the other hand, in the case of self-heating assumption, V_{OC} degrades with SC approximately above 100 suns, mainly due to the increase of the saturation current. As a first approximation the short-circuit current density J_{SC} grows linearly with SC. However, simulated data confirm a super-linear trend of J_{SC} at high SC. The simulated *FF*, equal to 83.7% at 1-sun, is reduced by the enhanced resistive power loss as the irradiance increases (Fig. 3) and it is approximately 74% at 100 suns. As a trade-off between the increase of both J and V_{OC} and the reduction of FF with increasing irradiance, a maximum efficiency of about 22.0% has been measured at about 60 suns (Fig. 3). At this concentration factor, the calculated effective cell temperature due to self-heating is 310 K. A plateau of the efficiency is observed within the SC -range 60–100 suns. In the case of constant temperature assumption, the maximum simulated efficiency is 22.9% at 84 suns as shown in Fig. 3. The optimum SC is higher than the one calculated under self-heating, since no V_{OC} degradation occurs at high light irradiation. [4-6]

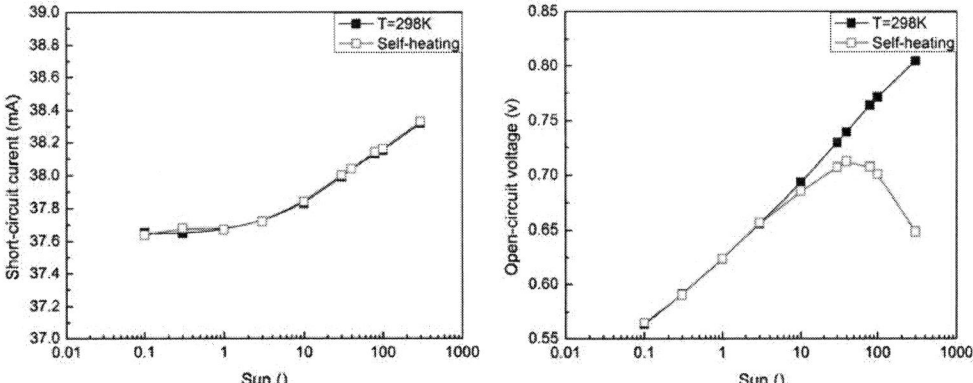

Fig. 2 Simulated short-circuit J_{SC} open-circuit voltage V_{OC} versus sun. Simulations have been performed in the case of constant temperature at T=298 K and under self-heating assumption.

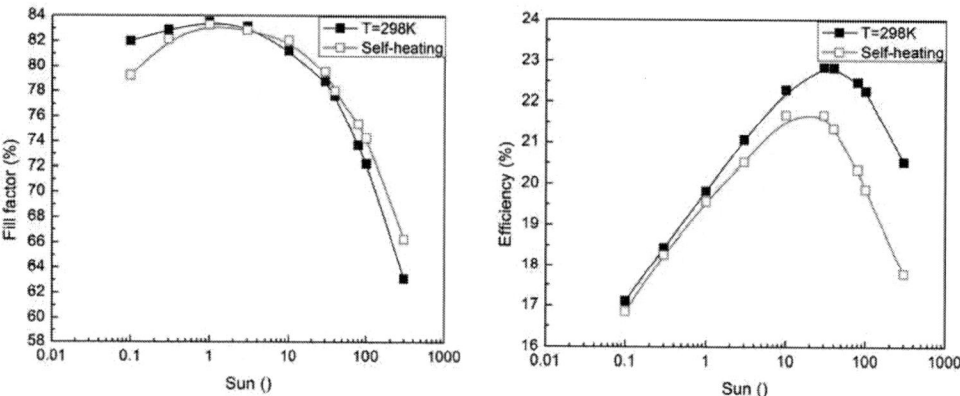

Fig. 3 Simulated fill factor *FF*, Efficiency η versus sun. Simulations have been performed in the case of constant temperature at *T*=298 K and under self-heating assumption.

5. Conclusion

This study attempts to analyze the temperature and self-heating effect by using 3-D simulation methodology and introduces a method to reduce the temperature-related negative effects on solar cells. Moreover, a new concentrator system for indoor cell characterization has been outlined. TCAD simulations of solar cells including accurate modeling of electro-optical properties were performed. Due to a device complexity many of them cannot be simulated in the full 3-D setup in the reasonable time. Therefore derived solutions based on the combined-mode setup coupling the 3-D TCAD model of the solar cell to a "classical" TCAD (2-D or 3-D) model of the active device are proposed. The latter methodology provides both good accuracy with respect to full 3-D modeling and short simulation time. Very good agreement between the new approach in the 3-D simulation and published results [2] confirms the validity of the proposed methodology. The advantages of the proposed method are in the high speed of simulation and simplicity of implementation for complete, high complexity structure analysis. The analysis of optical and electrical behavior can help during the design and optimization of parameters and geometry from semiconductor layers, metallization, package, and up to textured surface.

Acknowledgement

This work was supported in part by the ENIAC JU Project no. 621270/2013 eRamp and in part by grant APVV-14-0749 and VEGA 1/0491/15 supported by Ministry of Education, Science, Research and Sport of Slovakia.

References

[1] User Manual, Ver. I-2013.12, Synopsys TCAD Sentaurus, San Jose, CA, USA, 2013.
[2] M. Mikolašek et al., *Applied Surface Science, Vol.312*, 2014, pp. 145-151.
[3] P. Príbytný, et al., Simulation of Semiconductor Processes and Devices, Washington DC, USA, pp. 455 – 458, 2015.
[4] R. De Rose, *SciVerse ScienceDirect Energy Procedia* 27, 2012, pp. 197–202.
[5] F. Nallet et al., *Proc. 26th Int. Symp. Power Semicond. Devices IC's*, 2014, pp. 334-337.
[6] TCAD Sentaurus application note 2014, <https://solvnet.synopsys.com/retrieve/039602.html>.

ASDAM 2016, The 11th International Conference on Advanced Semiconductor
Devices And Microsystems, November 13-16, 2016, Smolenice, Slovakia

Polymer 3D photonic crystals for LED

M. Goraus, D. Pudis, P. Gaso, L.Suslik, D. Jandura

Department of Physics, University of Zilina,
Univerzitna 1, 01026 Zilina, Slovakia
e-mail: goraus@fyzika.uniza.sk

The properties of light emitting diodes (LEDs) are the best choice for future lighting due to their superior energy efficiency. Also three-dimensional (3D) photonic crystal (PhC) structures have the great potential to improve their optical properties. We present promising technology for creating 3D PhC structures based on polymers which can be directly applied on the LED chip. For the fabrication of 3D structures, we used direct laser writing (DLW) lithography. The structures were filled and covered with polydimethylsiloxane (PDMS) for creating the thin polymer membrane with 3D PhC. Prepared structures were experimentally investigated by scanning electron microscope (SEM) and optical microscope. The effect of 3D PhC on radiation pattern modification was investigated by goniophotometer measurements.

1. Introduction

After the light emitting diodes (LEDs) revolution and mass deployment of this technology, there is a great demand on improvement of their light extraction efficiency and modification of radiation pattern. Many groups used implementation of photonic crystal (PhC) in the LED surface to change the optical properties. Generally, the current technologies try to pattern the LED surface by employment of the combination of lithography and surface etching. Another interesting way is using the thin membranes, which can be applied on the LED surface.

In our recent research, we tried to prepare simple polymer membranes, which we applied on LED chip to enhance the properties of the light emission. We prepared simple polydimethylsiloxane (PDMS) polymer membranes with two-dimensional (2D) PhCs. These membranes were embossed from patterned photoresist masters, which were prepared by laser interference lithography. For square and triangular symmetry PhCs, we showed the modification of a far-field pattern [1]. Similarly, we prepared the PhC with higher symmetry known as photonic quasicrystals (PQCs) structures [1, 2]. PQC structures showed more isotropic diffraction in the far-field radiation pattern [3, 4]. The patterned PhC and PQC membranes diffract light from the LED chip and modify the Lambertian radiation diagram of conventional LEDs.

This paper presents the fabrication of three-dimensional (3D) polymer PhC structures fabricated in PDMS in combination with IP-Dip photoresist. Prepared polymer 3D PhC were applied on the LED chip. We created 3D PhC structures using commercial Nanoscribe system, which is based on the direct laser writing (DLW) method. The 3D structures were prepared by two-photon absorption in IP-Dip photoresist [5]. These structures were filled and covered with PDMS creating the thin polymer membrane with 3D PhCs inside. The prepared membrane with 3D PhCs was applied on the LED chip.

978-1-5090-3084-2/16 $31.00 © 2016 IEEE

Prepared structures were experimentally investigated by scanning electron microscope (SEM) and optical microscope. The effects of a 3D PhC PDMS membrane on the far-field radiation pattern of the LED were investigated by goniophotometer measurements.

2.   Structure design

In our experiment, we proposed the 3D PhC structure for the used LED diode with central wavelength at 635 nm. For the estimation of a photonic band gap (PBG), we use approximation of 3D structure by multilayer system based on effective refractive index using transfer matrix model. PhC structure is composed from spheres with diameter of app. 4.1 µm arranged in simple cubic lattice with period of app. 8.2 µm. Detailed dimensions of structure are shown in Fig. 1.

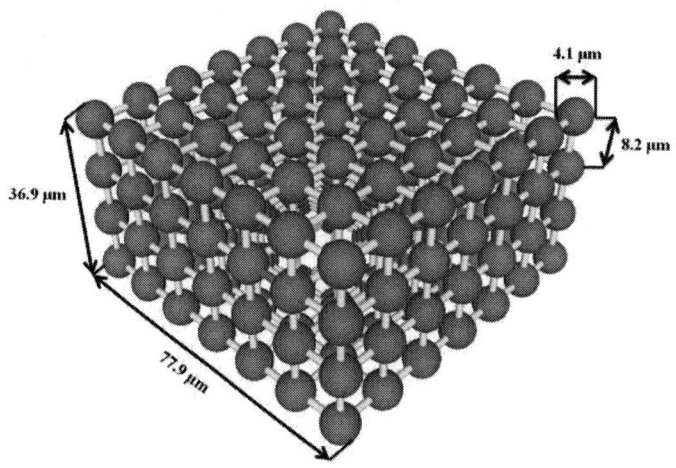

Fig. 1. Schema of proposed 3D PhC and detailed dimensions.

Unique optical and elastic properties of PDMS were used for application of polymer membranes with 3D PhC structures on the conventional LED chip. PDMS is thermally stable, allowing elongation more than 120%, in the visible and near-infrared range of spectrum shows high transparency and is good formable by embossing techniques [6]. We use PDMS to fill the structure prepared by DLW lithography. Arrangement of PDMS membrane with 3D PhC applied on the LED chip is shown in Fig 2.

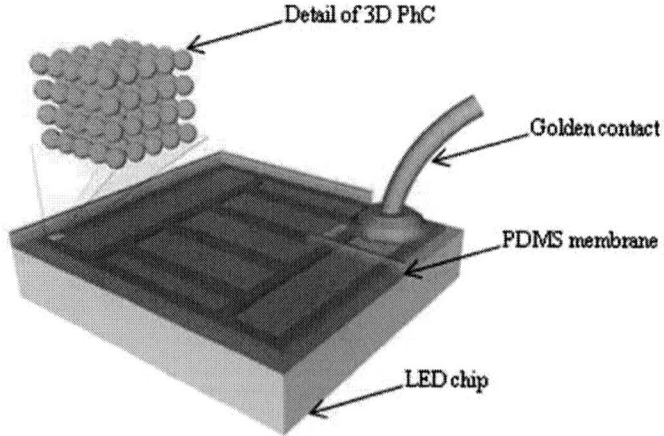

Fig. 2. PDMS membrane with 3D PhC structures applied on the conventional LED chip.

978-1-5090-3084-2/16 $31.00 © 2016 IEEE   54

3. Experimental

We prepared 3D PhC structures using commercial Nanoscribe system. System is based on the DLW method with two-photon absorption in IP-Dip photoresist. Limitation of exposure area for galvo system is app. 200 x 200 μm^2. For better results, we designed 3D PhC with dimension app. 78 x 78 μm^2. To cover the entire LED chip, we fabricated area of 3 x 3 structures as is documented in Fig. 3b.

Fig. 3. 3D PhC structure fabricated in IP-Dip photoresist. a) SEM detail of the structure and b) optical microscope image of structure in PDMS, where inset image shows final membrane applied on the LED chip.

After exposure and developing, we confirmed the quality of prepared 3D PhCs by SEM as is documented in Fig. 3a. The structures were then filled with liquid PDMS of Sylgard 184 (Dow Corning) with curing agent in ratio 1:10 and cured at 50°C at 8 hours. After this procedure, the membrane was mechanically separated from the glass substrate and placed on the LED chip with upward oriented structures. The 635 nm wavelength LED far-field pattern was measured by 3D distribution of optical field with resolution of 5° by goniophotometer with Si-detector placed app. 80 mm above the LED chip surface.

4. Results

PDMS membrane with 3D PhC structure was created and applied on the LED chip. Effect of 3D PhC PDMS membrane on diffraction properties of LEDs was investigated. Far-field radiation pattern diagram of LED with PDMS membrane was measured by goniophotometer at LED driving current 2 mA. This system enables intensity measurement in spherical coordinates by use of precise movable Si-detector in azimuthal and elevation angle. Measurement was performed in complete 3D space around the LED chip with 5° resolution. In measured radiation pattern of LED with 3D PhC structure, different shape of optical field can be clearly distinguished (Fig. 4a). While the unmodified conventional LED shows typical Lambertian radiation pattern, the 3D PhC structure in Fig. 4a shows evident modification of the radiation pattern, if the PDMS membrane with 3D PhC was used. Especially, the cut of radiation pattern in xz plane (Fig. 4b) clearly documents the character of light distribution, where the evident dip is in the central emission region of the LED chip, where typically LED shows the highest intensity of light radiation. This measurements document the effect of 3D PhC structure on radiation pattern.

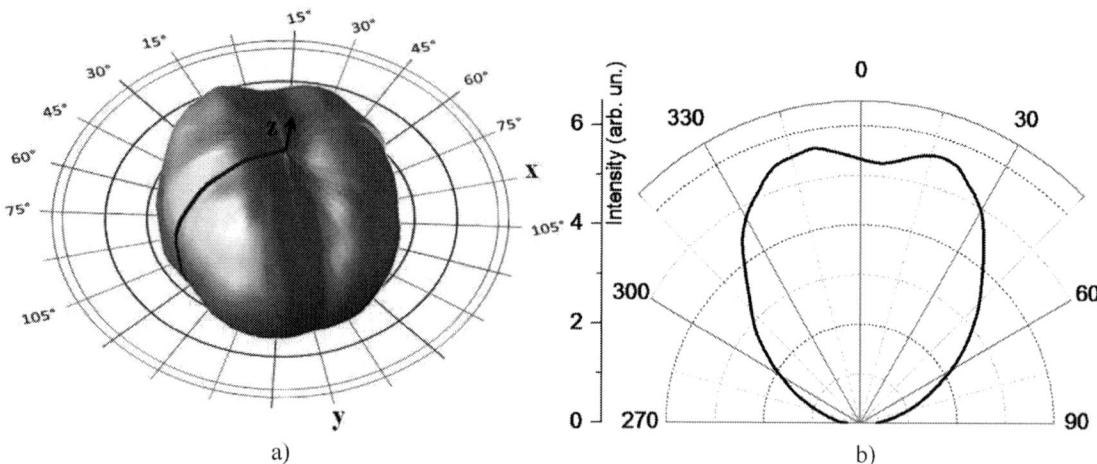

Fig. 4. Far-field radiation pattern of red LED with PDMS membrane with 3D PhC structure inside shown as a) 3D radiation pattern and b) cut in *xz* plane.

5. Conclusions

In this paper, we presented new technique for modification of radiation pattern of LEDs. We designed and prepared polymer PDMS membrane with 3D PhC structures. We showed considerable effect on radiation properties the LED emitting at wavelength of 635 nm. 3D PhCs were prepared by DLW lithography process and were analyzed by SEM and optical microscopy. These analysis confirmed the high quality of the prepared structures. Presented results in this paper document the possibilities of 3D PhC membranes, what may be highly attractive for improvement of optical properties of LEDs.

Acknowledgement

This work was supported by the Slovak National Grant Agency under the projects No. VEGA 1/0491/14 and 1/0278/15 and the Slovak Research and Development Agency under the project No. APVV 0395-12. This work was co-funded from EU sources and European regional development by project ITMS 26210120021.

References

[1] D. Pudis, L. Suslik, R. Nolte, P. Schaaf, J. Kovac, P. Gaso, in *Proceedings of the ASDAM 2014 Conference*, Smolenice, Slovakia, 2014, pp. 323-326.

[2] M. Goraus, D. Pudis, in *Proceedings of the ELEKTRO 2016 Conference*, Strbske Pleso, Slovakia, 2016, pp.612-615.

[3] L. Suslik, D. Pudis, M. Goraus, R. Nolte, J. Kovac, J. Durisova, P. Hronec, P. Schaaf, In: *Progress in Applied Surface, Interface and Thin Film Science 2015*, in press.

[4] L. Suslik, D. Pudis, M. Goraus, J. Durisova, J. Kovac. In *ADEPT 2016 Conference*, Tatranska Lomnica, Slovakia, 2016, pp. 159-162.

[5] P. Gaso, D. Jandura, D. Pudis, In *ADEPT 2016 Conference*, Tatranska Lomnica, Slovakia, 2016, pp. 163-166.

[6] Dow Corning, Sylgard 184 Silicone Elastomer Kit, [online], (16.10.2016), http://www.dowcorning.com/DataFiles/090276fe80190b08.pdf

ASDAM 2016, The 11th International Conference on Advanced Semiconductor
Devices And Microsystems, November 13-16, 2016, Smolenice, Slovakia

GaAsSb/InAs QDs structures for advanced telecom lasers

M. Zíková, A. Hospodková, J. Pangrác, J. Oswald and E. Hulicius

Institute of Physics, Czech Academy of Sciences, v.v.i.,
Cukrovarnická 10, Prague 6, 162 00, Czech Republic
e-mail: zikova@fzu.cz

Preparation and properties of InAs/GaAs quantum dots (QDs) prepared by MOVPE technology covered by GaAsSb strain reducing layer (SRL) with long emission wavelength suitable for telecommunication applications will be presented. Shift of the emission wavelength was achieved by introduction of GaAsSb SRL. SRL with high Sb concentration preserves QD size (which is about 15 nm wide at the base and 5 nm high), decreases the strain inside InAs QDs and decreases the barrier height in valence band. All these phenomena increase the photoluminescence (PL) wavelength. Different antimony content profile can significantly change the PL properties of such QD structures. Furthermore, high content of antimony leads to a creation of type II heterostructure for which a redshift of the PL wavelength with decreased PL intensity is typical. On this kind of structure, extremely long (record) emission wavelength at 1.8 μm was achieved. However low PL intensity may complicate light emitting applications.

1. Introduction

Unique properties of quantum dots (QDs) are given by three-dimensional quantum confinement and discrete density of states result in suitability for many applications. QDs in laser active region decrease the threshold current and improve temperature stability of lasers. The combination of InAs and GaAs materials provides suitable band alignment (deeper electron quantum well than those of the phosphide-based structures) for better temperature and power characteristics. Moreover, InAs/GaAs QD structures can be grown on cheaper GaAs substrates and exploit all advantages of well-established GaAs-based technology. That is why the topic of the InAs/GaAs QD based structures emitting or absorbing in the range of particular wavelengths is still current. While 1.3 μm has already been reached and implemented in industrially produced QD lasers, 1.55 μm and longer wavelengths are still difficult to achieve for InAs/GaAs QDs. For this purpose GaAsSb strain reducing layer (SRL) can be used.

2. Experimental

All considered samples were prepared by low pressure metal-organic vapor phase epitaxy in AIXTRON 200 apparatus on semiinsulating (100) GaAs substrate using Stranski–Krastanow growth mode. For the structure growth trimethylgallium, triethylgallium, trimethylindium, arsine, tertiarybutylarsine and triethylantimony were used. The structure was grown at a total pressure of 7 kPa, the total flow rate through the reactor was 10 slpm. The growth temperature was 650 °C for the first GaAs buffer layer, then the temperature was lowered to 510 °C for the growth of the rest of the structure: second buffer layer, InAs QDs, GaAsSb layer and GaAs capping layer. The formation and development of InAs QDs, as well as the growth of GaAs and GaAsSb layers, was monitored during the growth by the

978-1-5090-3084-2/16 $31.00 © 2016 IEEE   57

reflectance anisotropy spectroscopy measurement at 2.65 eV using EpiRAS 200 TT (LayTec) which also helped to check the structure quality.

Photoluminescence (PL) of the structure with QDs was excited by a semiconductor laser (670 nm) and was detected by a Ge detector using standard lock-in technique. All PL measurements were performed at a room temperature.

The HRTEM measurement was performed by 200 kV JEOL 2011 microscope with point resolution of 0.19 nm and the Cs correction equal to 0.5 mm.

For the simulations the Nextnano software [1] was used. First structure parameters such as materials, thicknesses or compositions are defined. Then, using a conjugate gradient method the total elastic energy, is minimized. From this a local strain tensor is obtained and then the band offsets and light/heavy holes splitting are computed. Finally multi-band Schrödinger and Poisson equations are solved using the GaAs, InAs, InGaAs and GaAsSb material parameters taken from [2].

3.   Results and Discussion

As was already mentioned, the main task of a SRL is to reduce strain inside a QD due to a suitable lattice constant being between the InAs and GaAs value enabling the strain relaxation inside QD, see a schematic drawing in Fig. 1 for illustration. For example the lattice constant of InAs is 6.058 Å, of GaAs is 5.653 Å and of $GaAs_{0.85}Sb_{0.15}$ (computed by Vegard´s law) is 5.719 Å. Lattice constant values were taken from [3].

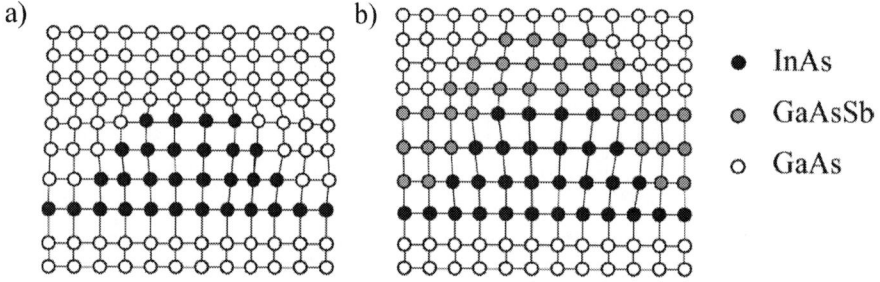

Fig. 1: Schematic view of difference of lattice constants inside InAs QD and in its vicinity a) covered only by GaAs, b) covered by GaAsSb SRL and consecutively capped by GaAs

The strain reduction inside InAs QD helps to redshift the PL by decreasing the band gap and thus the recombination energy in comparison to strained InAs. Another redshifting aspect of GaAsSb SRL is preserving the size of InAs QD by preventing their dissolution and larger (specifically higher) QDs emit on longer wavelength compared to smaller QDs.

In following chapters the influence of composition and composition gradation on the QD structure properties is described.

3.1 Different composition of GaAsSb SRL

Changing the GaAsSb composition leads to the change of structure band alignment. The shift of band edges was computed and results are shown in Fig. 2. Four structures with different GaAsSb composition of 0 %, 3 %, 13 % and 19 % were simulated, when 0 % GaAsSb corresponds to GaAs. The percentage used in article represents the aliquot part of the elements of the V.A group of the periodic table of elements.

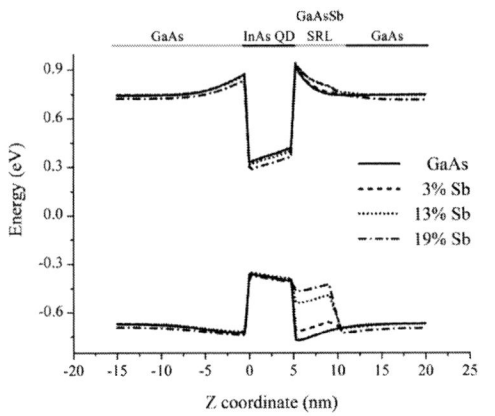

Fig. 2: Computed band edges of InAs QDs covered by GaAsSb SRL with different amount of Sb, 5 nm high QD with 5 nm thick GaAsSb SRL was considered

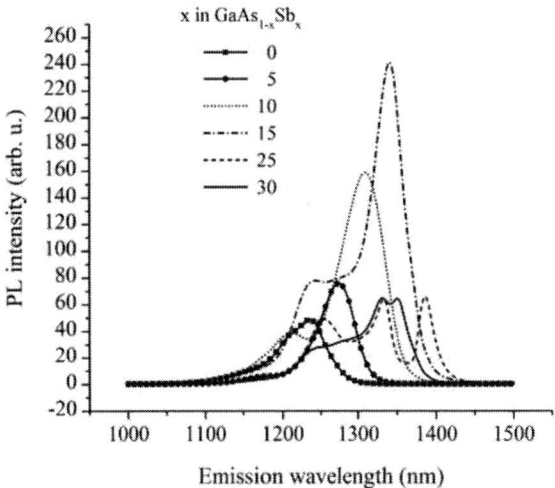

Fig. 3: PL spectra of samples with different composition of GaAsSb SRL

In Fig. 2 we can see that changing the composition of GaAsSb SRL strongly influences mainly the valence band edge, whereas the conduction band remains almost unchanged. We have also experimentally prepared structures with various GaAsSb composition and measured their PL spectra. Results are shown in Fig. 3.

For higher Sb content the sample PL is redshifted. The most intense PL was obtained for a sample with 15 % of Sb in GaAsSb. For this composition the heterostructure is type I with intense PL with long PL wavelength. The PL spectra for 25 % and 30 % have decreased PL intensity, because those structures are type II with decreased electron and hole overlap therefore the radiative recombination fades. Record wavelength obtained on such structure was 1.8 μm for a type II heterostructure [4]. For 25 % there are still some larger QDs that are type I and have redshifted PL, this is the peak at 1400 nm. The other two peaks having lower emission wavelength correspond to excited states (revealed by other PL measurement with different excitation intensity). For 30 % of Sb in GaAsSb SRL, there is only one significant peak around 1.35 μm, which is from the first excited state, which is still type I; the ground state PL peak is of the type II and is not observed at all, because its PL intensity is negligible.

3.2 Different gradation of GaAsSb SRL

What makes the technological sample preparation harder is the surfacing behavior of Sb atoms when growing GaAsSb layer. This was for example observed by reflection anisotropy spectroscopy measurement in [5]. This natural gradation was also revealed from the HRTEM measurement of our prepared sample, see Fig. 4. The determined composition shows higher concentration of Sb atoms on the top of flat part of GaAsSb SRL, although the composition according to the growth technology should be homogeneous.

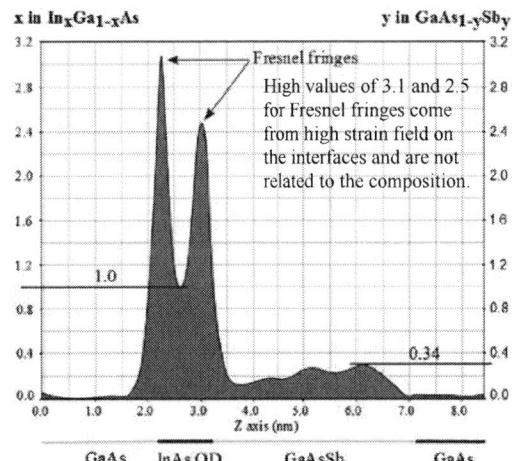

Fig.4: Determined unintentionally graded composition of a flat part of GaAsSb SRL

The natural gradation of the composition led us to the experiment concerning the influence of the layer gradation on the PL spectrum. Three samples with different amount of Sb in GaAsSb layer were prepared: increasing with low Sb content on the bottom of the layer and high Sb content on the top of the layer, decreasing with reversed concentration of Sb and constant with constant GaAsSb composition. The PL measurement results are shown in Fig. 5.

The PL spectrum shows the maximum wavelength of 1300 nm for constant composition of GaAsSb SRL. The composition was around 10 % of Sb in GaAsSb. For the SRL with decreasing amount of Sb, the valence bands of InAs and GaAsSb fuse together and the quantum trap effectively broadens, which results in a shift of ground state energy lever nearer to the valence band edge and the recombination energy decreases, therefore the PL wavelength redshifts. The sample with increasing gradient of the amount of Sb in GaAsSb SRL has also longer emitted PL which is probably due to the higher maximal content in the SRL. These results were also supported by theoretical simulations.

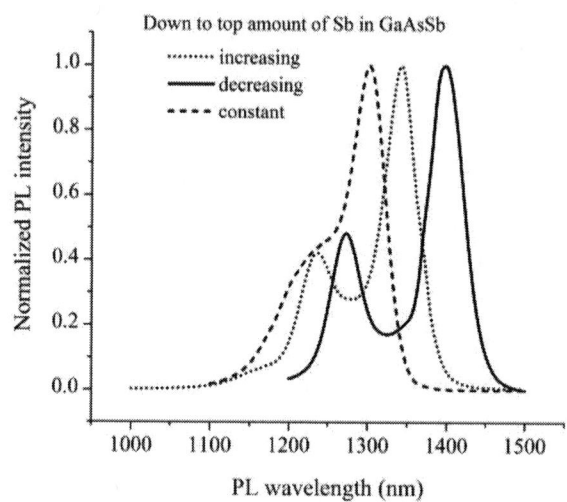

Fig. 5: PL spectra of three samples with different composition gradation of GaAsSb SRL: increasing, decreasing and constant

4. Conclusions

We have prepared several structures differing in the GaAsSb SRL composition or composition gradation. We found out that higher content in the GaAsSb SRL causes long wavelength PL emission from the sample. Although the longest PL wavelengths are obtained for Sb content over 20 %, the intensity is significantly decreased, because these structures become type II with spatially separated electrons and holes. For the applications using telecom wavelengths it is better to have balanced structure with PL wavelength as long as possible keeping the heterostructure to be type I to have intense PL. Further improvement can be done using the gradation of the concentration profile of Sb atoms in the GaAsSb layer. Both increasing and decreasing gradation profiles help to redshift the PL, but in the increasing one a triangular barrier in the valence band is formed, which can help to maintain type I heterostructure even for higher Sb content in GaAsSb.

Acknowledgement

This work was supported by the MEYS NPU project no. LO1603 – ASTRANIT.

References

[1]   S. Birner et al., *Acta Phys. Pol. A* **110**, 111, 2006.
[2]   I. Vurgaftman, J.R. Meyer, L.R. Ram-Mohan, *J. Appl. Phys.* **89**, 5815, 2001.
[3]   O. Madelung, *Semiconductors: Data Handbook*, Springer, Berlin, 2003.
[4]   M. Zíková et al., *J. Cryst. Growth* **414**, 167, 2015.
[5]   A. Hospodková et al., *J. Cryst. Growth* **414**, 156, 2015.

ASDAM 2016, The 11th International Conference on Advanced Semiconductor
Devices And Microsystems, November 13-16, 2016, Smolenice, Slovakia

Minimization of Self-Heating in SOI MOSFET Devices with SELBOX Structure

M. R. Narayanan and Hasan Al Nashash

Department of Electrical Engineering, College of Engineering,
American University of Sharjah, 26666 Sharjah, United Arab Emirates
e-mail: mnarayanan@aus.edu and hnashash@aus.edu

The SOI MOSFET devices offer excellent performance advantages over bulk MOSFET devices. The structural features of the SOI MOSFET devices introduce several undesirable effects which are not normally present in the bulk MOSFET devices such as kink effect and self-heating effect. The work in this paper is focused on minimizing the self-heating in the SOI MOSFET device with the use of back oxide at selected regions below source and drain instead of continuously as in SOI MOSFET devices. TCAD simulation results shows that the modified structure is effective in minimizing the self-heating in the SOI devices. It is observed that through proper selection of gap parameters the self-heating in the device can be controlled.

1.   Introduction

Silicon on Insulator (SOI) MOSFET devices were introduced in 70s in order to improve the radiation hardness of the circuits using MOSFET devices [1]. It was a need of the time as there was a widespread use of the MOS devices in the space and satellite communication equipment which were functioning in the radiation prone environments. These devices are identical to the MOSFET devices with external contacts namely source, drain and gate. The primary difference is; the active device region in the SOI MOSFET devices is provided with a dielectric isolation between the channel and substrate [2]. The SOI devices are found to have several advantages over bulk MOSFETs resulting from the structural features. These devices exhibit improved radiation hardness. The dielectric isolation of the active device region from the substrate leads to reduction in the parasitic capacitances which improves the high frequency response of the device.

In spite of these advantages, the SOI MOSFET devices introduce several issues not commonly known in the bulk devices. The dielectric material used for the isolation is most often silicon dioxide and it is known as buried oxide. Thermal conductivity of silicon dioxide is low; 1.4(W/m-K) as compared to silicon which is 148(W/m-K) at room temperature. Consequently, the buried oxide insulates the active device region from the bulk substrate electrically and thermally. The inefficient heat transfer from the channel to the lower substrate results in the accumulation of the heat in the active region. This leads to a rise in the device temperature known as 'self-heating' effect. Self-heating in the devices has several adverse effects in the performance aspects such as reduced carrier mobility, threshold voltage shift, degradation in the switching performance [3]. Additional issues in SOI devices due to the self-heating include electromigration and increase in the device failure rate [4]. Reliability studies show that for a rise in the device temperature by 10^0C, the device failure rate is doubled [5].

978-1-5090-3084-2/16 $31.00 © 2016 IEEE   61

2. Methodology

Early work for minimizing the self-heating in the SOI devices are mostly related to the use of materials with better thermal conductivity like silicon-over-insulator-multilayer (SIOM), alumina and aluminum nitride etc. [6] [7]. These methods involve introducing different materials in place of silicon dioxide buried layer which involves complex processes which are expensive too. An effective method for minimizing the self-heating can be realized through structural modification of the SOI MOSFET. In the modified device structure, the buried oxide is used at selected regions below the source and drain and not continuously as in the SOI devices [8] [9]. In the structural details shown in Figure 1, there is isolation between the two buried oxide segments. As the buried oxide covers selected regions below the source and drain, the device is named as SELBOX MOSFET [8]. It is found that the device temperature is maximum in the channel and especially close to the drain [10]. In SOI MOSFETs the heat transfer to the base substrate through the buried oxide with low thermal conductivity is inefficient. The material present in the region between two silicon dioxide segments in Figure 1 shown as W_{gap} will be silicon with higher thermal conductivity. This silicon segment will acts as an effective medium for heat transfer to the lower substrate. The enhanced heat transfer to the substrate leads to a minimization in the self-heating effect.

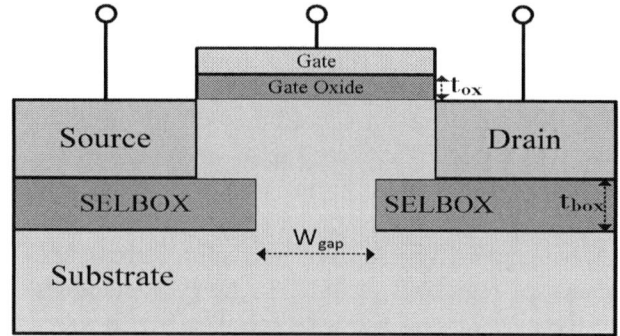

Figure 1: Selective Buried Oxide (SELBOX) MOSFET Structure

Device parameters

*Gate oxide thickness t_{box} ~10nm.
*Chanel length ~0.12nm.
*Gap length w_{gap} ~100nm.
*SELBOX thickness t_{ox} ~100nm.
*W/L ratio = 1
*Device structure length = 0.6μm.
*Substrate doping conc.
  ~2×10^{16} cm^{-3}.
*Source/Drain doping conc.
  ~1×10^{19} cm^{-3}.

3. Results

Figure 1 shows the structure of the device used for the thermal studies. Simulation of fabrication of the device is carried out using Silvaco Athena and Silvaco Atlas is used for device simulation [11]. Parameters of the device used for simulation are indicated in Figure 1. The heat conduction in the MOSFET device is analyzed by subjecting the devices for thermal simulation using Atlas Giga with the activation of lattice temperature mode. Based on literature for the SOI devices, the following models are used for device simulation [11]. Mobility models: field dependent mobility model and arora; recombination models: concentration dependent recombination model conmob and auger; carrier statistics model for bandgap narrowing bgn and impact ionization model from Selberherr are used for the simulation. Numeric methods chosen for simulation are Gummel and Newton methods. The software employs the heat conduction equation; equation 1 to compute the temperature distribution in the semiconductor device [11]. In equation 1, ρ is the material density, c_s is the specific heat, k is the thermal conductivity, T_L is the local lattice temperature, t is the time, P is the space and time dependent power density generated in the device and ∇T_L is the gradient of temperature

978-1-5090-3084-2/16 $31.00 © 2016 IEEE

$$\rho c_s \frac{\partial T_L}{\partial t} = \text{div}(k \, \nabla T_L) + P \tag{1}$$

The thermal profile of the devices under identical simulation conditions are given in Figure 2 and 3. The applied drain and gate voltages for simulation are V_{DS} = 3.0 V and V_{GS} = 2.0 V respectively. The peak temperature from simulation are $T_{MAX(SOI)}$ = 354.4K, $T_{MAX(SELBOX)}$ (gap = 0.04µm) = 334.8K. For bulk MOSFET with identical geometry $T_{MAX(Bulk)}$ = 311.1K. The peak device temperature reduces as we move from SOI MOSFET to SELBOX MOSFET and further with increase in the gap length between SELBOX segments. These simulation results demonstrate the ability of the SELBOX MOSFET in minimizing the self-heating effect [12].

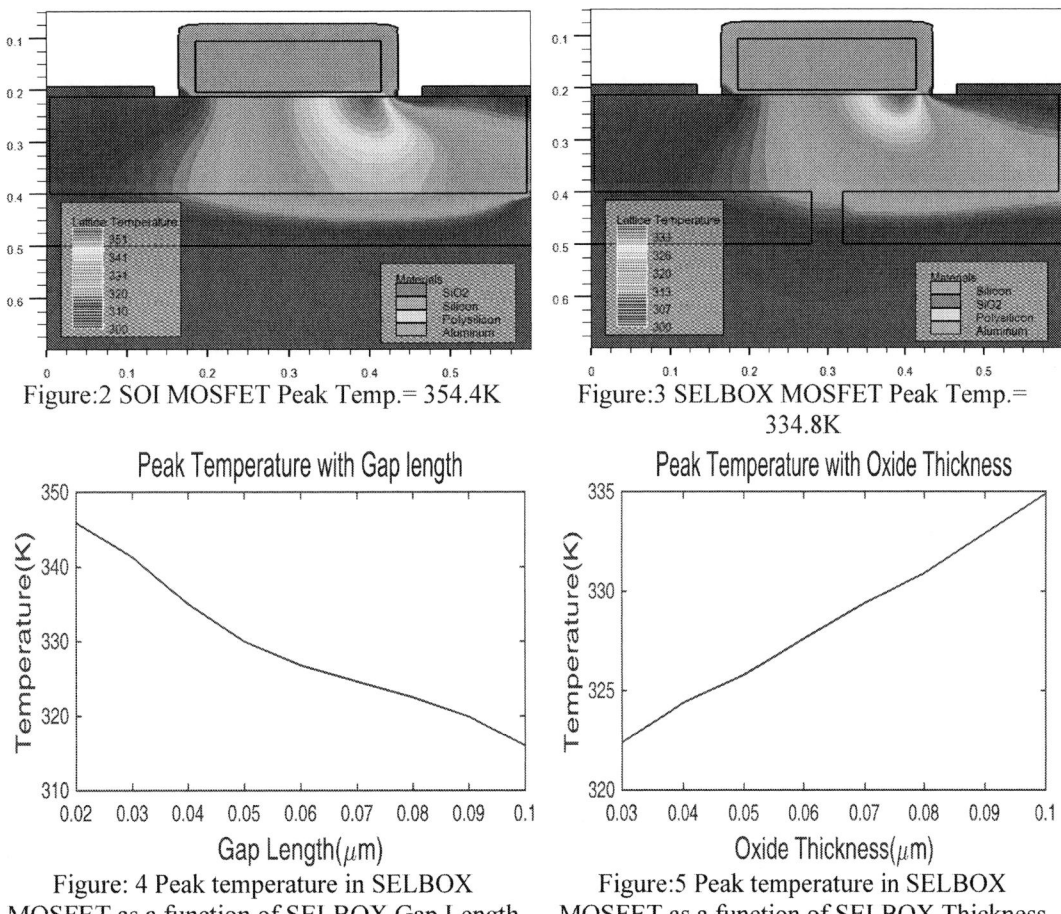

Figure:2 SOI MOSFET Peak Temp.= 354.4K

Figure:3 SELBOX MOSFET Peak Temp.= 334.8K

Figure: 4 Peak temperature in SELBOX MOSFET as a function of SELBOX Gap Length

Figure:5 Peak temperature in SELBOX MOSFET as a function of SELBOX Thickness

Figures 4 and 5 show the peak temperature variation as a function of the gap length and SELBOX thickness in the SELBOX MOSFET. With increase in the gap length, the area normal to the heat flow to the substrate increases. Thermal resistance of the silicon in the region between the SELBOX segments can be found with equation 2. Here, R_θ is thermal resistance, 'A' is the area normal to the heat flow which increases with 'W_{gap}'. With increase in W_{gap} the R_θ reduces as more volume will be covered by silicon than silicon dioxide.

$$R_\theta = \frac{t_{box}}{k*A} \qquad (2)$$

4. Conclusion

The self-heating in SOI MOSFET devices leads to reliability issues which need to be properly addressed. The SELBOX MOSFET with a narrow gap in the oxide offers an effective solution for minimizing the heating effect in the devices. The self-heating effect, the associated limitations of the SOI devices and the potential of the proposed structure in minimizing the rise in temperature have been analyzed through simulation. It was noticed that a narrow gap in the order of 0.05 μm can reduce the temperature due to self-heating by 20^0C which will have a reasonable impact on the reducing the device failure rate.

References

[1] G. K. Celler and S. Cristoloveanu, "Frontiers of Silicon-on-Insulator," *Journal of Applied Physics,* pp. 4955-4978, 1 May 2003.

[2] J. P. Collinge, Silicon-on-Insulator Technology: Materials to VLSI, 3 ed., New-York: Springer, 2004.

[3] Z. X. Zhang, Q. Lin, M. Zhu and C. L. Lin, "A New Structure of SOI MOSFET for Reducing Self-Heating Effect," *Elsevier Ceramics International,* vol. 30, no. 7, pp. 1289-1293, July 2004.

[4] D. Vasileska, K. Raleva and S. Goodnick, "Modelling Heating Effects in Nanoscale Devices," *Journal of Computational Electronics,* vol. 7, no. 2, pp. 66-93, 2008.

[5] M. C. Cheng, F. Yu, L. Jun, M. Shen and G. Ahmadi, "Steady State and Dynamic Thermal Models For Heat Flow Analysis of Silicon-on-Insulator MOSFETs," *Microelectronic Reliability,* vol. 44, no. 3, pp. 381-396, 2004.

[6] O. Rayssac, H. Moriceau, M. Oliver, I. Stoemenos, A. M. Cartier and B. Aspar, "From SOI to STOIM Technology: Application for Specific Semiconductor Processes," *Proceedings of the Tenth International Symposium on SOI Techniques and Devices,* 2001.

[7] C. Men, Z. Xu, Z. An, P. K. Chu, Q. Wan, X. Xie and C. Lin, "Fabrication of SOI Structure with AlN Film as Buried Insulator by Ion-Cut-Process," *Applied Surface Science Elsevier,* vol. 199, pp. 287-292, 2002.

[8] N. R. Madathumpadical, H. Al Nashash, B. Mazhari, D. Pal and M. Chandra, "Analysis of Kink Reduction in SOI MOSFET Using Selective Back Oxide Structure," *Active and Passive Electronic Components Hindawi Publishing Corporation,* vol. 2012, pp. 1-9, May 2012.

[9] Y. Dong, M. Chen, J. Chen, X. Wang, P. He, X. Lin, L. Tian and Z. Li, "Patterned Buried Oxide Layers undera Single MOSFET to Improve the Device Performance," *IOP Semiconductor Science and Technology,* vol. 19, pp. L25-L28, 2004.

[10] T. Sadi, R. W. Kelsall and N. J. Pilgrim, "Investigation of Self-Heating Effects in Submicrometer GaN/AlGaN HEMTs using an Electrothermal Monte Carlo Method," *IEEE Transactions on Electron Devices,* vol. 53, no. 12, pp. 2892-2893, 2006.

[11] Silvaco, "Atlas User's Manual Device Simulation Software," Silvaco International, Santa Clara California, 2004.

[12] N. R. Madathumpadical, H. Al Nashash, D. Pal and M. Chandra, "Thermal Model of MOSFET with SELBOX Structure," *Journal of Computational Electronics ,* pp. 1-9, 18 July 2013.

ASDAM 2016, The 11th International Conference on Advanced Semiconductor
Devices And Microsystems, November 13-16, 2016, Smolenice, Slovakia

Highly reliable long-term operation of AlGaN/GaN/AlN HFETs grown on silver substrate

A. Fox [a,b)], M. Mikulics [a,b)], M. Marso[c)], M. Kočan [d)], Z. Sofer [e)], P. Kordoš [f)],
H. Lüth [a,b)], J. Schubert [a,b)], D. Grützmacher [a,b)], and H. Hardtdegen [a,b)]

[a)] *Peter Grünberg Institute (PGI-9), Forschungszentrum Jülich, 52425 Jülich, Germany*
[b)] *JARA – Fundamentals of Future Information Technologies*
[c)] *Faculté des Sciences, de la Technologie et de la Communication,
Université du Luxembourg, L-1359 Luxembourg*
[d)] *School of Electrical, Electronic and Computer Engineering, University of Western
Australia, Crawley, WA, Australia*
[e)] *Dept. of Inorganic Chemistry, Institute of Chemical Technology, Prague, Technická 5,
166 28 Prague 6, Czech Republic*
[f)] *Institute of Electronics and Photonics, Slovak Technical University, SK-81219 Bratislava,
Slovak Republic*

*We developed a novel "combined" two-step epitaxial procedure based on
MOVPE and MBE for an optimized growth of group III-nitride layers on
silver-metallic substrates. The AlGaN/GaN/AlN heterostructures were used
for the fabrication of HFETs. The electrical properties as well as the long-
term operation properties were systematically studied and compared with
conventional devices. The improved heat dissipation in heterostructures
deposited directly on a silver substrate leads to a significant decrease in
channel temperature (-60% at 7W/mm) and affects the long term stability of
the drain current (± 2%) favorably in the whole range under investigation (up
to ~1000hours) compared with conventional AlGaN/GaN layers grown on
sapphire substrates. The results presented demonstrate the great potential of
the novel material-device concept.*

1. Introduction

The heat management in semiconductor devices represents a serious problem not only
for high power devices. Even though the efficiency of CMOS based circuits has greatly
increased the past decades, their high integration density leads to a non-negligible increase in
heat generation. Alternative device architectures or novel material systems are called for to
decrease heat accumulation. One possible solution is the growth and deposition of
semiconductor thin films on metallic substrates. There have been a number of experimental
reports: the growth of GaAs on Mo employing liquid phase epitaxy [1], the growth of Si on
Ni, W and steel [2,3], the MBE growth of GaN on Mo, Ta and Nb [4-6] and the growth of
GaN on Ag substrates and successful fabrication of a Schottky diode [7]. In addition, an
improvement in thermal stability and the avoidance of electrical and optical degradation
effects in photodetectors based on GaAs/Al substrate was observed [8].

Here, in this work we present a first long-term operated HFET based on
AlGaN/GaN/AlN heterostructure grown on a silver substrate in a two-step growth procedure
using MBE for the AlN buffer and MOVPE for AlGaN/GaN growth (Fig. 1). The HFETs
were fabricated conventionally using standard optical photolithography (Fig. 2). The ohmic
contact metallization is based on a Ti/Al/Ni/Au multilayer and was annealed at 850°C for 30 s

978-1-5090-3084-2/16 $31.00 © 2016 IEEE

in N_2 ambient. The source-drain spacing was 3 µm. The gate contacts consisted of standard Ni/Au metal layers. The designed gate lengths varied from 200 nm to 2 µm and were defined by e-beam and/or optical lithography.

2.   Device characterization

A typical DC characteristic of a Ag-metal substrate based HFET with a 200 nm gate length is shown in Fig. 3. Our device exhibits ~ 0.65 A/mm maximal drain current. This value is fully comparable with values previously reported on sapphire substrate [9]. Similarly, extrinsic transconductances evaluated from the DC output characteristics exhibited values about 160 mS/mm for 200 nm L_g in comparison to ~120 mS/mm for the device with 2 µm gate length. Furthermore, two HFET devices - a conventional AlGaN/GaN/sapphire device and our novel Ag-metal substrate based device - were tested in DC regime for long-term operation. We achieved very stable drain-source (I_{ds}) currents for the AlGaN/GaN/AlN/silver device during the about 1000 hours of operation in comparison to ~ 10 % decrease in I_{ds} for a conventional device (Fig.4).

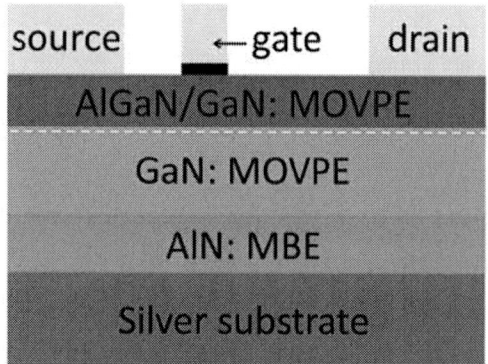

Fig. 1. Schematic cross-section of the two-step deposited AlGaN/GaN/AlN/silver substrate- material system and HFET structure.

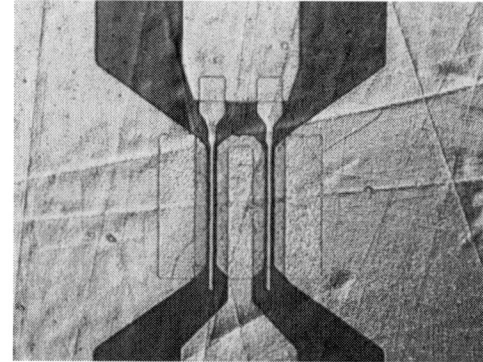

Fig. 2. SEM micrograph of an HFET structure fabricated from an AlGaN/GaN/AlN/silver-metallic substrate material system.

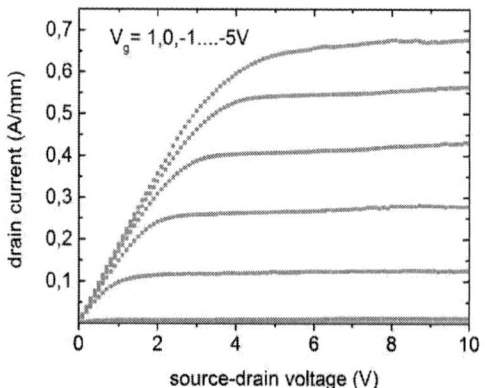

Fig. 3. DC output characteristic measurements on an HFET/ novel silver-substrate with a gate length of 200 nm exhibiting comparable maximal drain currents as previously reported for structures on sapphire substrates [9].

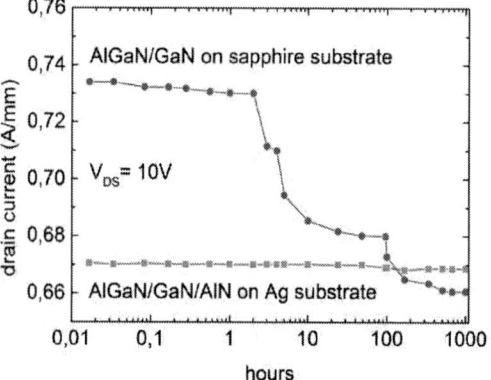

Fig. 4. Long-term operation-reliability measurements performed on conventional AlGaN/GaN/sapphire substrate and novel silver-substrate based devices. The improved long-term stability of the drain current for the metallic substrate is clearly visible.

978-1-5090-3084-2/16 $31.00 © 2016 IEEE

We explain this result by a significantly improved internal thermal heating management at the interface between the "active" transistor layers responsible for the current transport and the effective heat dissipation by means of the metallic substrate.

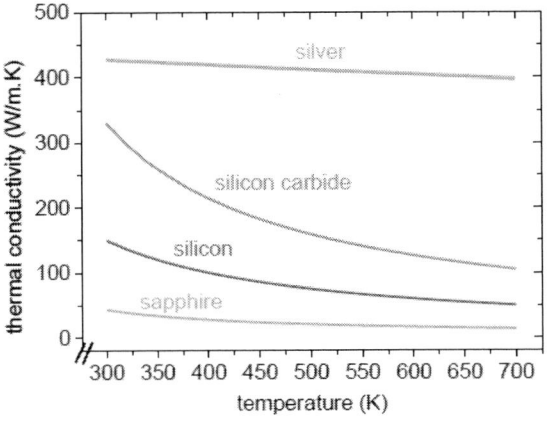

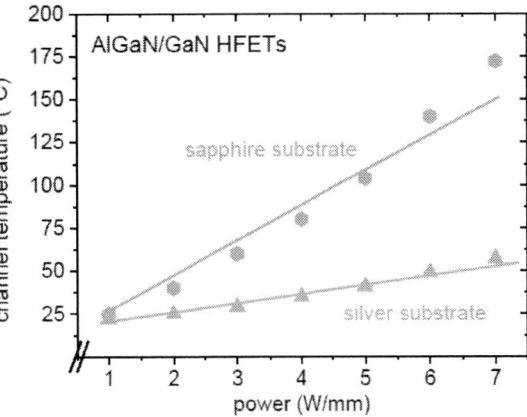

Fig. 5: Thermal conductivity as a function of temperature for different materials - sapphire, silicon and SiC conventionally used in III-nitride technology in comparison to the Ag-substrate. The thermal conductivity is higher at and above room temperature for the Ag-metallic substrate [10].

Fig. 6: Channel temperature as a function of dissipated power for both HFET devices – a conventional AlGaN/GaN heterostructure grown on sapphire and one on the novel silver-substrate. The channel temperature is unambiguously larger for HFET structures on sapphire substrates indicating less-efficient heat dissipation by the substrate.

3. Channel Temperature characterization

Conventional Raman thermography [11,12] was used to study the thermal management of AlGaN/GaN HFET devices. High lateral resolution measurements ($\sim 0.5 \mu m$) indicate a strongly non-uniform distribution in HFET devices based both on a sapphire and a silver substrate, respectively. For the sake of statistics, only the average values of the evaluated temperature were plotted in Fig.6. Since the thermal conductivity of the materials is temperature dependent (Fig.5) the relation between dissipated power and the temperature is non-linear and the measured temperatures deviate from the linear behavior. This tendency is more evident in the case of sapphire based devices, whereas a first increase in HFET channel temperature is observed at higher dissipated power levels for the Ag-substrate (Fig.6). This effect could be clearly attributed to the thermal conductivity decrease at higher operation temperatures and the increasing ineffectiveness of the heat dissipation process inside the HFET device. In analogy to our reliability measurements presented in Fig. 4, the channel temperature was monitored for ~10 hours for both devices under constant ~10V source–drain bias operation conditions. A temperature increase of up to 30% was determined for the conventional sapphire substrate based device whereas only an increase of ~5% was observed for the silver based HFET. This indicates that the observed reduction in drain-source (I_{ds}) currents for the AlGaN/GaN/AlN/sapphire and the improved long-term stability of the drain-source (I_{ds}) currents for the AlGaN/GaN/AlN/silver device can be attributed to the difference in channel temperature i.e. to the difference in heat dissipation by the respective substrate.

4.   Conclusions

AlGaN/GaN/AlN heterostructures were deposited by a two-step growth procedure on Ag-metallic substrates. HFETs were fabricated and characterized with respect to their DC characteristics, their long-term operation reliability and their channel temperature. The novel devices with a 200 nm gate length exhibit a maximal drain current ~ 0.65 A/mm and extrinsic transconductances of about 160 mS/mm. These values are fully comparable with those previously reported for HFETs on sapphire substrates [9]. The unusual long-term stability of the drain-source (I_{ds}) currents (± 2%) compares favorably with that of conventional AlGaN/GaN layers grown on sapphire substrates. It is attributed to the reduced channel temperature and is due to the more effective heat dissipation for a highly thermally conductive substrate at device operation conditions. Our current achievements demonstrate the strong potential for highly reliable transistor devices based on heterostructures grown on metallic substrates.

Acknowledgement

This work was partially supported by Czech Science Foundation (project No. 13-20507S).

References

[1]   J.M. Woodall, IBM Technical Disclosure Bulletin **21**, 2584 (1978).

[2]   G.W. Racette, and R.T. Frost, J. Crystal Growth **47**, 384 (1979).

[3]   D.E. Carlson, Third E.C. Photovoltaic Solar Energy Conference. Dordrecht, Netherlands: Reidel, 294 (1981).

[4]   K. Yamada, H. Asahi, H. Tampo, Y. Imanishi, K. Ohnishi, K. Asami, Proceedings of International Workshop on Nitride Semiconductors, Tokyo, Japan: Inst. Pure & Appl. Phys, 556 (2000).

[5]   K. Yamada, H. Asahi, H. Tampo, Y. Imanishi, Appl. Phys. Lett. **78**, 2849 (2001).

[6]   A.V. Andrianov, K. Yamada, H. Tampo, H. Asahi, Semiconductors **36**, 878 (2002).

[7]   M. Mikulics, M. Kocan, A. Rizzi, P. Javorka, Z. Sofer, J. Stejskal, M. Marso, P. Kordoš, H. Lüth, Appl. Phys. Lett. **87**, 212109 (2005).

[8]   M. Mikulics, R. Adam, Z. Sofer, H. Hardtdegen, S. Stanček, J. Knobbe, M. Kočan, J. Stejskal, D. Sedmidubský, M. Pavlovič, V. Nečas, D. Grützmacher, and M. Marso, Semicond. Sci. Technol. **25**, 075001 (2010).

[9]   M. Mikulics, H. Hardtdegen, Y.C. Arango, R. Adam, A. Fox, D. Grützmacher, D. Gregušová, S. Stanček, J. Novák, P. Kordoš, Z. Sofer, L. Juul, and M. Marso, Appl. Phys. Lett. **105**, 232102 (2014).

[10] M. Mikulics and H. Hardtdegen, submitted to Appl. Mater. Today (2016).

[11] M.S. Liu, L.A. Bursill, S. Prawer, K.W. Nugent, Y.Z. Tong, G.Y. Zhang, Appl. Phys. Lett. **74**, 3125 (1999).

[12] M. Kuball, J.M. Hayes, M.J. Uren, T. Martin, J.C.H. Birbeck, R.S. Balmer and B.T. Hughes, IEEE Electron Device Lett. **23**, 7 (2002).

ASDAM 2016, The 11th International Conference on Advanced Semiconductor
Devices And Microsystems, November 13-16, 2016, Smolenice, Slovakia

InGaN mesoscopic structures for low energy consumption nano-opto-electronics

H. Lüth[a,b], M. Mikulics[a,b], A. Winden[c], St. Trellenkamp[a,b], Z. Sofer[d], M. Marso[e],
D. Grützmacher[a,b], and H. Hardtdegen[a,b]

[a] Peter Grünberg Institute, Forschungszentrum Jülich, 52425 Jülich, Germany
[b] JARA – Fundamentals of Future Information Technology
[c] Robert Bosch GmbH, Reutlingen D-72760, Germany
[d] Dept. of Inorganic Chemistry, Institute of Chemical Technology, Prague, Technická 5,
166 28 Prague 6, Czech Republic
[e] Faculté des Sciences, de la Technologie et de la Communication,
Université du Luxembourg, L-1359 Luxembourg

Nano-LEDs based on mesoscopic structures are the key elements for future energy saving nano-opto-electronics as well as for fast and highly secure optical communication. We present first results using a vertical device layout in which nano-LED emitters based on InGaN mesoscopic structures deposited by metalorganic vapor phase epitaxy (MOVPE) were implemented. The nano-LEDs were integrated in a vertical device layout without degradation of their optical and electrical properties. The results presented demonstrate the great promise of the novel device concept and integration technology for low energy consumption nano-opto-electronics operated in the telecommunication wavelength range.

1. Introduction

There is a special need to develop nano-LED sources suitable for highly secure, ultrafast and efficient optical communication [1,2], which are compatible with established telecommunication systems [3,4]. Also alternative applications have been revealed for nano-LED sources such as lithography [4]. For most applications, the emitters should be singularly addressable in order to integrate them into a device circuit, which can stimulate "few"-photon emission. From a materials point of view, (InGa)N alloys can cover a large wavelength range since their band gaps exhibit values between ~ 0.7 and ~ 3.4 eV. They will therefore be able to emit light upon electrical stimulation in the range between ~ 365 and ~ 1770 nm i.e. also in the telecommunication wavelength range. Equally important is the fact that group III nitrides possess large band offsets in their heterostructures, leading to high exciton binding energies and strong quantum confinement effects [5]. Hence, there are bright prospects for "few" and single photon emission at room temperature. From a practical applications point of view, the fast transfer of information in the Gbit/s range calls for a device circuit and device layout, which is suitable for high-frequency operation. Group III-nitrides have already demonstrated their potential [6,7] to this end.

2. Material optimization

First, we started with the site-controlled growth of InGaN nanostructures via catalyst-free selective-area MOVPE [8,9]. The manufacturing process was optimized with respect to the mask pattern in order to be able to fabricate individually addressable InGaN nanopyramid

978-1-5090-3084-2/16 $31.00 © 2016 IEEE   69

based nano-LEDs. The starting point for growth were uniform and smooth 1.3 μm thick n-GaN layers deposited on sapphire (c-plane) substrates. They were masked with SiO_2.

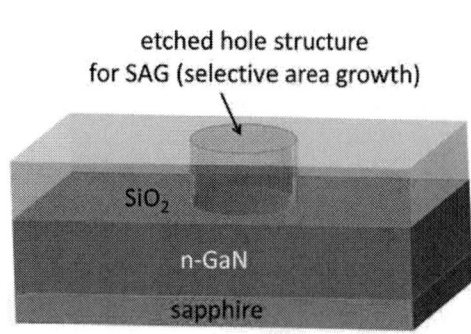

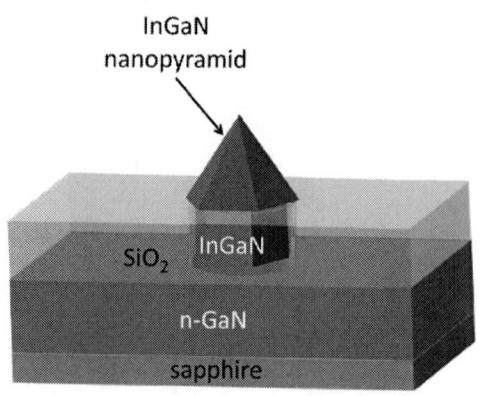

Figure 1a: 1.3 micrometer thick n-doped GaN layers on c-plane sapphire were employed as templates for selective area (SA) metalorganic vapor phase epitaxy (MOVPE) of group III-nitride nanostructures. These templates were masked with a 50 nm thick SiO_2 layer, which was patterned with a hexagonally arranged aperture array using a three-layer PMMA resist and e-beam lithography followed by CHF_3 reactive ion etching.

Figure 1b: After ex situ cleaning in H_2SO_4, InGaN nanopyramids with varying In content between 0 and 100 % were grown selectively at a temperature of 650 °C and a V/III molar flow ratio of $4.4x10^4$ optimized with respect to the highest selectivity. Trimethylindium (TMIn), triethylgallium (TEGa) and NH_3 were used as the precursors and N_2 as the carrier gas [10,11].

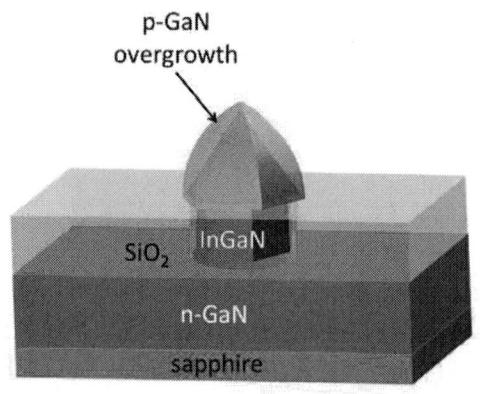

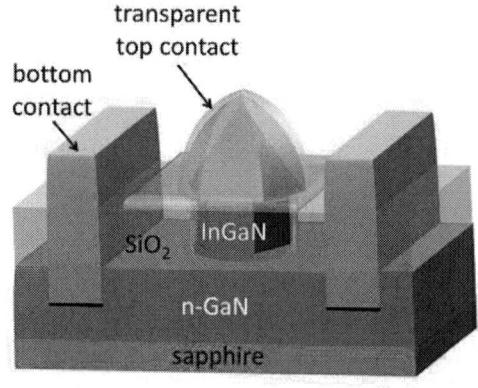

Figure 1c: After using a growth time which ensured fully developed nanopyramids, growth of the p-doped GaN cap layer was carried out by closing the TMIn source material and continuing growth with triethylgallium (TEGa) and the dopant precursor bis(cyclopentadienyl) magnesium (Cp_2Mg) for 30 min.

Figure 1d: bottom contacts (Ti/Al/Ni/Au) were defined with the help of conventional optical lithography and an Ar-IBE process. After the lift-off process bottom contacts were annealed in N_2 ambient at an optimized temperature preventing InGaN nanopyramid degradation. The transparent top contact (Ni/Au) was deposited on top of the p-GaN/InGaN nanopyramid and treated with an optimized laser micro annealing process [12].

Afterwards, the InGaN nanopyramids were integrated into a device layout for DC testing and future high-frequency operation. The fabrication process is described in figure 1. Additionally, the entire surrounding surface area (except for the mesa with bottom and top contacts) was coated with a 200 nm SiO_2 layer to prevent leakage currents. In the last step, the InGaN nanopyramid based nano-LED device was connected to contact pads (Ti/Au) for DC testing and future RF operation by employing optical lithography and a lift-off process (figure 2). In figure 2 SEM micrographs are presented of the prototype nano-LED device described above at different magnifications.

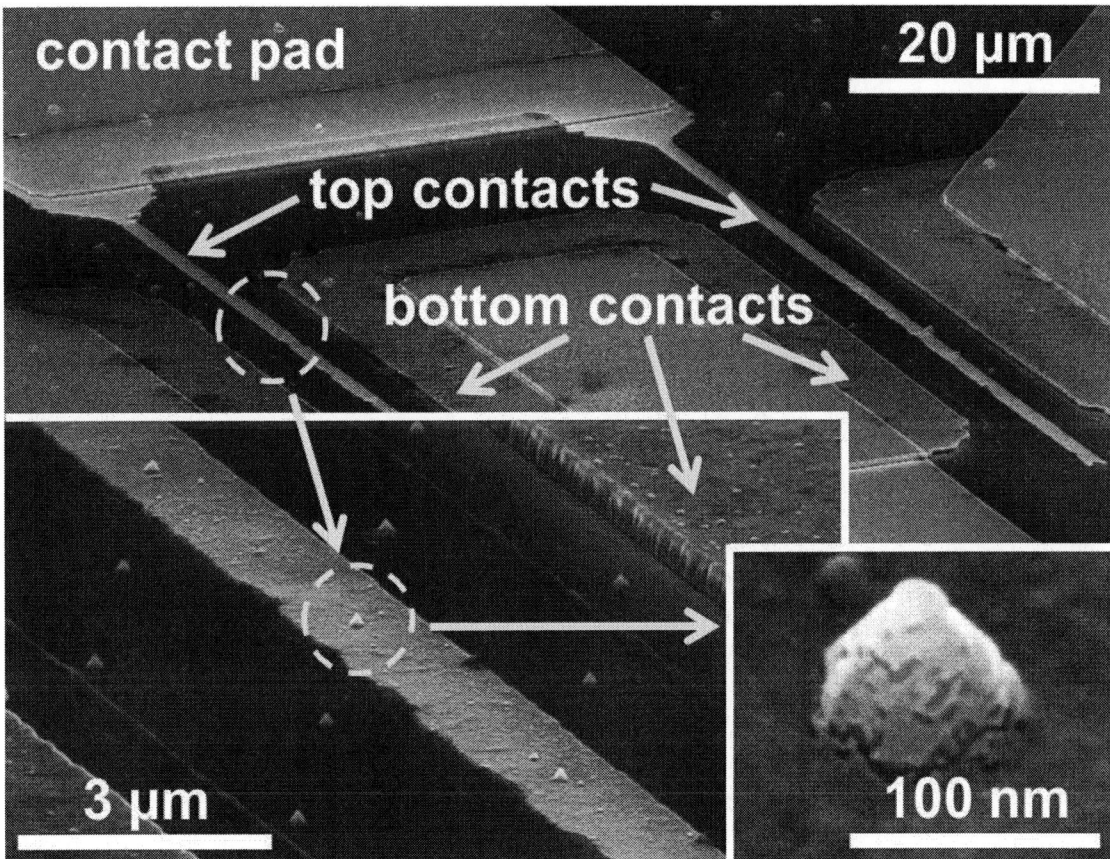

Figure 2: Scanning electron micrograph of the vertically integrated InGaN pyramids with mesoscopic size (prepared by SA-MOVPE) within hexagonally arranged arrays. The inset shows the top contact area region with a singularly addressable nano-LED structure in two different magnifications: the mesoscopic sized nanopyramid structure is covered with a transparent Ni/Au top contact. The device layout (presented in the figure-top-area) includes high frequency (HF) contact pads and vertically integrated nano-LEDs and was designed with respect to further DC testing and future HF operation.

3.   Micro photo (PL)- and electroluminescence (EL) characterization

At first micro PL studies were carried out on the core of the nano-LEDs: the InGaN pyramids with 100 nm size and with varying In content between 0 and 100 %. As expected, a systematic decrease in PL emission energy with increasing In composition is observed (figure 3). The recorded emission energy correlates with the room temperature band gaps of the InGaN alloys. The behavior relates to a very slight deviation from a linear interpolation between the respective band gaps of the binary compounds InN and GaN for the alloy. A nanopyramid with an In content of 90 % results in emission in the technologically relevant telecommunication wavelength range. This composition was therefore chosen to study the size-dependent EL of nano-LEDs. Representative spectra are presented in figure 4. We found that the band edge luminescence energy of the nano-LEDs red-shifts as the structure sizes

978-1-5090-3084-2/16 $31.00 © 2016 IEEE   71

decrease. We attribute this behavior to different degrees of strain originating from the interaction of the nano-LED with the surrounding SiO_2 isolation mask. Strain can therefore be exploited additionally to tune the wavelength emission of III-nitride nano-LEDs[13].

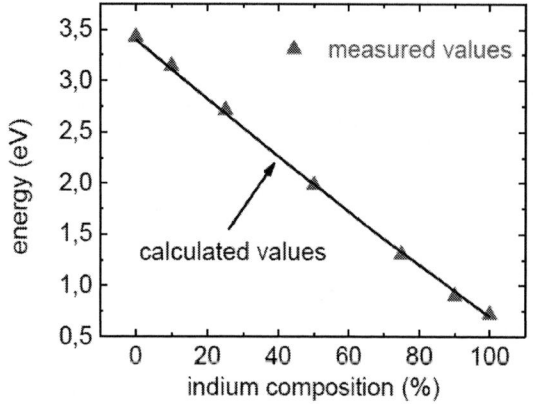

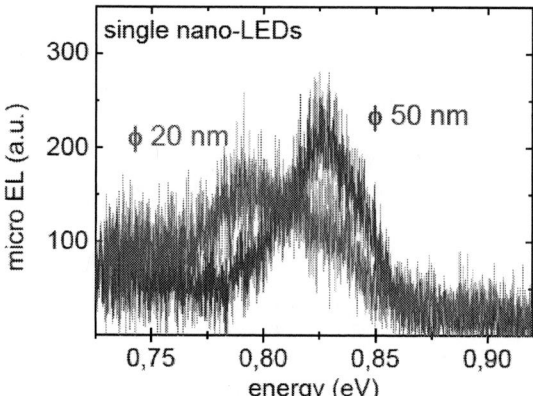

Figure 3: Dependence of the PL emission- band gap energy for InGaN nanopyramids (diameter: 100 nm) on In composition. When using an In composition of close to 90 %, emission is achieved in the technologically important telecommunication wavelength range.

Figure 4: Micro EL measurements for single 20 nm and 50 nm (diameter) vertically integrated InGaN nano-LED structures.The observed EL energy shift is attributed to differing size-related strain inducing interactions between nano-LED and surrounding isolation mask

4. Conclusion

A successful technological integration and material optimization of InGaN mesoscopic structures was presented. Our results reveal that these structures have a great potential as future low energy consumption and reliable nitride based emitting sources for an operation covering the UV, through the VIS and finally the telecommunication wavelength range.

Acknowledgement

This work was partially supported by Czech Science Foundation (project No. 13-20507S).

References

[1]   G.S. Buller, R.J. Collins, Meas. Sci. Technol. **21**, 012002 (2010).
[2]   Ł. Dusanowski et al., Appl. Phys. Lett. **105**, 021909 (2014).
[3]   M. Mikulics et al., Appl. Phys. Lett. **108**, 061107 (2016).
[4]   M. Mikulics, H. Hardtdegen, Nanotechnology. **26**, 185302 (2015).
[5]   A.F. Jarjour et al., Phys. Status Solidi. **206**, 2510–2523 (2009).
[6]   M. Mikulics et al., Appl. Phys. Lett. **86**, 211110 (2005).
[7]   M. Mikulics et al., IEEE Photonics Technol. Lett. **23**, 1189–1191 (2011).
[8]   A. Winden et al., Phys. Status Solidi. **9**, 624–627 (2012).
[9]   A. Winden et al., J. Cryst. Growth. **370**, 336–341 (2013).
[10]   H. Hardtdegen et al., J. Cryst. Growth. **124**, 420–426 (1992).
[11]   Y.S. Cho et al., Phys. Status Solidi. **3**, 1408–1411 (2006).
[12]   S. Riess et al., Jpn. J. Appl. Phys. **52**, 8–12 (2013).
[13]   M. Mikulics et al., Appl. Phys. Lett. 109, 041103 (2016).

ASDAM 2016, The 11th International Conference on Advanced Semiconductor
Devices And Microsystems, November 13-16, 2016, Smolenice, Slovakia

Electrical and optical characterization of freestanding $Ge_1Sb_2Te_4$ nano-membranes integrated in coplanar strip lines

M. Mikulics[a,b], M. Marso[c], R. Adam[a,b], M. Schuck[a,b], A. Fox[a,b], R. Sobolewski[d],
P. Kordoš[e], H. Lüth[a,b], D. Grützmacher[a,b] and H. Hardtdegen[a,b]

[a] Peter Grünberg Institute, Forschungszentrum Jülich, 52425 Jülich, Germany
[b] JARA – Fundamentals of Future Information Technology
[c] Faculté des Sciences, de la Technologie et de la Communication,
Université du Luxembourg, L-1359 Luxembourg
[d] Department of Electrical and Computer Engineering and Laboratory for Laser Energetics,
University of Rochester, Rochester, New York 14627-0231, USA
[e] Institute of Electronics and Photonics, Slovak Technical University, SK-81219 Bratislava,
Slovak Republic

We developed a transfer and integration technique for freestanding $Ge_1Sb_2Te_4$ nano-membranes. A so-called laser micro annealing process for the precisely-local formation of low resistance $Ge_1Sb_2Te_4/Ti/Au$ ohmic contacts was optimized. Ultrafast switching effects were studied on nano-membrane devices after their implementation into a group III-nitride based optoelectronic circuit. Highly resistive switching was observed without any noticeable structural deterioration. Charge transport measurements indicate that current densities exhibit values in the "READ resp. SET/RESET" regime from 10^{-6} down to $10^{-11} A/\mu m^3$ after ~100 switching cycles. Our study on freestanding $Ge_1Sb_2Te_4$ nano-membranes contributes to a better understanding of charge transport related physical phenomena in novel chalcogenide material systems with high potential for future low energy consumption data storage devices.

1. Introduction

Since the beginning of the 21st century, the concerted developing efforts especially in opto-electronics on the nano-meter scale and the progress in related scientific fields have brought about new insights in applied physics. They resulted in a number of novel applications for commercial electronics, energy harvesting, medicine as well as environmental friendly products and services. Finally yet importantly, they also reached so-called green ITs with a focus on efficient and low energy consuming data storage. In general, low dimensional nanometer sized objects exhibit unique material properties, which allow the fabrication of novel energy saving devices [1]. These potentially solve and overcome problems related to thermal degradation and device reliability issues associated with a high integration density. The next generation of such low energy consuming data storage devices are needed primarily for high information rate communication systems and should operate up to the THz frequency range. The currently used "fast" (up to ~ 100 MHz) phase change memories (PCM) mostly employ micro/nano trench cell structures based on chalcogenides [2]. However, the efficiency limits of such structures has already been reached and, therefore, further performance improvement of data storage devices can be obtained only by alternative material structures with picosecond or rather subpicosecond charge transport and switching dynamics. Here we explore the properties of freestanding $Ge_1Sb_2Te_4$ based nano-membrane structures as possible

978-1-5090-3084-2/16 $31.00 © 2016 IEEE 73

candidates for the application in next generation highly efficient and ultrafast data storage devices. This systematical study describes the fabrication process, the ohmic contact optimization / formation during the local laser micro annealing procedure and the device characteristics as well as the switching dynamics in long-term operation.

2. Device fabrication

The fabrication process starts with the Metal-Organic Vapor Phase Epitaxy (MOVPE) growth of planar $Ge_1Sb_2Te_4$ layers on undoped silicon (111) substrates employing an optimized growth procedure [3]. After MOVPE growth, thin (~ 100 nm) nano-membrane structures were removed from the "native" substrate, and an improved transfer technique was used for the nano-membrane positioning onto the "host" substrate. The entire technological process was described in detail in our previous work for the fabrication of micrometer sized picosecond and femtosecond photo-switches [4,5]. We improved the micromanipulation of the nano-membrane structures and increased the effectiveness of the fabrication processes by using a micrometer-sized pipette of quartz glass, which allows precise manipulation of our nanometer/micrometer sized structures. The schematics of the process is presented in Figure 1. After nano-membrane transfer to the host substrate, the structures were positioned with 5 - 10 µm accuracy on the desired location and bonded through van der Waals forces to the coplanar strip line.

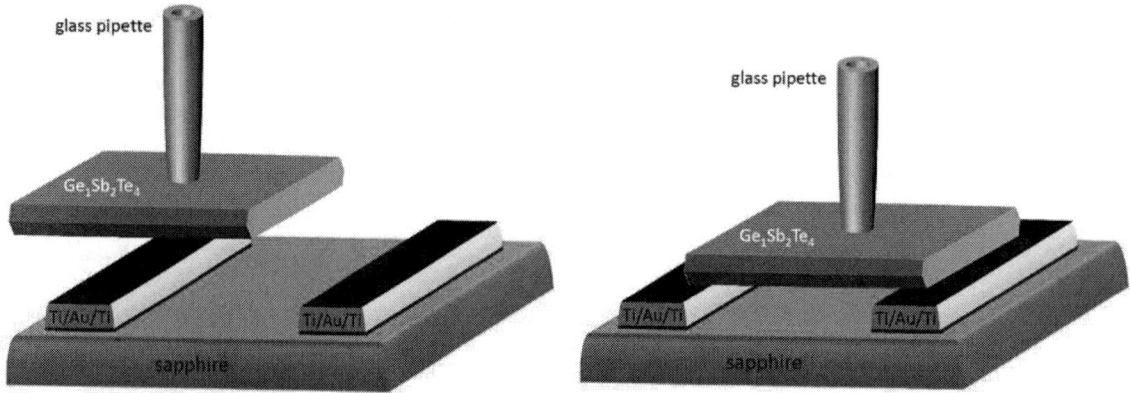

Figure 1: Transfer and integration scheme for a freestanding $Ge_1Sb_2Te_4$ nano-membrane. The nano-membrane is transferred to the non-native sapphire substrate using the glass pipette/micro manipulator assisted transfer technique [4,5].

3. Laser Micro Annealing (LMA) dynamics, micro PL and DC measurements

The nano-membranes integrated in CPS lines were locally annealed at various annealing times from 10 seconds to 15000 seconds in nitrogen atmosphere. The aim was to improve the adhesion between the $Ge_1Sb_2Te_4$ nano-membrane structure and the Ti/Au/Ti coplanar strip lines after the transfer process on the one hand and to eliminate possible defects on the side walls of the nano-membrane as well as to initialize ohmic contact formation on the other hand. The annealing process was performed by means of a focused laser beam (HeCd, 325 nm). The so-called laser micro annealing (LMA) procedure is shown schematically in Figure2. It should be noted that the laser beam was focused onto an area of ~ 5 µm in diameter at the interface between nano-membrane and coplanar strip lines. Subsequently, the nano-membranes integrated into the coplanar strip lines were optically inspected (see Figure 3) and their resistance determined as a function of annealing time. Figure. 4 presents this dependence. The laser input power was kept constant at ~ 1 kW/cm². The decrease in resistance in the first 2 regions is related to fast and slower diffusion processes. In the 3rd region, thermal effects start to dominate the resistance behavior and degradation of the nano-membrane sets in. In the 4th region, after ~ 3.5 hours, the nano-membranes start to melt and to decompose. The lowest resistance was achieved in the second region (slow diffusion). The values determined - ~ 0.4 - 0.6 kΩ/μm - are fully comparable with those previously reported [6].

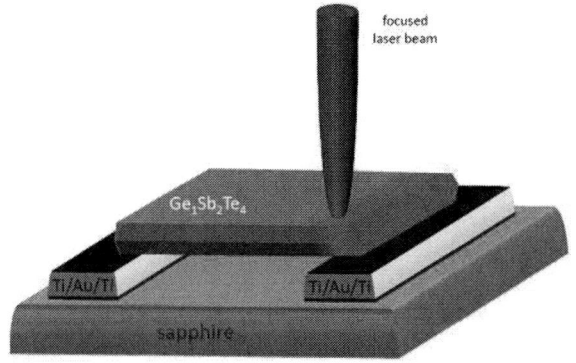

Figure 2: Schematics of the so-called laser micro annealing (LMA) process. The entire structure was annealed by using a focused HeCd laser beam.

Figure 3: Fully integrated $Ge_1Sb_2Te_4$ nano-membrane in CPS lines and locally formed ohmic contacts.

Micro photoluminescence (μ−PL) measurements (Fig.5.) were performed at 300 K using a continuous wave He-Cd laser as the excitation source (325 nm) directly before and after the annealing process as well as after "switching" characterization. The spot size of the laser was about 500 nm. The laser light was focused on the nano-membrane structure by a 100x UV objective, resulting in a lateral resolution of ~ 500 nm. The sample emission was analyzed by a spectrometer with ~ 0.1 nm resolution. The spectrum obtained before switching exhibits a narrow ~ 100meV emission peak at 0.57 eV (black curve). After annealing, the red spectrum was measured with a ~ 210 meV broad peak at 0.82 eV. The peak form and position are indicative of the band gaps [7] for crystalline trigonal and amorphous $Ge_1Sb_2Te_4$, respectively.

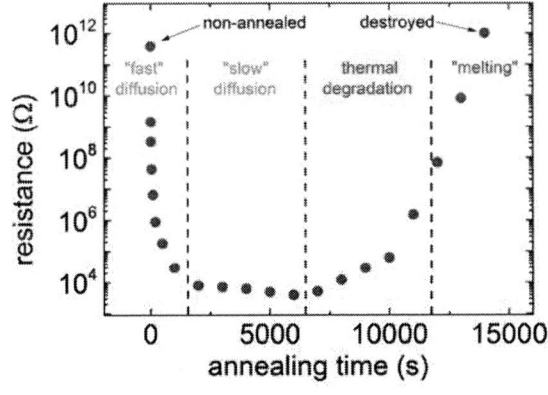

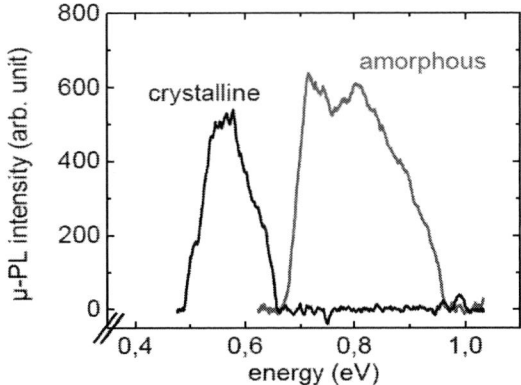

Figure 4: Total resistance (nano-membrane and contacts) as a function of annealing time.

Figure 5: Micro PL intensity measurements for $Ge_1Sb_2Te_4$ crystalline and amorphous phases.

4. Switching characteristics

Ultrafast switching effects were studied on nano-membrane devices after their implementation into a group III-nitride based optoelectronic circuit [8,9]. Figure 6 presents a representative current-voltage characteristic after ~ 100 switching cycles. Highly resistive switching was observed without any noticeable structural deterioration. Charge transport measurements indicate that current densities exhibit values in the "READ and SET/RESET" regime from 10^{-6} down to $10^{-11} A/\mu m^3$ after ~ 100 switching cycles (Fig.7.).

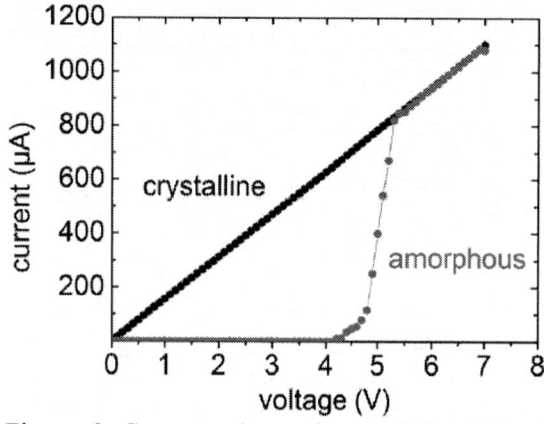

Figure 6: Current-voltage characteristic obtained after ~ 100 switching cycles demonstrates the threshold switching effect and the phase change behavior for the $Ge_1Sb_2Te_4$ nano-membrane.

Figure 7: Cycling performance of the SET and RESET state of a freestanding single $Ge_1Sb_2Te_4$ nano-membrane integrated in Ti/Au/Ti coplanar strip lines.

5. Conclusion

We fabricated and tested freestanding $Ge_1Sb_2Te_4$ nano-membrane devices transferred to a sapphire-host substrate and integrated them into CPS lines. The influence of micro laser annealing on total resistance was investigated and the annealing procedure optimized. Our study on freestanding $Ge_1Sb_2Te_4$ nano-membranes contributes to a better understanding of charge transport related physical phenomena in novel chalcogenide material systems with high potential for future low energy consumption data storage devices.

Acknowledgement

Funding from the European Union FP7 project "SYNAPSE" and the German Research Foundation within SFB 917 "Nanoswitches" is acknowledged.

References

[1]   R.E. Simpson et al., Nat. Nanotechnol. **6**, 501–505 (2011).
[2]   G.W. Burr et al., J. Vac. Sci. Technol. B **28**, 223 (2010).
[3]   H. Hardtdegen et al. J. Alloys and Compd. **679**, 285-292 (2016).
[4]   M. Mikulics et al., Appl. Phys. Lett. **101**, 031111 (2012).
[5]   M. Mikulics et al., Semicond. Sci. Technol. **29**, 045022 (2014).
[6]   T. Siegrist et al., Ann. Rev. Condens. Matter Phys. **3**, 215-237 (2012).
[7]   J.W. Park et al., Phys. Rev. B **80**, 115209 (2009).
[8]   M. Mikulics et al., IEEE Photonics Technology Letters, **23**, pp. 1189-1191 (2011).
[9]   M. Mikulics et al., to be submitted to Appl. Phys. Lett. in (2016).

ASDAM 2016, The 11th International Conference on Advanced Semiconductor
Devices And Microsystems, November 13-16, 2016, Smolenice, Slovakia

Hybrid Optoelectronics Based on a Nanocrystal/III-N Nano-LED Platform

M. Marso[a], M. Mikulics[b,c], H. Lüth[b,c], Z. Sofer[d], P. Kordoš[e] and H. Hardtdegen[b,c]

[a] Faculté des Sciences, de la Technologie et de la Communication,
Université du Luxembourg, L-1359 Luxembourg
[b] Peter Grünberg Institute, Forschungszentrum Jülich, 52425 Jülich, Germany
[c] JARA – Fundamentals of Future Information Technology
[d] Institute of Chemical Technology, Prague, Technická 5, 166 28 Prague 6, Czech Republic
[e] Institute of Electronics and Photonics, Slovak Technical University, SK-81219 Bratislava,
Slovak Republic

We fabricated and tested hybrid-III-nitride (CdSe nanocrystal/p-GaN/MQW/n-GaN/sapphire) based nano-LEDs and integrated them into a device layout suitable for DC testing and designed for future operation at high frequencies. Our studies provide clear evidence that our technological process for the vertically integrated nano-LED emitters is perfectly suited for long-term operation without any indication of degradation effects. This novel technology shows strong potential for a future single photon based OE circuit.

1.   Introduction

Recently we introduced a universal hybrid-optoelectronic platform [1,2] as an alternative to conventionally used and well-established technological solutions [3,4]. "Few" photon sources as well as single photon emitters operating also under harsh conditions and preferably at room temperature are proposed to be the key-technology for future low energy consumption nano-opto-electronics. Additionally, such photon sources based on hybrid device architectures are required for highly secure, ultrafast optoelectronics [5-7] and future spin-photonic applications. Therefore there is a strong need to develop such emitting sources covering a broad wavelength range - starting from conventional infrared wavelengths (~ 1.3÷1.7 µm) used for telecommunication, through the visible range for quantum cryptography and down to the UV range for optoelectronics [8]. III-nitrides with their well-known unique material properties are predestined to serve as a basis for such hybrid solutions. The basic idea behind our hybrid optoelectronic platform is the combination of electrically driven nano-LED (light emitting diodes) sources, used for primary optical excitation, and secondary emission from excited mesoscopic or nanocrystal structures. This hybrid solution overcomes technological difficulties associated with a contacting of only "few" nanometer sized objects. In addition, it contributes essentially to a decrease in production costs. Of course, there are still unanswered questions regarding the whole nano-LED/nanocrystal (NC) integration technology, the material and system compatibility, degradation effects as well as the suitability for a possibly required reliable long-term operation.

In this contribution, we will present a hybrid nano-LED device technology. We will start with the preparation of single vertically stacked p-GaN/MQW/n-GaN nano-LED-structures in well-positioned arrays and will continue with the device technology and subsequent nanocrystal structure integration.

2.   Device fabrication

978-1-5090-3084-2/16 $31.00 © 2016 IEEE

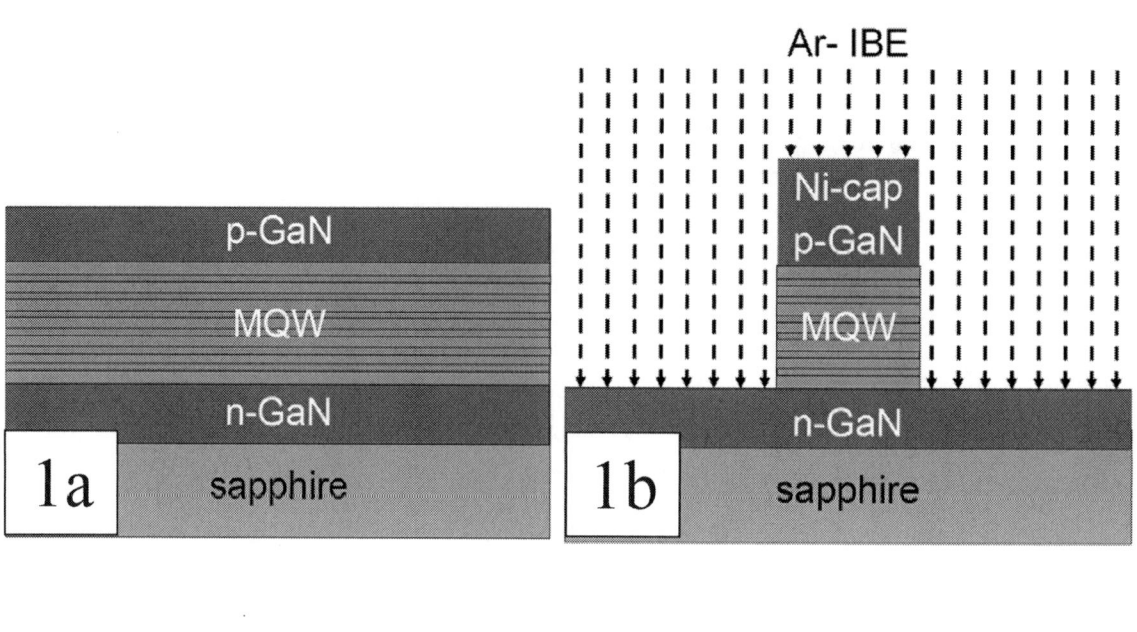

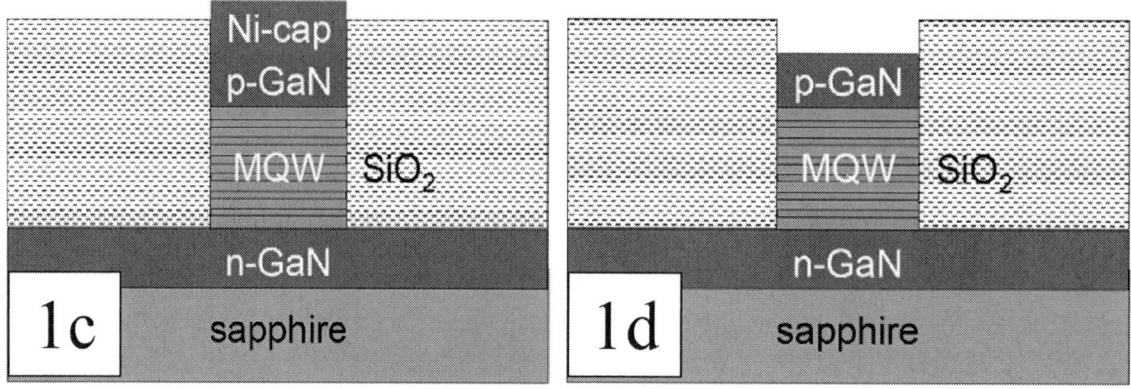

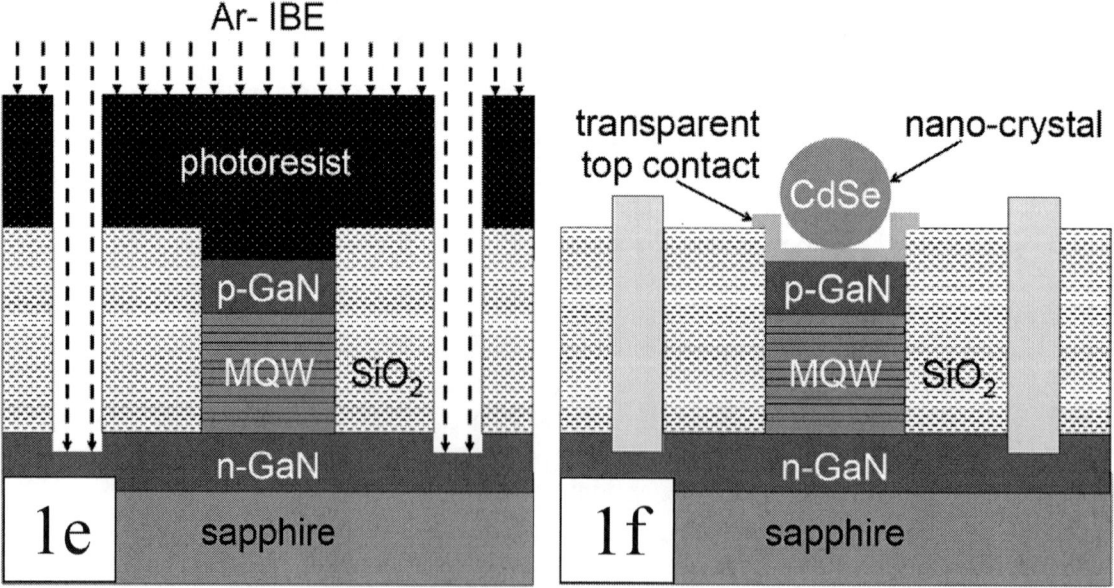

Fig. 1: Schematics-fabrication process of hybrid/III-nitride nano-LED structure integrated into a vertical device layout.

First, we started with the structuring of the III-nitride layer system (p-GaN/MQW/n-GaN/sapphire) grown by MOVPE (Fig. 1a) [9,10]. Nano-LEDs were defined using electron-beam lithography combined with Ar-ion beam etching through Ni-capping layers (Fig. 1b). The etching process needed to be optimized carefully [11]. All nanostructures were isolated from their neighbours using a SiO_2 thin film (Fig. 1c). The Ni-capping layer was removed by a $HCl:H_2O$ solution (Fig. 1d). The large bottom (Ti/Al/Ni/Au) and the top (Ni/Au) contacts (Fig. 1e and 1f) were defined using optical lithography. The CdSe nanocrystal was placed above the transparent top contact of the nano-LED. The entire fabrication process has been described in detail in [2]. Figure 3 presents the final hybrid nano-crystal/nano-LED device integrated into a device layout suitable for DC and HF characterization.

3. DC characterization and Micro electroluminescence measurements

First, our hybrid-nano-LEDs (see Fig.3) were tested by DC measurements (figure 2). The I/V characteristic shown in Figure 2 indicates a very low energy consumption of ~ 40nW at 5V bias, which may be around 6–7 orders of magnitude lower than for a conventional LED.

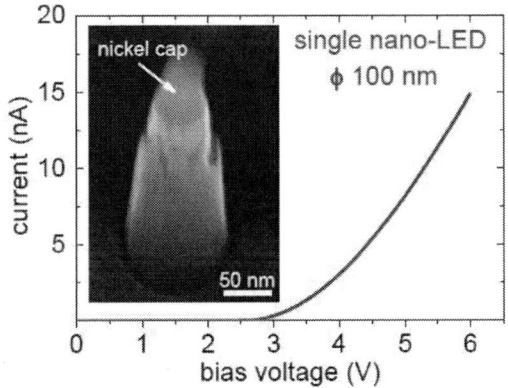

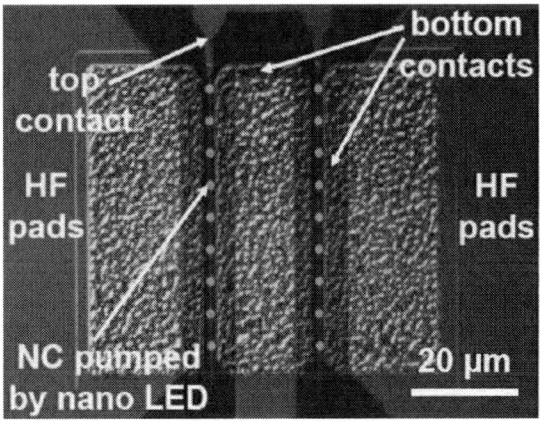

Fig. 2. IV characteristic of a single nano-LED (diameter 100 nm, SEM micrograph in the inset) [2].

Fig.3. Fully integrated NC/nano-LED structures (green dots, as a guide to the eye) in the device layout suitable for DC and HF characterization.

Representative optical spectra from a single nitride based nano-LED structure and secondary emission from an excited CdSe nanocrystal are presented in [2,12]. The pulse response for a single nano-LED and a secondary excited NC was recorded with the help of an ultrafast photodetector. The pulse FWHM (nano-LED) is about 0.9 ns at 400 nm and ~1.4 ns for the NC (at ~540 nm). This new nanoscale hybrid-LED device (Fig.3.) could theoretically achieve transmission speeds of 1 Gbps. Figure 5 presents the electroluminescence intensity for the nano-LED with a diameter of 100nm and the induced PL emission of the nanocrystal at DC "long-term" operation. I-V curve measurements and the LED emission indicate stable (~ 10%) optical characteristics for ~ 1000 hours. The slight decrease in EL intensity together with the increase in current are attributed to thermal effects. The photoluminescence emitted from the NC, however, is constant and not affected, since the LED`s (pump) emission is still sufficient for secondary NC emission. The choice of the semiconductor nanocrystal (and its size) can be employed to induce emission at the desired wavelength. Furthermore, we determined a photon-photon down conversion efficiency of 27%. Our results disclose the potential of the

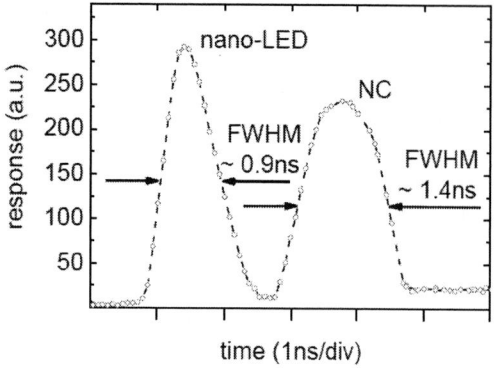

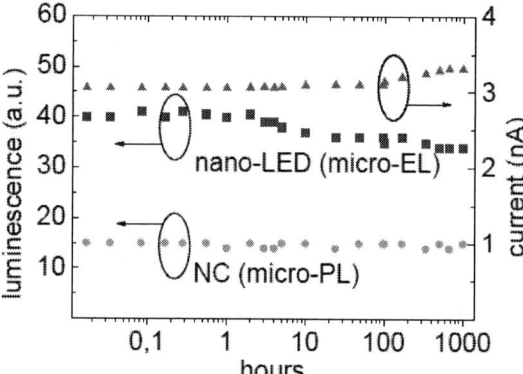

Fig. 4. Response measurements for a single ø 100 nm nano-LED (pump) and CdSe nano-crystal (NC) with a diameter of ~4 nm. Nano-LED source operates at a maximum repetition rate of 1 GHz.

Fig.5. Luminescence intensity and current as a function of time. LED emission as well as secondary NC emission are long-term stable. The slight decrease in LED intensity is due to thermal effects.

hybrid optoelectronic platform for a large range of low energy consumption reliable "few" photon sources.

4.   Conclusion

We successfully fabricated and tested a hybrid nano-optoelectronic platform based on a nanocrystal and a group III-N nano-LED. The nanocrystals were electro-optically pumped by the electrically driven LEDs and emit without any need to contact them. The reliable emission and its efficiency together with the wide variety of possible nanocrystals make this device concept an attractive basis for a large range of low energy consumption photon sources.

References

[1]   M. Mikulics and H. Hardtdegen, Germany, 12 December 2012 patent specification DE102012025088 (A1), WO2014094705 A1 (2013).

[2]   M. Mikulics et al., Appl. Phys. Lett. **108**, 061107 (2016).

[3]   M.A. Schreuder et al., Nano Lett. **10**, 573 (2010).

[4]   N. Narendran et al, in *Lightemitting Diodes: Research Manufacturing and Applictions IV Proc. SPIE*, San Jose, USA, 26 - 27 January 2000, edited by H.W. Yao, I.T. Ferguson, and E.F. Schubert, pp. 240–248.

[5]   B. Lounis and M. Orrit, Reports Prog. Phys. **68**, 1129–79 (2005).

[6]   N. Gisin, G. Ribordy, W. Tittel and H. Zbinden, Rev. Mod. Phys. **74**, 145–95 (2002).

[7]   G.S. Buller and R.J. Collins, Meas. Sci. Technol. **21** 012002 (2010).

[8]   M. Mikulics and H. Hardtdegen, Nanotechnology **26**, 185302 (2015).

[9]   H. Hardtdegen, M. Hollfelder, R. Meyer, R. Carius, H. Münder, S. Frohnhoff, D. Szynka, and H. Lüth, J. Cryst. Growth **124**, 420 (1992).

[10]   Y.S. Cho, H. Hardtdegen, N. Kaluza, N. Thillosen, R. Steins, Z. Sofer, and H. Lüth, Phys. Status Solidi **3**, 1408 (2006).

[11]   J. Moers, M. Mikulics, M. Marso, A. Haab, St. Trellenkamp, Z. Sofer, D. Grützmacher, and H. Hardtdegen, ASDAM 2016.

[12]   H. Hardtdegen and M. Mikulics, ASDAM 2016.

ASDAM 2016, The 11th International Conference on Advanced Semiconductor
Devices And Microsystems, November 13-16, 2016, Smolenice, Slovakia

Fabrication of UV sources for novel lithographical techniques: development of nano-LED etching procedures

J. Moers[a,b)], M. Mikulics[a,b)], M. Marso[c)], St. Trellenkamp[a,b)], Z. Sofer[d)],
D. Grützmacher[a,b)], and H. Hardtdegen[a,b)]

[a)] *Peter Grünberg Institute, Forschungszentrum Jülich, 52425 Jülich, Germany*
[b)] *JARA – Fundamentals of Future Information Technology*
[c)] *Faculté des Sciences, de la Technologie et de la Communication,*
Université du Luxembourg, L-1359 Luxembourg
[d)] *Dept. of Inorganic Chemistry, Institute of Chemical Technology, Prague, Technická 5,*
166 28 Prague 6, Czech Republic

The development of two different dry etching approaches – ion beam etching (IBE) and reactive ion etching (RIE) - is reported for the fabrication of nano-LEDs as UV sources. The IBE approach leads to nano-LEDs with higher emission intensity but with rougher side-walls and broader FWHM.

1. Introduction

During the last decades, light emitting diodes (LEDs) based on different material systems were implemented in a broad range of optoelectronic applications. Beside being applicable for ultra-fast optoelectronic circuits [1,2], they are suitable for emission at short wavelengths down to the ultraviolet (UV) range. Recently, we proposed that nano-LED arrays can be used as key elements for future nano-LED assisted lithography [3,4]. In this novel lithographical approach, a train of nano-LED-generated photons initializes substantial changes in photochemical molecules to produce the desired structures down to the molecular scale. Group-III-nitride-based nano-LEDs arranged in arrays are the essential part of this novel lithographical technique. The development and production of stable and reliable nano-LED sources plays an important role in the further progress in this field until this novel technique will be well established and suitable for mass production.

The formation of such UV nano-LEDs by etching nanowires (NWs) from layers, however, is still a challenge and one of the most critical steps in the nano-LED fabrication process. This work focuses on two different ion beam etching approaches: the first, based on chemical etching, is the reactive ion etching (RIE) approach; the second, based on physical etching, is argon ion beam etching (IBE). Both approaches and their influence on the nano-LEDs´ optical properties will be presented and their advantages and disadvantages evaluated and discussed in detail.

2. Sample preparation

First, we started with the deposition of p-GaN/MQW/n-GaN LED layer structures by Metalorganic Vapor Phase Epitaxy (MOVPE) [5,6]. Subsequently nano-LEDs were defined using electron-beam lithography for the creation of Ni-caps (Fig. 1 and Fig. 2) which served as the etching mask. The entire fabrication process has been described in detail in [4,7]. RIE experiments were carried out in an Oxford Plasmalab 100 reactor with an ICP180 source at a constant pressure of 4 µbar and a sample temperature of 5°C using chlorine:argon gas mixtures ranging between 20:0 and 0:20. The RF power was varied between 150 and 200 W and the ICP power from 500 to 2500 W. Similarly, Ar-IBE experiments were performed in an IBE-Oxford equipment. Different series of nano-LEDs (Fig.3.) were prepared under

978-1-5090-3084-2/16 $31.00 © 2016 IEEE

optimized ion beam etching conditions i.e. on etching parameters such as accelerating voltage (from 200V up to 500V) and beam current (from 20mA up to 88mA). Additionally the influence of incident beam angle (in the range ~5°÷30°) and its effect on the nano-LED/NW´s surface morphology as well as on their optical characteristics were studied systematically with the aim of preventing channeling effects and suppressing non-radiative recombination. The optical properties of the nano-LED structures processed were evaluated with the help of micro-photoluminescence (micro-PL) measurements (Fig.4.).

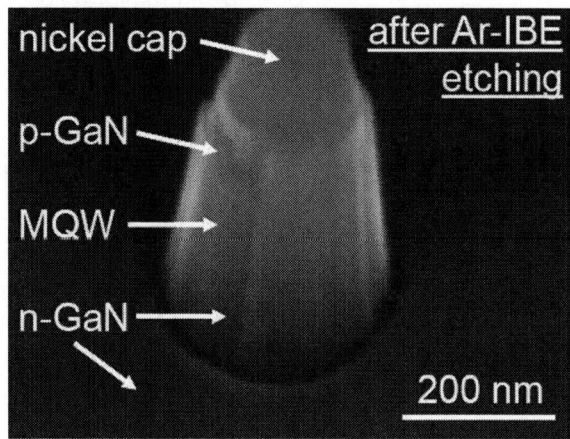

Fig. 1: Fabrication schematics of the group-III nitride based nano-LED defined with the help of Ar ion beam etching. A nickel cap serves as the protecting/etching mask [4,7].

Fig. 2: SEM image of nano-LED structure after the Ar-IBE process. The surface morphology is strongly influenced by the ion beam etching process conditions [4,7].

3. Micro photoluminescence measurements

Arrays of hexagonally arranged nano-LEDs were prepared with both approaches and with different diameters ranging from 100nm to 200nm and a pitch of 3µm. A representative scanning electron microscopy (SEM) image of an array fabricated using IBE is presented in Figure 3. Beside the hexagonally arranged nano-LED pillars, small "parasitic" nanostructures are observed. A micro PL mapping of the etched structures reveals, however, that only the nano-LED pillars are optically active. Such a mapping of the array performed at room temperature and at 405 nm wavelength is presented in Figure 4. The influence of structure size on micro PL emission energy and intensity was investigated for individual nano-LEDs and is displayed in Figure 5. The PL peak intensity and energy are linearly dependent on the structure size [4,7]. As a consequence of nanostructure formation by the Ar-IBE process, the nanostructures relax in a size-dependent way [8] leading to a red-shift of ~ 100meV for a diameter of 100 nm in comparison to its 200 nm etched nanostructure counterpart. This is a reproducible effect and can be used to tune the emission wavelength [4]. The PL intensity is primarily dependent on the (nano-LED) volume (see Figures 5 and 6). In addition, the intensity is strongly affected by the etching conditions. Channeling effects as well as surface roughening can initialize an increase in non-radiative recombination centers in the region close to the structure "wall". Figure 6 presents the PL intensity as a function of Ar- ion accelerating voltage (V_a) during the etching. A decrease in accelerating voltage from 500 V to 200 V causes a significant increase in PL intensity emitted by a single nano-LED by ~ 20 to 400 % for the largest and the smallest nano-LED, respectively. It is obvious, that a careful optimization of etching parameters is a prerequisite for a stable and reliable nano-LED source.

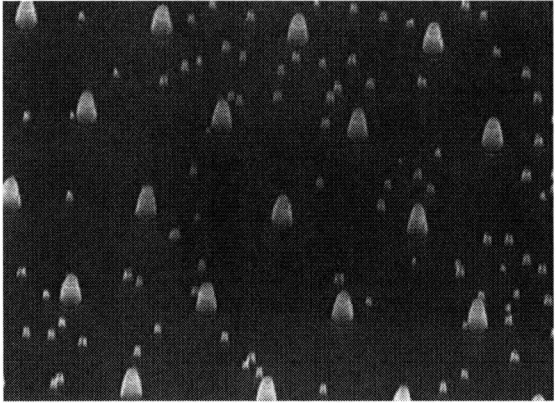

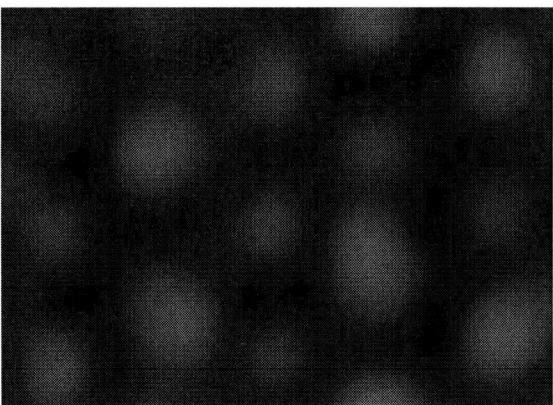

Fig. 3: An ensemble of III-nitride nano-LED structures arranged in hexagonal array. Parasitic "nanostructures" between "major" nano-LEDs are optically inactive.

Fig. 4: Lateral mapping of the micro photoluminescence from the nano-LED array performed at room temperature and 405 nm emission wavelength.

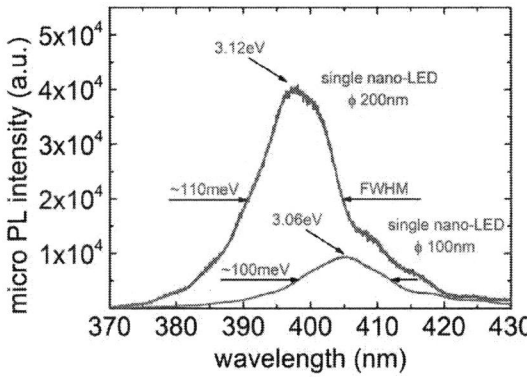

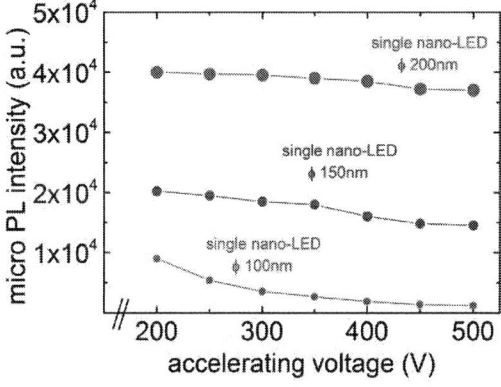

Fig.5: micro PL intensity recorded for nano-LEDs with 100 and 200nm diameter produced using an optimized Ar-IBE process. The size dependent nanostructure relaxation leads to a red-shift of the emission by ~ 100meV.

Fig.6: micro PL intensity of nano-LEDs with 100nm, 150nm and 200nm diameter as a function of Ar-IBE accelerating voltage (V_a). An up to fourfold intensity increase is observed as the V_a is reduced to 200V for the smallest diameter [4].

Similar arrays were also fabricated using RIE as the etching approach. Here also, a careful optimization of the process parameters was mandatory. The influence of inductively couple plasma (ICP) power on emission intensity was investigated for three different nano-LED diameters. We found, that the intensity increases at first as the power is increased but then drops again and that the optimal power increases with nano-LED diameter by a factor of ~ two in the range studied. At the optimal ICP power (and radio frequency (RF) power, not presented here) etching conditions, the influence of Cl_2/Ar gas etching mixture on emission intensity was studied. This ratio correlates with the influence of chemical versus physical etching in RIE. Here, a ratio of 16:4 is optimal independent of the nano-LED diameter.

The characteristics of nano-LEDs fabricated using both approaches were compared. The side-walls of the IBE etched nano-LED pillars are rougher than those prepared by RIE (not presented here). This could be detrimental for further device processing, when the nano-LEDs are embedded in a dielectric for device isolation. The intensity of the nano-LEDs is 5 – 10 times lower for the RIE process. It is evident, that the RIE process affects more of the emitting MQW region than the IBE process. The FWHM, however, is substantially larger for the IBE approach, which is attributed to the increase defect formation on the side walls of the nano-LEDs by channeling effects.

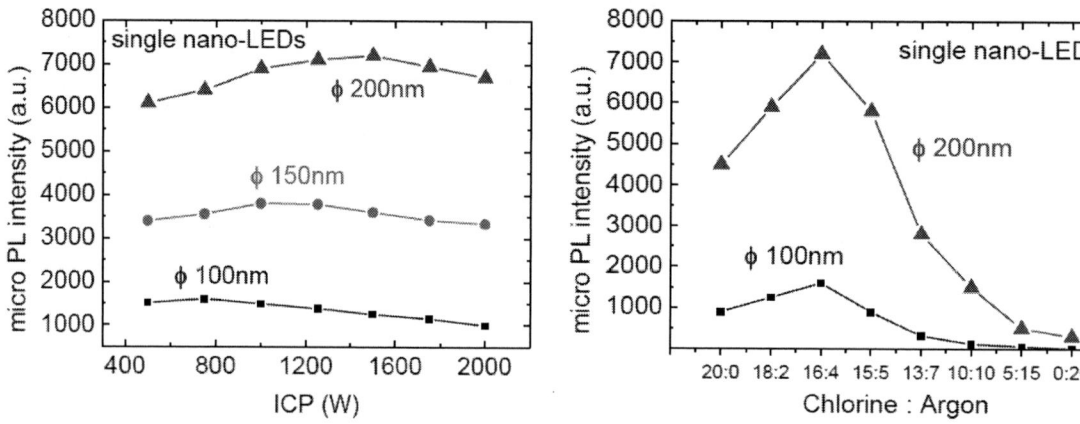

Fig.7: micro PL of nano-LEDs with 100nm, 150nm and 200nm diameter as a function of RIE inductively coupled plasma (ICP) power. The optimal ICP power increases with nano-LED diameter.

Fig.8: micro PL intensity for two different nano-LED diameters as a function of chlorine:argon etching gas ratio. A ratio of 16:4 is optimal for the etching of nano-LEDs in RIE.

4. Conclusion

We successfully optimized two different etching approaches for nano-LED fabrication using micro-photoluminescence spectroscopy as the characterization tool. The group-III-nitride-based nano-LEDs prepared by both approaches are suitable for the novel lithography technique, which has the potential to significantly simplify and accelerate future nano-lithographical processes in mass-production. The advantages and disadvantages of both etching approaches for fabrication of nano-LEDs are compared and discussed.

Acknowledgement

This work was partially supported by Czech Science Foundation (project No. 13-20507S).

References

[1]   M. Mikulics et al., Appl. Phys. Lett. **86**, 211110 (2005).
[2]   M. Mikulics et al., IEEE Photon. Technol. Lett. **23**, 1189 (2011).
[3]   M. Mikulics, and H. Hardtdegen, Patent DE20121016178 20120816, (2013).
[4]   M. Mikulics, and H. Hardtdegen, Nanotechnology **26**, 185302 (2015).
[5]   H. Hardtdegen et al., J. Cryst. Growth **124**, 420 (1992).
[6]   Y.S. Cho et al., Phys. Status Solidi **3**, 1408 (2006).
[7]   M. Mikulics et al., Appl. Phys. Lett. **108**, 061107 (2016).
[8]   N. Thillosen et al., Nano Lett. **6**, 704–708 (2006).

ASDAM 2016, The 11th International Conference on Advanced Semiconductor
Devices And Microsystems, November 13-16, 2016, Smolenice, Slovakia

The effect of process parameters and annealing on the properties of Ti/Pt films for miniature temperature sensors

I. Hotový[1], Š. Haščík[2], M. Predanocy[1], M. Mikolášek[1], V. Řehaček[1], I. Kostič[3], P. Nemec[3], A. Benčurová[3], D. Rossberg[4], L. Spiess[4]

[1]Institute of Electronics and Photonics, Slovak University of Technology,
Ilkovicova 3, 812 19 Bratislava, Slovakia
ivan.hotovy@stuba.sk

[2]Institute of Electrical Engineering, Slovak Academy of Sciences, Bratislava, Slovakia
[3]Institute of Informatics, Slovak Academy of Sciences, Bratislava, Slovakia
[4]Department of Materials Technology, Technical University of Ilmenau, D-98684 Ilmenau, Germany

We present a Pt film temperature element of the size of 1.5×1.5 mm² with the meander width of 20 µm. It was investigated the effect of Pt film thickness prepared on alumina ceramics and the post-deposition annealing process on structural and electrical properties. As-deposited Pt films are polycrystalline with strongly preferred orientation on the (111) plane and the grains size is about 20 nm. The peak strength of (111) increased after annealing for all investigated samples and also the grain sizes increased up to 67 nm. Annealed Pt films exhibit a lower resistivity in comparison with the as-deposited films.

1. Introduction

Platinum is the material of choice for metallic components in microelectronics, sensorics and microsystems. Pt exhibits long-term stability, reproducible resistance behaviour and chemical stability at high temperatures and is interested as material in resistance temperature detectors. Pt is also used for its ability to be patterned by microelectronic fabrication techniques [1]. Silicon substrates are often used for sensor application due to their compatibility with the electronic circuit process. In generally, there are problems with Pt adhesion towards SiO_2 insulation layers. Therefore, an intermediate layer between Pt and the substrate are often used to generate a blocking interdiffusion and to improve the adhesion. Hence, he elements and compound such as Ti, Ta, Cr, RuO_2, IrO_2 are investigated. There is a challenge to utilize the alumina substrate for its electrical insulated properties, a relatively high thermal conductivity; it is suitable to work at high temperatures and in biological environment. Typical application using alumina substrates we can find in chemical sensors, microheaters, hotplates, micro chemical emission, sensors, thermocouples and fuel cell systems [2-3]. So, it is a need to pay attention also to the research and development of miniaturized Pt elements fabricated on the alumina substrates using microelectronic and microsystem technology.

In this work we investigated the effect of Pt film thickness prepared on alumina ceramics and the post-deposition annealing process on electrical properties. We also describe development and fabrication of the temperature element and some its structural and electrical properties.

978-1-5090-3084-2/16 $31.00 © 2016 IEEE

2. Experimental details

For film deposition a dc magnetron sputtering system was used containing four targets whereby Ti and Pt targets were arranged side by side. A sputtering power for Pt deposition of 300 W was used. The total working sputtering gas pressure was kept at 0.5 Pa and adjusted by a piezoceramic valve. The Pt film thicknesses measured by Talystep were in the range of 600 ÷ 1200 nm for all examined samples. The deposition parameters for the Ti intermediate film were 0.8 Pa and 600 W resulting in a rate of 10 nm/min. The Ti film thickness was kept constant at 10 nm in all deposition cases and Pt was deposited on Ti without breaking the vacuum. The Ti/Pt thin films were prepared onto unheated alumina substrate. In order to stabilize the properties of Ti/Pt thin films, selected samples have been annealed in a furnace at 1100°C in nitrogen atmosphere for 2 hours.

The crystal structure was identified with a Theta-Theta Diffractometer D5000 with a Goebel mirror in the grazing incidence geometry with CuKα radiation. The average crystalline grain size of Pt particles was estimated from the integral breadths and the peak position of an XRD line broadened according the Scherrer formula. The surface morphology and the FIB-SEM cross section were observed by Zeiss FIB Auriga.

Finally, temperature sensor fabrication was realized by plasma etching process. The alumina substrate was covered homogeneously by sputtered Ti/Pt films with various Pt thicknesses in the range of 600 ÷ 1200 nm. A conventional lithography process with photoresist AZ 4562 was carried out to define a meander pattern of the width of 20 μm. After pattern transfer, the wafers were hardbaked at 140°C for 3 min and consequently the etching process was started. Dry etching through resist mask was performed in an Oxford Plasmalab System100 reactor utilizing a inductively coupled plasma (ICP) source operated at 13.56 MHz and separate RF (13.56 MHz) biasing of the sample electrode. Pt films were etched in Cl_2/Ar plasma. The etching process conditions were performed at 500-1500 W for ICP power, 135 W for RFpower, 10 mtorr for process pressure, 20°C for substrate temperature and the flow rates were 20 sccm for Cl_2 and 10 sccm for Ar. The size of chip was 1.5×1.5 mm^2.

Electrical resistivity, before and after annealing, was determined from the values of sheet resistance measured by using digital four point probe at room temperature and measured film thicknesses. The temperature coefficient of resistance (TCR) was calculated by measuring the change in the electrical resistance per degree of temperature, controlling the temperature using the furnace.

3. Results

3.1 Structural and surface analysis

From XRD diffractograms (Fig. 1) it was found there was strongly preferred orientation on the (111) plane for as-deposited Pt film. The grains size is near 20 nm. The peak strength of (111) increased after annealing for all investigated samples and also the grain sizes increased up to 67 nm. These sizes are in good agreement with SEM. In Bragg Brentano peaks of the alumina substrate are also to see, because 1 μm Pt is not enough to absorb all X-rays. But with GID, penetration of X-rays is smaller. The rough substrate is the reason, that also the (111) Pt-peaks is the strongest also in GID. The annealed samples shows in GID additional peaks, so at this incidence angles the penetration of X-rays is not enough to get peaks from corundum. Platinum oxides and titanium oxides have almost the same crystal structure like corundum. So we can explain the black regions in SEM came from thin PtO_x or TiO_x.

From SEM observation (Fig. 2a) was found that surface morphology of all samples was characterized by a rough and compact granular structure reflecting the alumina substrate morphology. An agglomeration of small grains with arbitrary shapes was created on every alumina grain. Pt grows not only on top of the alumina grains but also on grain sides and along their grain boundaries and it is filling the space between Al_2O_3 grains. It was seen that the average grain sizes in the lateral direction of as-deposited Pt films were smaller (about 20 nm) than for Pt annealed films (about 60 nm). FIB-SEM cross sectional images (Fig. 2b) showed Pt polycrystalline columnar structure.

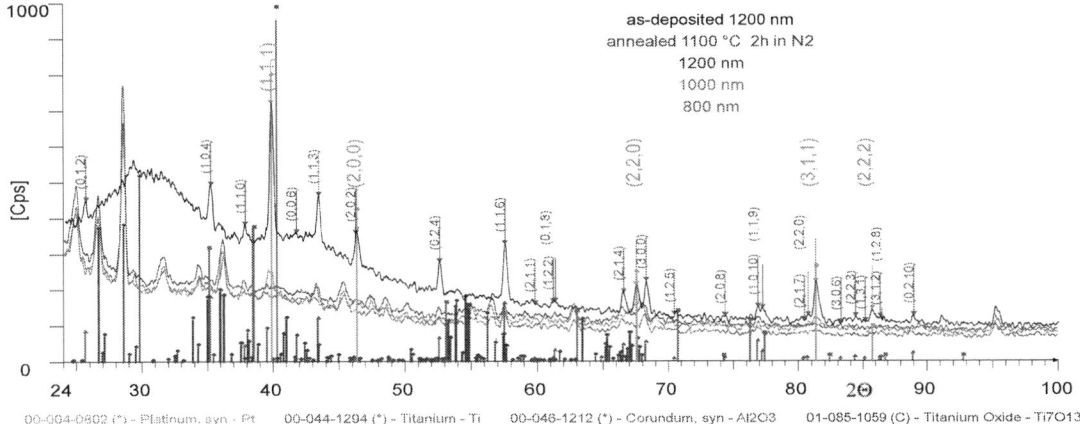

Fig. 1: XRD diffractograms of Ti/Pt films as-deposited 1200 nm, annealed 800, 1000 and 1200 nm in GID geometry.

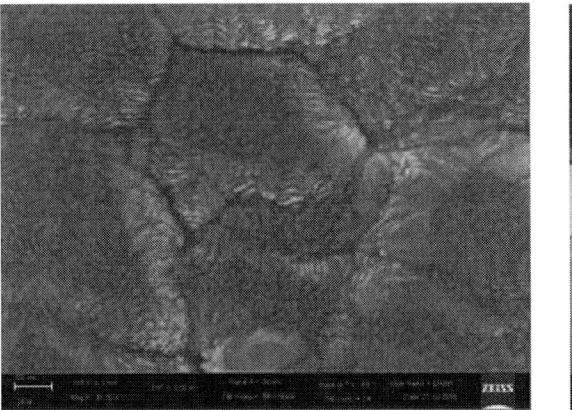 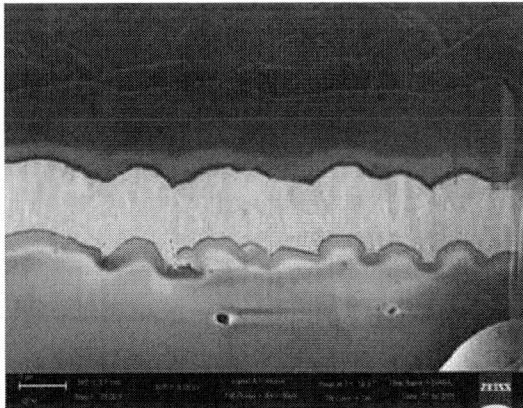

Fig. 2: SEM surface and FIB-SEM cross section image of as-deposited Ti/Pt films (10/1200 nm).

3.2 Electrical characterization

Table 1 displays the results of the resistivity tests performed at room temperature on as-deposited and annealed Ti/Pt films with different thicknesses. The as-deposited film resistivity values for all examined samples are higher than the reported values for Pt bulk material and these values slightly decrease with increasing Pt thickness [4]. After annealing all samples exhibit comparable values of resistivity. However, these values decreased due to the film microstructure which contains larger grains identified by XRD and SEM.

Table 1

Thickness of Pt [μm]	0.6	0.8	1.0	1.2
Resistivity [Ω·cm] (As deposited)	2.9×10^{-5}	3.1×10^{-5}	2.8×10^{-5}	2.2×10^{-5}
Resistivity [Ω·cm] (Annealed)	1.2×10^{-5}	1.0×10^{-5}	1.0×10^{-5}	0.9×10^{-5}
Initial resistance at 25°C [Ω/°C]	122.1	116.0	84.8	62.3
TCR [ppm]	3025	3078	3073	3050

Figure 3 represents developed Pt film sensitive element created with 20 μm line width and 20 μm spacing. Its resistance showed a good linear correlation with the temperature when it was increased up 300°C.

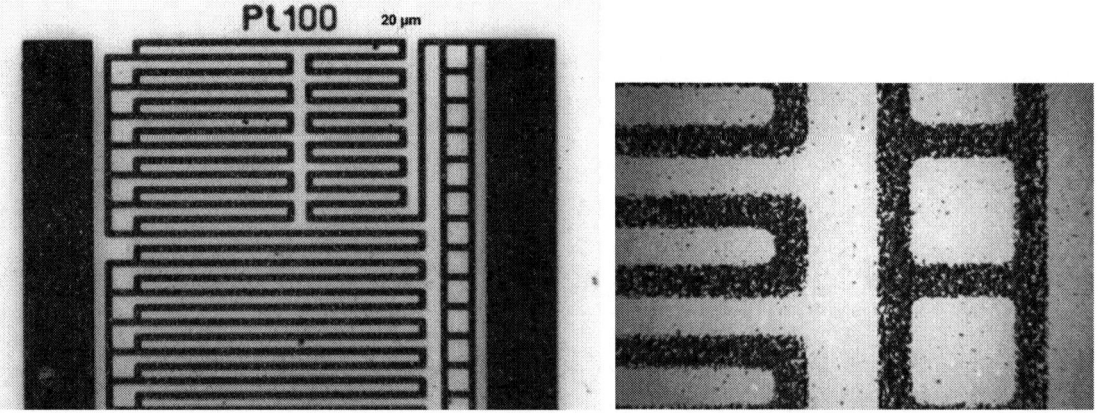

Fig. 3: Topology of Pt sensitive element and the detail of etched Pt film.

4. Conclusion

We successfully designed and developed a Pt film temperature element of the size of 1.5×1.5 mm^2 with the meander width of 20 μm and the thickness in the range of $0.6 \div 1.2$μm prepared on the alumina substrate. It was found that as-deposited Pt films are polycrystalline with strongly preferred orientation on the (111) plane and the grains size is about 20 nm. The annealed samples shows in GID geometry additional peaks probably belong to PtO$_x$. Their structure exhibits a larger gains size. Annealed Pt films have lower resistivity in comparison with the as-deposited films.

Acknowledgement

The work was supported by the Scientific Grant Agency of the Ministry of Education of the Slovak Republic and of the Slovak Academy of Sciences, No. 1/0828/16 and 2/0134/15, by project SAFESENS (agreement No. 621272-1 and German grant VDI 16ES0226) co-funded the ENIAC-JU and by PPP program of DAAD.

References

[1] J. Han, P. Cheng, H. Wang at al: *Materials Letters* **125** (2014), 224.

[2] N.H. Al-Hardan, M.J. Abdullah, A.A. Aziz, Z. Hassan: Materials Science in Semiconductor Processing 13 (2010), 199.

[3] G. Schmidt, J. Dellith, E. Kesser, U. Schinkel: *Applied Surface Science* **313** (2014), 267.

[4] E. Ciftyurek, K. Sabolsky, E. Sabolsky: *Sensors and Actuators B* 181 (2013), 702.

ASDAM 2016, The 11th International Conference on Advanced Semiconductor
Devices And Microsystems, November 13-16, 2016, Smolenice, Slovakia

Multi-bit Adder Design using ME-MTJ technology

Nishtha Sharma [1], Andrew Marshall [1], Jonathan Bird [2] and Peter Dowben [3]

1. The University of Texas at Dallas, Richardson, TX
2. University at Buffalo, the State University of New York, Buffalo, NY
3. University of Nebraska, Lincoln, Lincoln, NE

Nishtha.sharma@utdallas.edu

The inherent advantages of ME-MTJ devices are extremely low power consumption and the device level memory capability. We demonstrate the complex circuit design required for Magneto-Electric Magnetic Tunnel Junction (ME-MTJ) based logic. The clocking required is readily achievable, permitting large-scale designs to be considered. Compact models for the ME-MTJ based devices have been previously proposed. Using the models and circuits, we present simulation results of a 5-bit adder along with a complex clocking scheme.

1.   Introduction

Voltage-controlled spin electronics is considered a likely path for continued progress in information technology. This technology aims at reduced power consumption, increased integration density and enhanced functionality where non-volatile memory is combined with high speed logical processing. Promising spintronic device concepts use the electric control of interface and surface magnetization [1]. The ME-MTJ device first proposed by Binek and Doudin in [2] is one such beyond CMOS device, which displays the electric control of interface. As the CMOS process nodes extend to sub 10nm, there is a potential need for an alternate technology. The ME-MTJ device is one such technology with promising performance and area advantages. A MATLAB based model has been used to evaluate the performance of ME-MTJ based devices, and the results are compared with CMOS based devices [3]. We have extended this by developing a set of SPICE models written in VerilogA and hence validated the methodology with a 1-bit adder simulation [4]. Here, we extend the methodology with a 5-bit full adder simulation introducing a complex clocking scheme for larger circuit design.

2.   The ME-MTJ device and compact model

The basic ME-MTJ device is a three terminal voltage controlled device [1][5], with logic and memory capabilities. It consists of an antiferromagnetic layer (chromia) exchange biased with the free ferromagnetic layer (FM) of the (Magnetic Tunnel Junction) MTJ cell as shown in Fig. 1. It can be modified to derive more complex logic gates such as Majority/Minority gate, Inverter/buffer and the XNOR/XOR gate as shown in Fig. 1a. We have previously shown models using MATLAB to analyze these ME-MTJ devices and estimate the energy and delay terms [3]. Fig. 1b shows the comparison results compared to the CMOS based devices.

978-1-5090-3084-2/16 $31.00 © 2016 IEEE

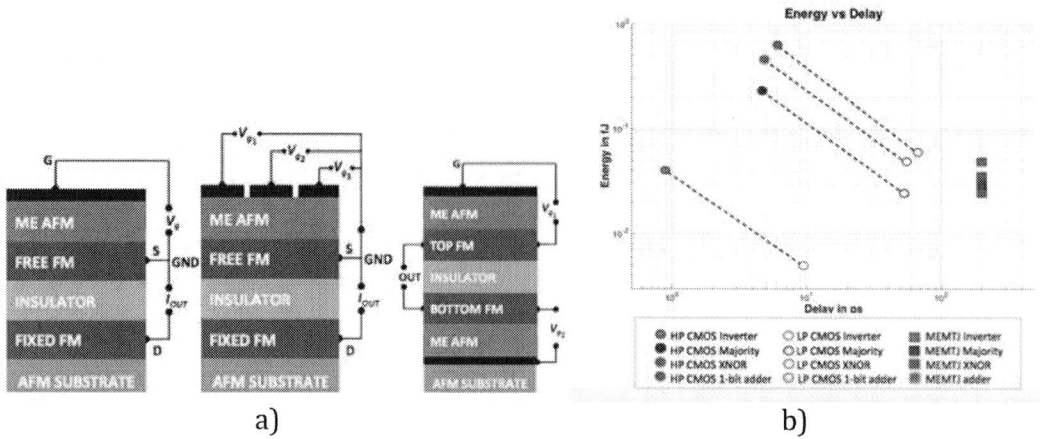

Fig. 1 Basic ME-MTJ, Majority Version and XNOR gate. Energy-Delay graph of the ME-MTJ based devices compared to CMOS based devices. For simpler circuits like an inverter, CMOS process has a lot of benefits, but with increasing circuit complexity, the CMOS based circuits degrade in performance, but the ME-MTJ device performance stays consistent.

MATLAB does not have the capability to analyze circuit details of large circuits, and as a result, in order to analyze larger, more complex circuits, we developed models that are compatible with circuit level simulators such as SPICE or SPECTRE [6]. VerilogA is an industry standard modeling language that can be read by circuit level simulators [7]. The compact models provide device behavior when used in a Spectre circuit simulation environment. They include ME-MTJ physical attributes incorporating write delay and charging effects, and reading bias effect on Tunneling Magneto-Resistance (TMR) [3][8]. Based on the polarity of voltages applied across the input, the tunneling magneto-resistance (TMR) of the MTJ cell changes. The ME-MTJ based devices can be used for IC design of hybrid ME-MTJ/CMOS systems as well as standalone ME-MTJ based devices.

3. Simulation Results

The schematic of the 5-bit adder is shown in Fig. 2. Transient simulations have been performed, ensuring model efficiency. The read voltage is assumed to be 0.11V and the threshold is set at +/-50mV for spin vector switching. The basic element of the 5-bit adder, i.e. a 1-bit adder is shown in Fig. 2 along with the clocking scheme consisting of a three-phase clock signal and global reset in Fig. 3. The 1-bit full adder consists of three ME-MTJ based majority gate and two inverters. Clk0, Clk1 and Clk2 are the pull up signals. The set of signals applied till stage 4 are unique and they are repeated at stage 5. In other words, the clock signals applied to stage n and stage n+3 are the same with a time lag of one cycle.

We have included a CMOS clocking and reset scheme. Reset transistors are used at each stage to reset the device memory, since the device performs a dual logic/memory function. This resets the logic state of the ME-MTJ device to a known logic state.

978-1-5090-3084-2/16 $31.00 © 2016 IEEE

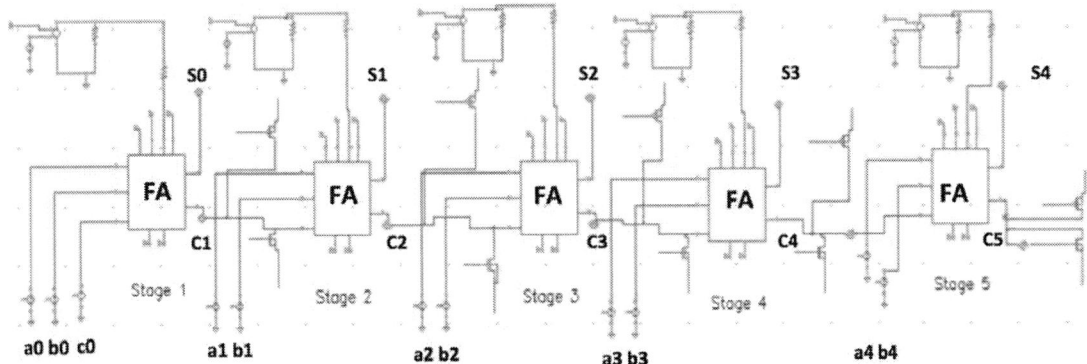

Fig. 2: Circuit schematic of ME-MTJ based 5-bit adder in Cadence spectre.

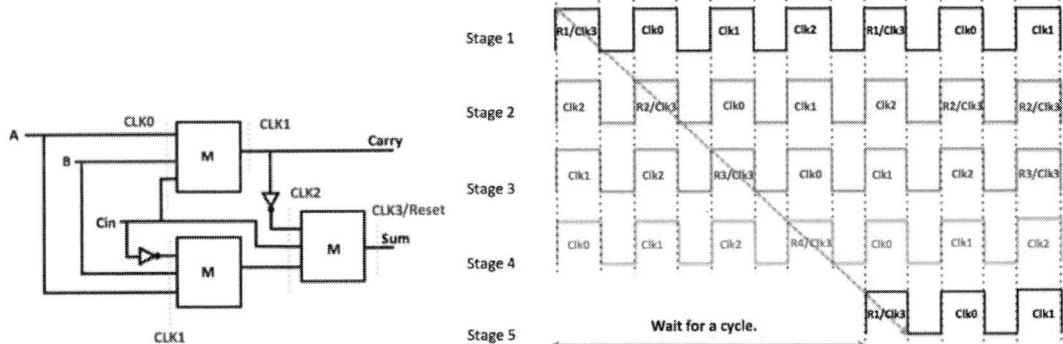

Fig. 2: a) ME-MTJ based 1-bit full adder shown with the clocking scheme. b) Clocking scheme of the ME-MTJ based 5-bit adder.

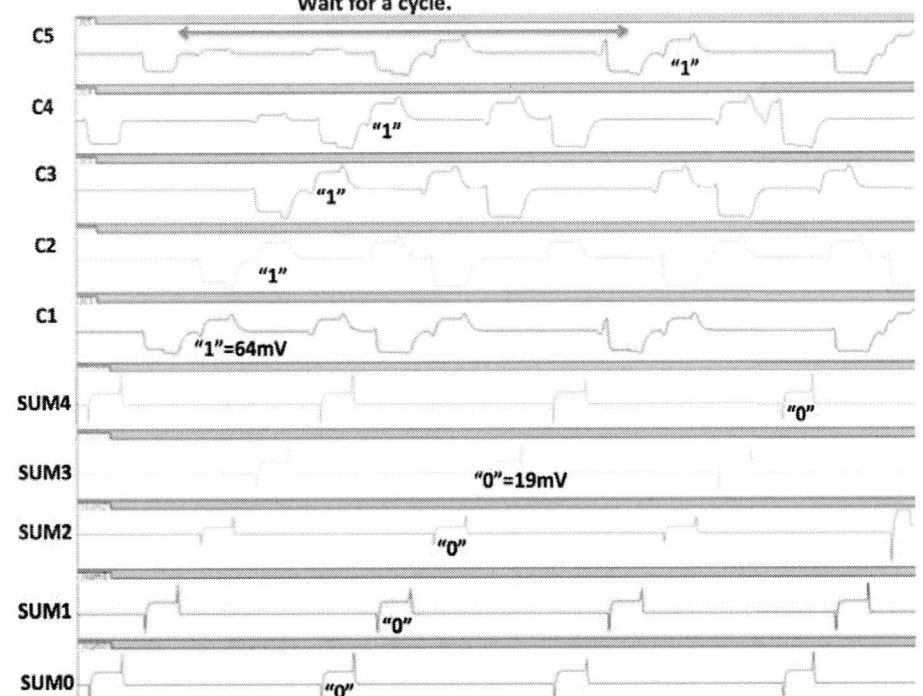

Fig. 4: Transient simulation of 5-bit adder. The input bit stream are a4a3a2a1a0=**00000** and b4b3b2b1b0=**11111** and input carry, c0=**1**. These input are selected to observe maximum delay in the system.

978-1-5090-3084-2/16 $31.00 © 2016 IEEE   91

For the CMOS clocking scheme, a pull-up device is used to enable sensing of the output node electrically. It consists of a PMOS transistor with a clocked input. The output can thus be read either asynchronously or synchronously depending on the requirement. The simulation results are shown in Fig. 4. The input bit streams are $a_4a_3a_2a_1a_0=00000$ and $b_4b_3b_2b_1b_0=11111$ and input carry, $c_0=1$. These input are selected to observe maximum delay in the system. Each sum and carry stage has a pullup and a reset signal applied to it. After the carry is propagated to stage 4, there is a forced lag of one cycle and then the clock signals at the stage 5 go high. The sum and the carry output obtained is $S_4S_3S_2S_1S_0=00000$ and $C_5C_4C_3C_2C_1=11111$. Simulation results indicate that the ME-MTJ models can be used for more complex circuit design. Furthermore, results indicate that appropriate clocking and reset functions can be generated for larger circuit design. Stage 5 is the time-delayed version of stage 1, and further stages can also be added in a similar way.

4. Conclusion

We have validated an ME-MTJ based 5-bit full adder using the compact models developed in [3]. In addition, we proposed a complex clocking scheme for the adder, which can be extended, to any multi-bit adder. This analysis paves a way for the complex circuit design using the ME-MTJ based devices.

Acknowledgement

This project was supported by the Nanoelectronics Research Corporation (NERC), a wholly-owned subsidiary of the Semiconductor Research Corporation (SRC), through the Center for Nanoferroic Devices (CNFD), an SRC-NRI Nanoelectronics Research Initiative Center under Task ID 2398.001."

References

[1] He, Xi, et al. "Robust isothermal electric control of exchange bias at room temperature." Nature materials 9.7 (2010): 579-585.

[2] Binek, Ch. & Doudin, B. "Magnetoelectronics with magnetoelectrics." Journal of Physics: Condensed Matter 17.2 (2004):

[3] L39. Sharma, N., et al. "Compact-device model development for the energy-delay analysis of magneto-electric magnetic tunnel junction structures." *Semiconductor Science and Technology* 31.6 (2016): 065022.

[4] Sharma Nishtha, et al. "VerilogA based Compact model of a three-terminal ME-MTJ device." 2016 16th IEEE Conference on Nanotechnology, IEEE NANO 2016.

[5] Bibes, M., and Agnès B. "Towards a magnetoelectric memory." Nature Materials 7 (2008): 425-426.

[6] Virtuoso User Guide, Cadence 5.1.41 Spectre guide documents, 2008.

[7] McAndrew, Colin C., et al. "Best Practices for Compact Modeling in Verilog-A." Electron Devices Society, IEEE Journal of the 3.5 (2015): 383-396].

[8] Zhang, Yue, et al. "Compact modeling of perpendicular-anisotropy CoFeB/MgO magnetic tunnel junctions." IEEE Transactions on Electron Devices 59.3 (2012): 819-826.

ASDAM 2016, The 11th International Conference on Advanced Semiconductor
Devices And Microsystems, November 13-16, 2016, Smolenice, Slovakia

Technology of AlSb/GaSb based LED nanostructures for high temperature superlinear luminescence

E. Hulicius[1], A. Hospodková[1], J. Pangrác[1], M. Zíková[1]

[1]Institute of Physics, CAS, v.v.i., 162 00, Cukrovarnická 10, Prague 6, Czech Republic
e-mail: hulicius@fzu.cz

The superlinear (SL) electroluminescence (EL) of the MOVPE structures based on AlSb/InAs$_{(1-x)}$Sb$_x$/AlSb deep quantum wells (QWs) grown by MOVPE on n-GaSb:Te substrates was measured. Preparation technology of these structures is described in more detail.

Dependences of the EL spectra and optical power on driving current of nano-heterostructures with a deep AlSb/InAs$_{(1-x)}$Sb$_x$/AlSb QW for 77 – 300 K temperature range are presented. Intensive two-band SL EL in the 0.5 - 0.8 eV photon energy range and optical power enhancement with the drive current at room temperature caused by the contribution of the additional electron-hole pairs, generated at AlSb/InAs interface, due to the impact ionization were found.

Study of the SL EL temperature dependence at 90 – 300 K range, based on our previous work, enabled us to define the role of the first and second heavy hole levels in the radiative recombination process.

1. Introduction

AlSb/InAsSb/AlSb based heterostructures and nanostructures with quantum wells (QWs) grown on GaSb are promising materials for optoelectronic devices for near- and mid-IR spectral regions (LEDs, laser diodes and detectors) for ecological monitoring, gas analysis and medical diagnostics [1, 2]. For several applications it is necessary to broaden their operating temperature over 300 K. However, optical power and quantum efficiency of the light emitting devices based on the narrow bandgap semiconductor compounds (InAsSb, InGaAsSb) are limited by nonradiative Auger recombination. This Auger process can be suppressed using deep QWs [3]. Earlier we have proposed a method [4] to increase the optical power in bulk narrow bandgap heterostructures with high potential band offsets and later [5] in GaSb-based nanostructures with a deep QW by using the effect of impact ionization on the heterojunction with high band offset.

2. Experimental

Structures consisting of 17 nm AlSb/ 5 nm InAs$_{1-x}$Sb$_x$/17 nm AlSb QW and 0.4 µm p-GaSb cap layer were grown on (100) n-GaSb:Te epiready (SWI) substrate in Laytec EpiRAS 200TT equipped AIXTRON 200 machine by low-pressure MOVPE. Before the growth, GaSb substrates were held under 2 mM/min TESb flow at 520 °C for 12 min to remove oxides and Sb desorption from the surface. Reflectance anisotropy spectroscopy (RAS) measurements for the deoxidation process of GaSb substrate is shown in Fig. 2 a). The oxide-free surface can be recognized by RAS spectrum [5]. The growth of the structure was carried out in H$_2$ atmosphere (10 slpm total flow) under a total pressure of 150 hPa and growth temperature of 560 °C. Triethylgallium TEGa, trimethylindium TMIn, tris(tertiarybutyl)aluminium TtBAl, triethylantimony TESb and tertiarybutylarsine tBAs were used as precursors. Samples with three different Sb concentrations x in InAs$_{(1-x)}$Sb$_x$ QW were

978-1-5090-3084-2/16 $31.00 © 2016 IEEE

prepared. In contrast to [6], AlAs-like interfaces between AlSb barrier and InAsSb QW were avoided by special switching sequence of metalorganics. We have found in-situ measurement technique RAS as very important tool for recognition of successful high quality structure growth. Examples of AlSb/InAs$_{(1-x)}$Sb$_x$/AlSb structure growths with different composition of InAsSb QW are shown in Fig. 2 b).

3.   Results and Discussion

MOVPE prepared QW structures with different InAsSb composition with intensive two-band electroluminescence (EL) were measured in the photon energy range of 0.5 - 0.8 eV under 20-200 mA drive current at temperatures 77 K and from 22 °C up to 180 °C (~300 - 450 K). Scheme of the structure energy band diagram is shown in Fig. 1. The aim was to prepare structure with large discontinuity of conduction band between barrier and QW layer. In such structure the excess energy of electron, coming from barrier to QW, can be used for generation of additional electron-hole pair in QW. We call suggested process "interface impact ionization". High quality of heterointerface is required to observe this phenomenon.

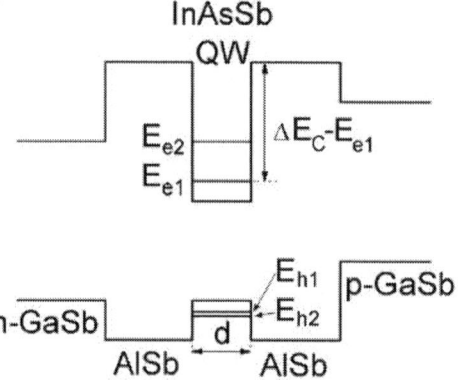

Fig. 1. The type I heterostructure - energy band diagram with deep AlSb/InAs$_{1-x}$Sb$_x$/AlSb QW grown on n-GaSb substrate, d ~5 nm.

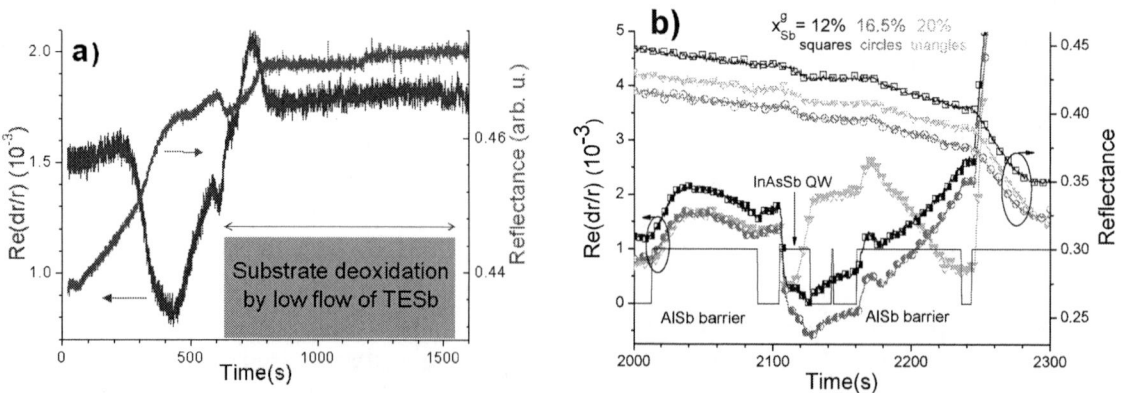

Fig. 2. RAS time resolved measurements during a) the deoxidation process of GaSb substrate; b) the AlSb/InAsSb/AlSb structure growth with different composition of the InAsSb QW, in the gas phase there was 12 % (squares), 16.5 % (circles) and 20 % (triangles) of antimony given by
$$x^g_{Sb} = x^g_{TESb} / (x^g_{TESb} + x^g_{tBAs})$$

978-1-5090-3084-2/16 $31.00 © 2016 IEEE   94

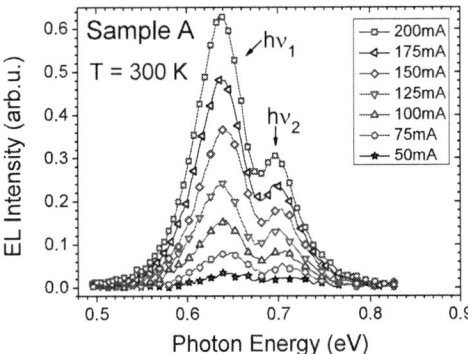

Fig. 3. EL spectra for different currents. Spectra at 300 K consist of two bands with peaks at $h\nu_1 = 0.635$ eV and $h\nu_2 = 0.695$ eV. Energy difference was ~ 60 meV. See [5].

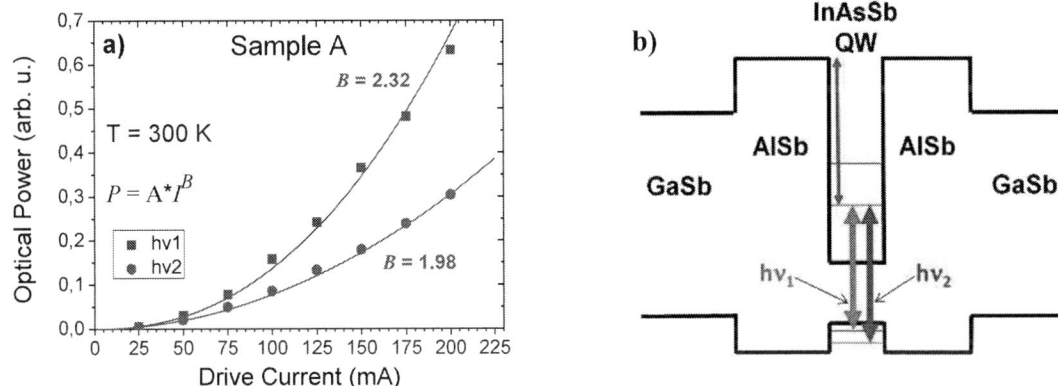

Fig. 4. a) Optical power P vs drive current I for two EL bands (T = 300 K). Points - experimental data, solid curve - according to $P = A \cdot I^B$, see [5]. b) Different transitions in InAsSb deep QW.

Our earlier results [5] of the stronger superluminescence (SL) for $h\nu_1$ peak, which is shown in Fig. 3 and 4 a), can be explained by a higher probability of impact ionization to the 1st hole level $h\nu_1$ with the lower energy - see Fig. 4 b).

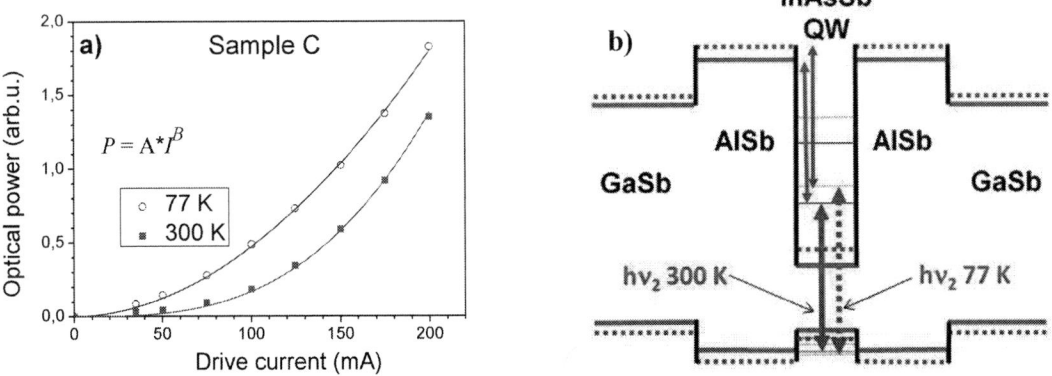

Fig. 5. a) Stronger increase of EL intensity at higher temperature. Points - experiment, solid curve - approximation according to $P = A \cdot I^B$ (A is a fitting parameter, B is the exponent value). b) Transitions for different temperatures.

Optical power dependence on the drive current at 77 K and 300 K is shown in Fig. 5 a). Due to decreased InAsSb QW E_g, the impact ionization became more probable for higher temperatures; see Fig. 5 b). Optical power P depends on the current I by $P = A \cdot I^B$, where A is a fitting parameter and B is the exponent value. *B index increases with T: B = 1.5-2 at 77 K, B = 2-3 at 300 K in $P = A \cdot I^B$. E_g (77 K) = 0.294 eV, E_g (300 K) = 0.246 eV.*

4 Conclusions

We have successfully prepared GaSb-based structures with a deep AlSb/InAsSb/AlSb QW. Unusual superluminescent behavior was observed. It means that the optical power P enhancement of EL with a drive current has exponential formula $P = A \, I^B$ and the exponent B is higher than 1 in the region above the threshold. For our structures the exponent B was in the range of 1.5 - 3 for the temperature range of 77 - 300 K.

This enhancement of optical power is explained by so called "impact ionization on the interface", where the excess of the electron energy on the interface with big conduction band discontinuity is used for generation of additional electron-hole pair in the InAsSb QW.

Several observed phenomena prove this hypothesis. First, stronger enhancement of EL and higher coefficient B was observed for higher temperatures ($B = 1.5 - 2$ at 77 K, $B = 2 - 3$ at 300 K). Second, the stronger enhancement was observed for transition with lower transition energy $h\nu_l$ from the lowest electron to the lowest heavy hole energy level. Third, the electroluminescent band caused by the transitions to the second heavy hole level exists only for temperatures above 190 K.

Presented results for the LEDs with deep InAsSb QW pave the way for the efficient mid-IR devices operating in wide temperature range from -200 °C up to $+200$ °C.

Acknowledgement

The authors acknowledge support from NPU LO1603 - ASTRANIT.

References

[1] M. B. Frish, R. T. Wainner, M.C. Laderer, B.D. Green, M.G. Allen, *IEEE Sensors Journal* **10**, 639-646, 2010.

[2] M. W. Sigrist, R. Bartlome, D. Marinov, J. M. Rey, D. E. Vogler, H. Wächter, *Appl. Phys. B - Lasers & Optics* **90**, 289-300, 2008.

[3] L. V. Danilov, G. G. Zegrya, *Semiconductors* **42**, 550-556, 2008.

[4] K. V. Kalinina, M. P. Mikhailova, B. E. Zhurtanov, N. D. Stoyanov, Yu. P. Yakovlev, *Semiconductors* **47**, 73-80, 2013.

[5] K. Möller, L. Töben, Z. Kollonitsch, Ch. Giesen, M. Heuken, F. Willig, T. Hannappel, *Appl. Surf. Sci.* **242**, 392-398 2005.

[6] M. P. Mikhailova, E. V. Ivanov, L. V. Danilov, K. V. Kalinina, N. D. Stoyanov, G. G. Zegrya, Yu. P. Yakovlev, E. Hulicius, A. Hospodková, J. Pangrác, M. Zíková, *J. Appl. Phys.* **112**, 023108 2012.

ASDAM 2016, The 11th International Conference on Advanced Semiconductor
Devices And Microsystems, November 13-16, 2016, Smolenice, Slovakia

3D energy harvester with tunable resonant frequency

V. Janicek

Department of Microelectronics, FEE CTU in Prague,
Technicka 2, 166 27 Prague 6, Czech Republic
e-mail: janicev@fel.cvut.cz

This paper describes the design and function upgrade of 3D electrostatic energy harvester. The structure consisting of simple finger layout works on the principle of an electrostatic converter and converts a non-electric energy into electrical energy by periodical modification of the gap between the electrodes. The mechanical structure is modeled as a 3D silicon-based MEMS. The basic structure reaches a low resonant frequency of 100 Hz. Add-on circuit makes it possible to tune the resonant frequency to the desired value to be able to resonate with the ambient mechanical vibrations. This makes the energy harvesting process more efficient.

1. Introduction

The massive onset of wireless technologies and IoT equipment puts an increased emphasis on quality of the power parts. Conventional electrochemical batteries require from time to time service intervention to be replaced. Mobile devices need some infinite energy source which could supply power for a extended period of time. There are two ways how to reach this goal – design with very low power consumption and sleep modes. However, this makes it impossible to use these devices in real-time mode applications where a non-interrupted flow of measured data is needed. The second way is to use an ambient non-electrical energy to be changed into electricity. These power converters are called energy harvesters. There are many types of ambient energies like the flow, light, heat or mechanical energy [1,2,3,4]. Using this kind of conversion makes it possible to get an infinite power source for the powered device. The proposed generator is based on beam structure with fixed natural resonant frequency. The layout and the model are based on three-dimensional silicon based MEMS. Compared to previously published papers this article focuses on the concept of the structure with tunable resonant frequency. This approach would lead to broader usage field and better yield of provided ambient energy.

2. Energy harvester principle

The designed energy equipment is using a combination of the electrostatic and piezoelectric generator (required as a start-up power source) in the form of MEMS structures. The structure has two electrodes working together as a capacitor. The moving mass located on one of the moving (floating) electrode should fine tune the resonant frequency. The design was optimized to reach very low natural resonant frequency of about 50-100 Hz. Because of production technology the higher (108 Hz) was reached. Electrostatic generator [5] uses the forces generated between the opposite charges on the plates of a charged capacitor. Separation of charge Q on the electrodes depends on the potential difference V between them according to equation $Q = CV_{VAR}$. C_{VAR} capacity is a function of geometry (topology) and electrode properties of materials that surround them. When moving a mass m in the range of $z(t)$, the capacity changes between C_{MAX} and C_{MIN}. The mechanical movement forces the floating

978-1-5090-3084-2/16 $31.00 © 2016 IEEE

electrode to move between its two limit positions and increases and decreases periodically the capacity value.

2.1 Topology design of a fixed resonant frequency harvester

Topology and the simulations were done with CoventorWare and Ansys. The layout consists of three parts (see Fig. 1a) - movable bomb electrode – part A, fixed electrodes – part B and spring suspension – part C.

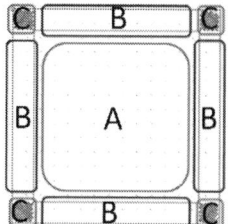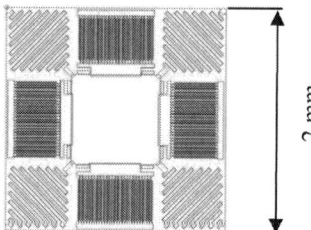

Figure 1: Basic comb structure topology (a), and final topology with suspension springs (b)

Fig. 1b shows the topology with a complete set of both types of electrodes, long wrapped spring suspensions and mechanical stops (rectangular blocks right under the movable electrode). These mechanical stops limit the movement excitation of the floating electrode to avoid mechanical damage to the structure and short-circuit between the electrodes.

3. Analysis

The first four modal frequencies (state of the system in equilibrium when the mechanically undamped (lossless) system reacts to external motion excitation with an unlimited deflection) I got from simulations showed very low resonance of the structure. The first moving (see Fig. 2a) the structure in the x-axis (lowest) reached 108 Hz, which makes the structure perfect for applications in the field of mechanical motors or rotating actuators working on power frequency of 230 V/50 Hz. The second frequency (moves the structure in Y-axis which is perpendicular to the X-axis of the first frequency movement). The third modal frequency moves the structure in the Z-axis (see Fig. 2b). All of these three movements are causing a small change in the capacitance between the fixed and floating electrode. The next (higher) natural frequencies have only negligible effect on the change of the capacitance, therefore, I do not suppose to use the harvester at so high frequencies.

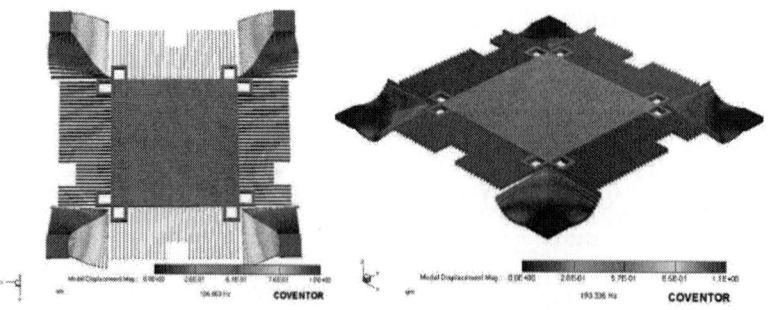

Figure 2: Movement of the electrodes at the modal resonant frequencies
(The scale of deflection is due to small shifts multiplied by the real and solid electrodes are not shown.)

Fig.3 illustrates the graph with the dependence of the movement excitation at different modal frequencies. It is clear that the fourth modal frequency reaches approx. 310 Hz which is too high for getting such a high ambient energy somewhere in the real industry environment. The maximum deflection of the floating electrode depends on the frequency of ambient energy mechanical oscillations.

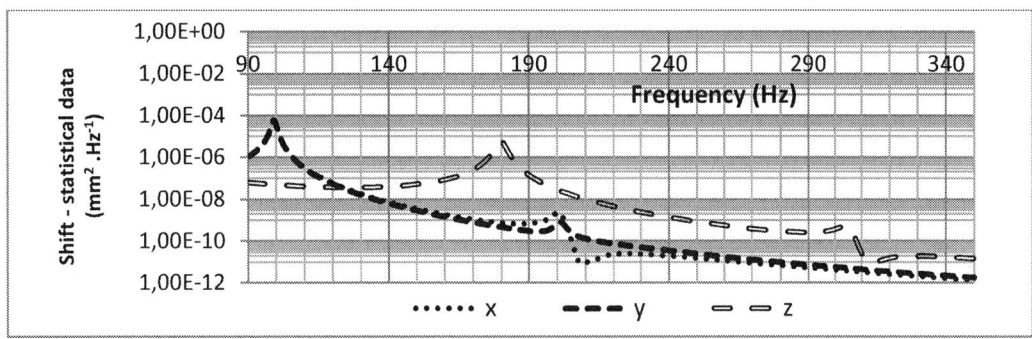

Figure 3: Random vibrations excitation in all three axes with a uniform statistical distribution from 90 Hz to 350 Hz. Acceleration 3500 (mm. $s^{-2}.Hz^{-1}$)

This frequency but also depends on many other aspects like the weight of the floating electrode, stiffness and form of the spring suspensions or electrical field between the electrodes. The first two (mechanical based) ways are quite hard to be adjusted during the service of the harvester. Adjusting the mass by adding or reducing mechanical mass located at the middle of the whole structure will reduce the natural frequencies and increases the deflection and internal stress. Only the last one – electrical field – can be adjusted in real-time during the harvester is in the process of harvesting energy. It also has a big advantage that we can fit the modal (resonant) frequency of the harvester depending on the frequency of the mechanical oscillations. Changing the mechanical stiffness of the structure through the last method can be done just with the help of one more adjustable voltage source. And this is a crucial aspect when fine tuning the modal frequency.

4. Tunable energy harvester concept

The Fig.4 shows the principle concept of a tunable energy harvester. Because of the isolation of each finger capacitor we can use one pair of capacitors to be controlled by an external voltage source (V_{ADJ}) which is controlled by a detection circuit (AFD). This circuit consists of an accelerometer which is controlling the power supply V_{ADJ}. The dependence progress is quadratic – see equation 1.

$$F_e = \frac{\varepsilon S U^2}{2d^2} \tag{1}$$

The damping effect caused by the charge deposited at the plates of the floating electrode and electrical force will influence the resonant frequency of the structure. The detection circuit (AFD) monitors the ambient conditions and increases or decreases the voltage V_{ADJ} to damp the movement of the floating electrode to reach the same resonant frequency like the ambient energy provides.

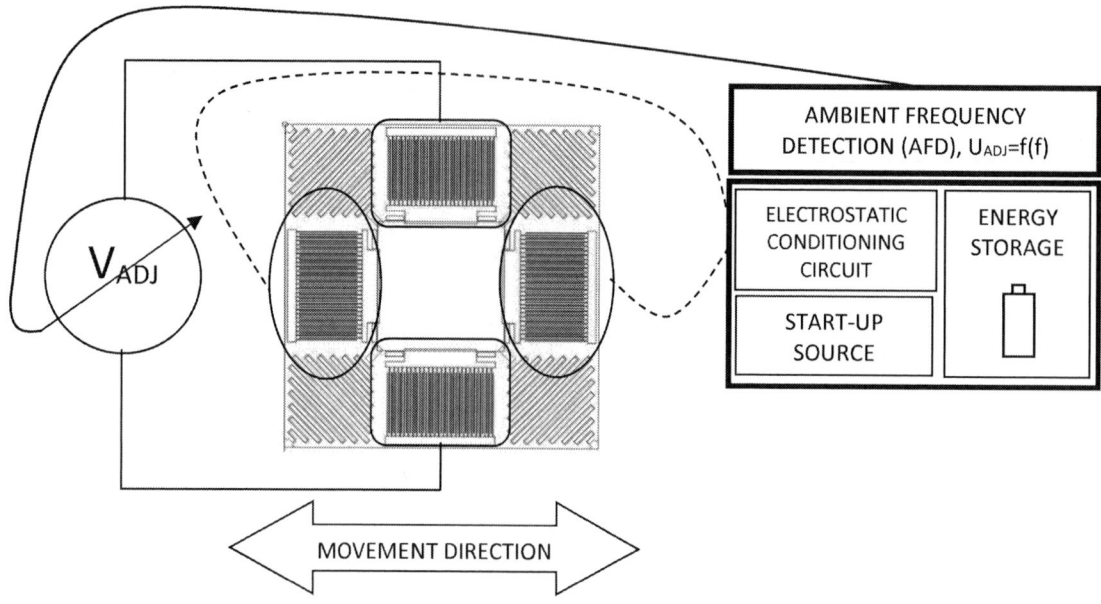

Figure 4: Electronically tunable energy harvester with automatic setting of resonant frequency

5. Conclusions

The proposed generator provides a new ability to set set the resonant frequency of the structure electronically. This will adjust the stiffness of the system (floating electrode and spring suspensions) to the desired value do the higher efficiency can be reached. The system is autonomously controlled, and it doesn't need any external services. This is a perfect combination for the broad product field of Internet of Things stuff.

Acknowledgement

This work has been supported by the grant of The Ministry of Interior of the Czech Republic No.VG20102015015.

References

[1] Magno, Michele, et al. "Kinetic energy harvesting: Toward autonomous wearable sensing for Internet of Things." *Power Electronics, Electrical Drives, Automation and Motion (SPEEDAM), 2016 International Symposium on.* IEEE, 2016.

[2] Ponnusamy, Vasaki, et al. "Energy Harvesting Methods for Internet of Things."*Biologically-Inspired Energy Harvesting through Wireless Sensor Technologies* (2016): 51.

[3] Shaikh, Faisal Karim, and Sherali Zeadally. "Energy harvesting in wireless sensor networks: A comprehensive review." *Renewable and Sustainable Energy Reviews* 55 (2016): 1041-1054.

[4] Ramya, R., G. Saravanakumar, and S. Ravi. "Energy Harvesting in Wireless Sensor Networks." *Artificial Intelligence and Evolutionary Computations in Engineering Systems.* Springer India, 2016. 841-853.

[5] S. Roundy et al, "Microelectrostatic vibration-to-electricity converters," *Proc. IMECE,* 2002.

ASDAM 2016, The 11th International Conference on Advanced Semiconductor
Devices And Microsystems, November 13-16, 2016, Smolenice, Slovakia

Reducing Crosstalk in the Internal Structures of Integrated Circuits

J. Novak and J. Foit

Department of Microelectronics, Czech Technical University in Prague
Technicka 2, CZ16627 Prague, Czech Republic
e-mail: novakj2@feld.cvut.cz and foit@feld.cvut.cz

The advent of novel sub-micron technologies of IC fabrication led to such a decrease in lead-to-lead separation that it is not possible any more to neglect the influence of these leads on the reliability of the system operation. Both the small lead separation and the application of multilayer interconnecting systems cause parasitic electromagnetic couplings; in the case of a unipolar CMOS technology, the capacitive coupling is the dominant effect. It is impossible to measure direct the rapid variations voltage between leads inside the IC.

1. Introduction

Equations for finding the primary electrical parameters of interconnecting systems can be defined on the basis of analytical solutions of electromagnetic fields [1]. These equations were defined under certain simplifying conditions and therefore they are only valid for simple configurations of interconnecting systems.

When trying to solve the electromagnetic fields in particular interconnecting systems inside integrated circuits, we can't avoid the numerical modeling of these fields. In cases of more complex structures, no general analytical solution is possible, due to the simplifications that eliminate local properties of the systems, important for calculations of the interconnecting system electrical parameters.

The numerical solution is based on an approximation of the solved area by a network of nodal points and its accuracy depends on the number of nodal points within that area [2]. The better accuracy is required, the larger number of nodal points must be present in the area solved. The networks of nodal points are created according to the spatial layout of the interconnecting system. The models represent only certain elements of interconnecting systems, describing the relevant properties of these systems. The networks in interconnecting systems are designed as gradual, with higher nodal point spatial density in the immediate vicinity of the interconnecting leads [3].

2. Spatial element of conductor

A rectangular block was used as the basic spatial element in designing the network. A "modified block" appeared as advantageous in the case of a line-type interconnecting grid where there do not appear geometrical alterations of the system in one dimension. This element type makes possible to create a very complex grid structure in a two-dimensional surface, and the spatial element is created by a simple expansion into the third dimension.

2.1 Three-lead connecting network

This system consists of three connecting leads in the same metallization layer. In solving this system we are first of all interested in the coupling capacitance between the edge leads (No. 1 and 3) of the system [4]. The solution of capacitance between the neighboring leads 1-2 and 2-3 corresponds to the twin.-lead system The lead dimensions correspond to the M3 metallization layer metallization layer in the AMI C05M-A 3M/2P/HR [5] process technology. The situation is described by a cross-section through the spatial model shown in Fig. 1.

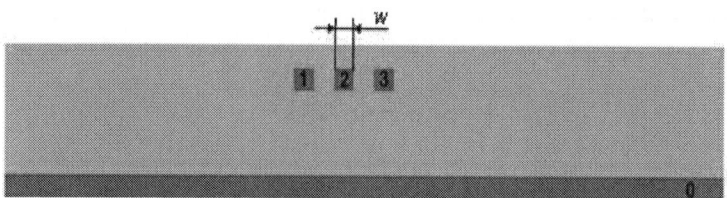

Fig. 1: Cross-section of a three-lead system.

From the electric field intensity distribution it is possible to deduce an extremely slight influence on the lead No. 3. This is in conformity with the low mutual capacitance C'_{13} between the peripheral leads of the system. If the central leads No. 2 is connected to the ground plane, it creates electrical shielding against unwanted effects caused by the lead No. 3 electric field. A more pronounced cut in the C'_{13} coupling capacitance can be obtained by increasing the central lead No. 2. width, its effect is shown in Fig. 2. This electrical shielding is only effective up to a width of the central lead of $w = 3.3$ μm, further increases do not bring about any marked cut in the C'_{13} coupling capacitance, it only increases the occupied area of the integrated circuit. The shielding effects of the central conductor in a three-lead system can be generalized to a condition that the shielding conductor width should not exceed triple the width of the signal conductors.

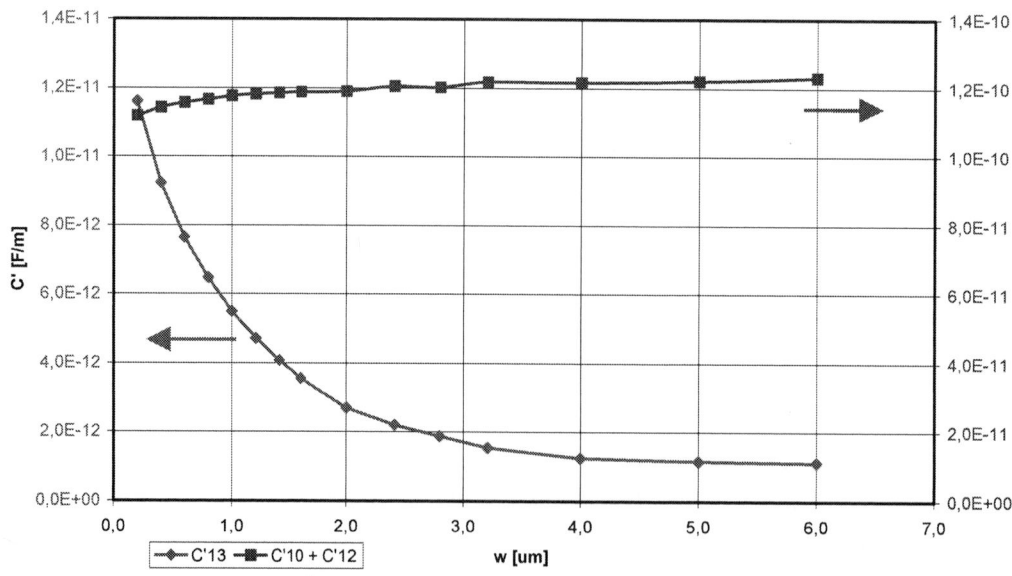

Fig. 2: Coupling capacitances versus central conductor width w

2.2 Maximum shielding in the integrated circuit

An example of such a system with a "protected conductor" is shown in Fig. 3. The separation of the interfering conductor from the edge screening conductor is d = 1 - 5 μm. In this example of system using a protected conductor, the matter of interest is the coupling capacitance between the signal conductors No. 1 and 2, the remaining leads No. 3, 4, 5, 6 are just shielding conductors.

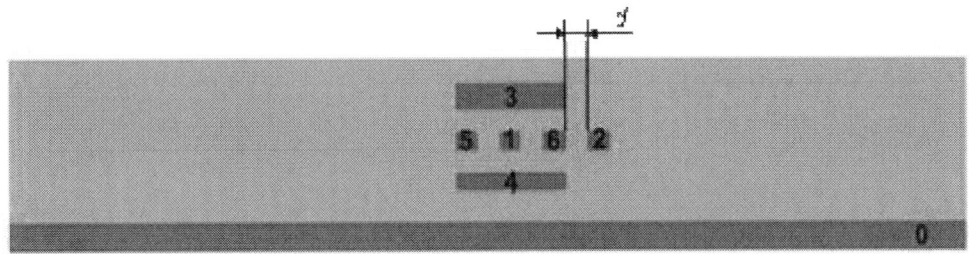

Fig. 3: Cross-section of a shielded conductor system
with a "protected conductor".

By combining the direct and indirect screening it is possible to achieve an extremely small coupling capacitance between the signal leads (Fig. 3). In cases when even this cut of coupling capacitance is not sufficient, the screening conductors in different layers can be interconnected by means of conductive connections and so to create rings surrounding the protected conductor. In this way we can design a structure around the protected conductor, resembling a Faraday´s cage (Fig. 4), or a coaxial cable. The dimensions of the feed-through contacts (VIA1, VIA2) are firmly determined by the integrated circuit fabrication technology and they cannot be changed. In the case of a spatial model of the lead system (Fig. 4), identical dimensions of the feed-through contacts VIA1, VIA2 were used.

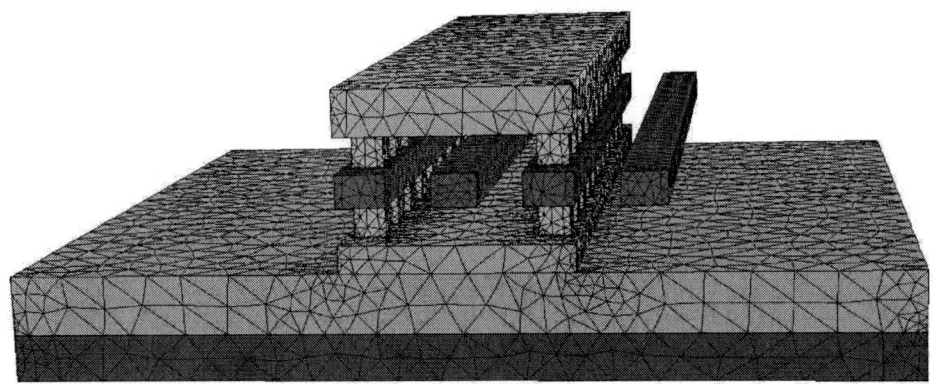

Fig. 4: Cross-section of a shielded conductor system with a "protected conductor".

The three-dimensional spatial model has its base size 20 μm x 20μm. The feed-through contacts form rows of identical vertical connections repeating in 1 μm intervals. Since the three-dimensional model contains a large number of these connections, this segment of the lead system can be considered as homogeneous and the system can be assumed to behave as a connecting line system with well defined capacitances per unit length (Fig. 5).

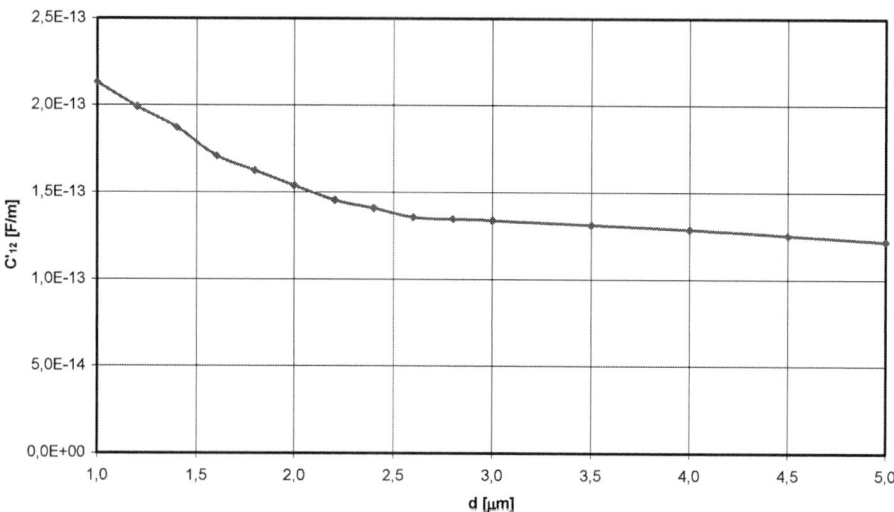

Fig. 5: Coupling capacitance C'_{12} versus conductor spacing d in a protected system

3. SUMMARY

The coupling capacitance between signal conductors can be limited by the application of electric shielding. From the point of view of maximum suppression of the signal conductor coupling capacitance, the protected conductor interconnection system appears to be very efficient (Fig. 4). This type of interconnection system is of most interest in the design of mixed-signal integrated circuits where the low parasitic coupling between digital and analog parts of the circuit is of paramount importance.

The accurate calculation of parasitic capacitances of the signal interconnections helps us to design detailed circuit models of the interconnecting systems. Critical spots in an electronic system, where parasitic couplings cause failures, can be found by simulation of the circuit models. Then, it is possible to take appropriate measures to decrease the parasitic capacitances of signal conductors, like adjusting the integrated circuit layout or by applying shield conductors.

Acknowledgement

This research has been supported by the following research programme of the Czech Technical University in Prague.

References

[1] C. S. Walker: "Capacitance, Inductance and Crosstalk Analysis", Artech House, 1990
[2] J. M. Min, The finite element method in electromagnetism, New York, J. Wiley 1993
[3] B. W. Boast, Principles of Electric and Magnetic Fields, New York, Harper and Brothers, 1986
[4] J. Novak, J. Foit, Analysis and measurement of capacitive coupling in integrated circuits, ASDAM'08, Smolenice, Slovakia, pp. 223 – 226, 2008, ISBN 978-1-4244-2325-5
[5] AMIS, C05M-A - 3M/2P/HR process, Design Rule Manual, 1999

ASDAM 2016, The 11th International Conference on Advanced Semiconductor Devices And Microsystems, November 13-16, 2016, Smolenice, Slovakia

Magnetically Levitated and Guided Systems

Florian Puci

Department of Microelectronics, Faculty of Electrical Engineering, CTU in Prague,
Technická 2, 166 27 Praha, Czech Republic
e-mail: puciflor@fel.cvut.cz

The paper describes fundamentals of the magnetic levitation technology. A general background of the magnetic levitation is given, followed by its applications, comparison with other types of technologies, the current stage of its development, etc. Further, two main types of magnetically levitated systems, within their subgroups, are compared, in general and details characteristic basis. A comparison between the AC and DC power supplies for these systems, including the pros and cons of each type, is also provided in the paper.

1. Introduction

The next-generation of lithography requires a high precision stage, which is compatible with a high vacuum condition and for this, a magnetic levitation stage with two or more degrees-of-freedom is considered state-of-the-art technology. Nowdays the size of wafer is moving up to 12 inch in order to enhance the efficiency of a batch process. Therefore, a high resolution process and a large operating range concurrently with the application into a super-clean environment are the requirements for the related micro-actuators. The noncontact characteristic of magnetic levitation technology enables high precision positioning as well as no particle generation. The manufacturing process of a recent semiconductor IC, imposes very severe constraints on not only the processing accuracy but also the working environment. However, the heat is inevitably generated, while using electromagnetic actuators for levitation, which deforms the structures and degrades accuracy of the stage, and though a gravity compensator is required. Though, a fully operational system with magnetic levitation technology is required to have the following characteristics [1]:

- **Gravity compensation with no power consumption**
- **Large and homogenous force density**
- **Zero stiffness between the translator and stator part**
- **Position-independent dynamic force**
- **Enough workspace for operation in all directions applications**

Between the parts with relative motion of new type of wafer transporters, represented for example by belt conveyers or articulated robots, there are mechanical contacts. Generally, in order to minimize and reduce the friction and enhance smooth operation, lubricants (oils) are used in the contact regions. However, due to the rubbing activity, there are always generated a lot of tiny solid particles which are mainly absorbed by the lubricant. But during the production process, it can happen that some particles can still flee the wafer transporter. Eventually, some of the particles will drop back onto the surface of wafers, damaging the delicate and newly manufactured integrated circuit (IC) device. In fact, in the nowdays microelectronic industry, the wafer transporter is ranked the number one in the list of particle source extrinsic to the fabrication process. The presence of lubricant in the contact regions has

978-1-5090-3084-2/16 $31.00 © 2016 IEEE

a negative contribution to the process cleanness and this results from the volatility of lubricant. The surface properties are strongly affected by the migration and the diffusion of the gaseous elements into the surface of wafers.

Though a solution can be that the rate of evaporation can be enhanced by the application of a vacuum environment, because the undesirable gaseous elements are evaporated to the processing environment. Obviously, reduction or if possible, total elimination of the contact between relatively moving components, is the best way to eliminate wafer transport contamination. In this manner, a particle-free and oil-free environment can be obtained and consequently both particle generation and lubricant evaporation will disappear simultaneously. For example, in a wafer transporter, mechanical contacts usually occur and are continuously present in the region between the component that carries the wafer, so the wafer carrier, and the transporter base that supports the carrier. By separating the wafer carrier from the transporter base, these mechanical contacts can be eliminated. This can be achieved by means of levitation, where non-contact physical forces are applied between the carrier and the base, so the carrier is separated from the transporter base.

In the known physical world, there exist three main methods that the free-floating levitation can be achieved:

a) air-bearing   b) electrostatic levitation   c) magnetic levitation (maglev)

In order to determine the most appropriate and effective method for our application, it required a general physical understanding of each method's working theory including some basic knowledge of integrated circuit fabrication technology. An air-bearing system basically requires two very important components: a special filter used to exclude particles generated by the air pump from entering the transporter and an additional circulation system used to enclose the air. In comparison with the previous system, in the electrostatic levitation system, the particle residues which are generated in the first installation or accidentally generated during wafer loading and unloading process will be attracted by the electrostatic field. Consequently, this can cause possible particle contamination, with the particles being adhere to the transporter due to the electrostatic field.

Given the above arguments and in order to achieve various performance targets, the precision stage using a novel contact-free planar actuator based on the magnetic levitation technology is highly recommended and this paper consider it the best solution for IC fabrication. Generally, as the MAGLEV eliminates the friction due to a mechanical contact, it has a wide range of applications requiring super-cleanness environment, including here the semiconductor wafer transfer. The manner of functionality and construction of MAGLEV eliminates the problems caused by using the above technologies. It creates a stable state without any mechanical contact when the gravitational force is solely counterbalanced by magnetic forces. This system includes levitating tracks, stabilizing tracks, and propulsion coils [2].

2.   Types of planar actuators

There exist two types of planar actuators, referred as either with moving coils or moving magnets. The actuator from the first type consists of moving air-core coils and stationary magnets. The main advantage of such system is the usage of a small number of coils and their amplifiers, as the stroke force can be easily increased by adding a few more magnets in the magnet array. Moreover, the simpler design of these actuators allows control of the torque on the translator part by using different coil topologies. A big disadvantage for this class is that there should be a cable to connect the translator and the stator part, as the coils require power and cooling. The following Figure 1 shows an example of these planar actuators [3].

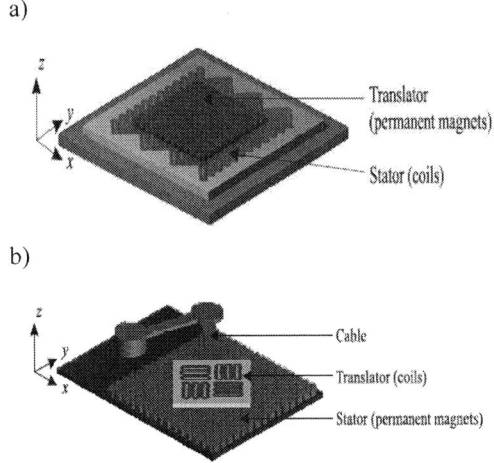

Figure. 1. Planar actuators: a) moving magnets, b) moving coils [3]

The second type of actuators with magnetic levitation consists of moving magnets and stationary air-core coils. In contrary with the previous class of actuators, it doesn't require a cable for connecting the translator with the stator part, which is really a big advantage from the design point of view. Coils are located and powered in the stator platform, which means smaller amount of disturbances to the translator. But, the torque decoupling as a function of position is more complex than in the moving-coils planar actuators. In manufacturing processes, working environment affects the quality of the precision products.

Different types of conventional transportation systems such for example: belt-type conveyors or articulated robots generate dusts and pollution due to the mechanical friction or lubrication, and are inadequate and inappropriate to satisfy the environmental demands. The magnetic levitated carrier system for the transportation systems has the advantages of being contact free, can eliminate the mechanical components e.g., gears, guide etc., reduce the mechanical alignment and maintenance cost, hence it satisfies the environmental demands. Therefore, research on contact-free type transportation system and actuator has been actively performed by worldwide researchers. Modern applications of levitation in equipments like magnetic bearings and magnetically levitated vehicles have given renewed impetus to research efforts in the direction of electromagnetic levitation [3] [4].

3. MAGLEV supply source types

There exist two approaches for designing a controlled force magnetic levitation system: one uses attractive forces interacting between an electromagnet and a soft magnetic flux closure, and the other is based on repulsive forces between electromagnets and appropriately magnetized permanent magnets.

Electromagnetic levitation technology, where the attractive forces are used, there is either an AC or DC source to drive the electromagnets. Although there have been built and tested several experimental systems which use as the supply the AC sources, this method of stabilization is appropriate for applications where mass of the levitated object, translator, is relatively small. The losses caused by the effect of the eddy currents and the complex circuit design and control of the power modulation, makes the AC method of stabilization inappropriate for heavy payloads.

In contrast with the above AC method, the explicit DC method, generally known also as the electromagnetic levitation system (EMLS), is characterized by simpler circuitry configuration and favourable power requirement. In the circuits using the DC EMLS, a switched mode power amplifier is mainly part of the design and used to control and utilize the current as well as the attraction force of the electromagnets. A simple electromagnetic levitation system consists of four main components: (i) Actuator and Rail, (ii) Position Sensor, (iii) Controller, (iv) Power amplifier [5].

The electrodynamic system uses the repulsive forces to levitate the object. In contrast with the first system, the electromagnetic system, produces levitation of the object based on the attractive forces between electromagnets and the levitated ferromagnetic objects. As part of this system, a position transducer is used to sense and measure and the gap between the magnet pole-face and the ferromagnetic object. Then the output signal from the transducer is fed back to a comparator. The process continues with output signal of the comparator being applied to a position controller, giving in this way the reference current for the current loop. The actual current supplying the coil is sensed and compared with the reference current by the current sensor. A very important element in this process is also the power amplifier, which produces necessary currents in the actuator coils after receiving a command from the current controller. Prior to this, the current error process task is completed in the current controller. The currents from the power amplifier going through coils generate requisite magnetic forces which are the key for levitating the translator.

Applications of electromagnetic levitated and guided systems, are for example, in the field of transportation vehicles, frictionless bearings and conveyor systems [5].

Acknowledgement

Research described in the paper was supervised by Prof. Ing. Miroslav Husak, CSc., FEE CTU in Prague and has been supported by the CTU project No. SGS14/195/OHK3/3T/13 „Development of Smart Devices and Systems in the Field of Microelectronics, Nanoelectronics, Optoelectronics and Micro-nano-optoelectronic structures, elements and systems".

References

[1]   Y. M. Choi, M. G. Lee, D. G. Gweon, J. Jeong, *A new magnetic bearing using Halbach magnet arrays for a magnetic levitation stage*, South Korea, 2009.

[2]   H. Park, S. K. Lee, J. H. Yi, S. H. Kim, Y. K. Kwak, I. A. Wang, *Contactless magnetically levitated silicon wafer transport system, Journal of Mechatronics, Chatsworth,* 1999.

[3]   C. M. M. van Lierop, *Magnetically levitated planar actuator with moving magnets: Dynamics, commutation and control design*, Eindhoven, The Netherlands, 2008.

[4]   J. W. Jansen, *Magnetically levitated planar actuator with moving magnets: Electromechanical analysis and design*, Eindhoven, The Netherlands, 2007.

[5]   P. K. Biswasa, S. Banerjeeb, International Journal of Applied Science and Engineering, *Design and ANSYS Software Based Simulation of U-I Type Actuator and Rail Used in Electromagnetic Levitation System*, Durgapur, West Bengal, India, 2014.

978-1-5090-3084-2/16 $31.00 © 2016 IEEE

ASDAM 2016, The 11th International Conference on Advanced Semiconductor
Devices And Microsystems, November 13-16, 2016, Smolenice, Slovakia

Model of the Triboelectric Generator

M. Husak and A. Bily

Department of Microelectronics, Faculty of Electrical Engineering, CTU in Prague
Technicka 2, 166 27 Prague 6, Czech Republic
e-mail: husak@fel.cvut.cz

The core of the article solves the microsystem model of the electric generator based on the triboelectric effect. The aim was to verify the possibility of using triboelectric effect, verifying properties using macromodel, determine the essential characteristics, finding the optimum load, output voltage and output power achieved. Attention is given to the three basic principles of activities - the vertical sliding, the lateral sliding, the shift of one electrode with the free triboelectric layer. The attention is focused on information about the structure design, information of the measured parameters with free triboelectric layer, where have been achieved the best results. Different structures have been designed, realized and measured, the example of structure with the free triboelectric layer was realized by mechanical machining of the cardboard paper, aluminum foil and teflon (PTFE). The output voltage reached amplitude of 40 V at the load 1 MΩ.

1. Introduction

A significant change occurring in present life in the last twenty years is the search and development of modern portable, personal and wearable electronics. A character of this modern technology trend is the great increase in the electronic devices or systems, each of that requires a mobile power supply source. The traditional way is to use batteries to power electronics. The battery has a limited power, replacing the battery for each device becomes a great problem, especially, who will replace the battery source, how do we know it is out of power, when should we replace the battery etc. This is not a problem if the number of batteries is limited, but the situation changes if the number of battery power supply to be replaced becomes huge. With the use of great amount of battery power supply, recycling of these energy sources become a great challange, because batteries will necessarily cause environmental issues if the chemicals for making the batteries leak out [1].

The trend in the development of electronics is toward low power electronics, which enables it to use the energy harvested from the environment energy of the device to power the device, in this time is called as self-powered systems or energy harvesting systems [2] for widely used application in different areas as example: security systems, smart buildings, biomedical systems, automotive, portable/wearable personal electronics, ultrasensitive chemical sensors, nanorobotics, MEMS, remote and mobile environmental sensors, sensor nets etc. For energy harvesting can be used the new technologies that can pick-up energy from the environment as micro or nanopower sources. It is the newly emerging field of nanoenergy, which is linked with the application of smart materials and nanomaterials as well as nanotechnology for harvesting energy to micropower micro or nanosystems [3]. One type of the energy harvesting generator work on the triboelectric effect [4]. Potential applications as a new energy technology and in self-powered active sensors are expressed in [5].

978-1-5090-3084-2/16 $31.00 © 2016 IEEE

The material with the triboelectric effect can obtain electrical charge after it contacts a different material by friction mechanism. After two different materials come into contact, a chemical bond called adhesion is formed between parts of the two surfaces, the electric charges move from one material to the other to equalize their electrochemical potential. The moving charges can be electrons or ions or molecules. When separated, some of the bonded atoms have a trend to hold extra electrons, and some trend to give them away, possibility of creating the triboelectric charges on the surfaces. The triboelectric charges on dielectric surfaces can be a force for driving electrons in the electrode to flow to balance the electric potential drop created. Based on such principle, four different modes of triboelectric generators can be introduced [1].

Vertical contact separation mode. Two different dielectric layers face each other, with electrodes deposited on the top and the bottom surfaces of the stacked structure. A contact between the two dielectric layers generates oppositely charged surfaces. The layer surfaces are separated by a small distance by the lifting of an external force, a potential drop is created. If the two electrodes are electrically connected by a load, free electrons in one electrode would flow to the other electrode to create an opposite potential in order to balance the electrostatic field. Once the distance is closed, the triboelectric charge created potential disappears and the electrons flow back [6].

Lateral sliding mode. Two dielectric layers are in contact, a relative shifting in parallel to the surface generates triboelectric charges on the two surfaces [7]. A lateral polarization is thus introduced along the shifting direction, which drives the electrons on the top and bottom electrodes to flow to fully balance the field created by the triboelectric charges. A periodic shifting apart and closing generates an AC output. The shifting can be a planar motion, a cylindrical rotation or disc rotation.

Single electrode mode. Vertical contact-separation mode and lateral sliding mode have two electrodes connected to a load. Sometimes, the part of the Triboelectric generator cannot be electrically connected to the load (for example for a mobile object, such as a human walking on a suitable surface). To harvest energy from such a case, can be used a single electrode triboelectric generator. The electrode on the bottom part of the generator is grounded. An approaching or departing of the top object from the bottom one would change the local electrical field distribution. There are electron exchanges between the bottom electrode and the ground to maintain the potential change of the electrode. This energy harvesting strategy can be in both contact-separation mode and contact-sliding mode.

Freestanding triboelectric layer mode. The generator consists from a pair of symmetric electrodes under a dielectric layer. The size of the electrodes and the distance between the two are of the same order as the size of the moving object. The object's approach to and/or departure from the electrodes generates an asymmetric charge distribution, which causes the electrons to flow between the two electrodes to balance the local potential distribution [8]. The oscillation of the electrons between the paired electrodes generates electrical charge. The moving object does not have to touch directly the top dielectric layer of the electrodes so that, in rotation mode, free rotation is possible without direct mechanical contact.

2. Design and realization of the triboelectric system

Different materials were used for model realizations. Teflon (PTFE) was used for the contact surfaces of dielectric-conductor. Electrodes for charge conduction were made from Cu layer. Al layer can serve as second triboelectric layer (combination with teflon). Al layer cans serves as counterpart teflon layer. Material 4.4'-oxydiphenylene-pyromellitimide (kapton) can be used for realization of interface surfaces of the dielectric/dielectric. Material PMMA (polymethylmethacrylate or plexiglass) can be used for second dielectric realization. Other dielectric materials used for the realization triboelectric systems are: FEP (Fluorinated ethylene propylene), PDMS (Polydimethylsiloxane), PET (Polyethylentereftalát). Conductive materials Cu, Ag, Al can be used for contacts realization. PMMA, PET and cardboard paper can be used for realization of the construct elements of the triboelectric systems. Micro-springs or conductive foam material (molitan) can be used for the realization of separators. Several versions of the structures for operation in various operating modes of triboelectric system were designed and realized [9]. Teflon, paper and aluminum were used to the realization of the generator structures.

Triboelectric system with vertical contact separation of dielectric layer from the conductive electrode was first proposed system. This principle was chosen for the verification of the triboelectric effect. The structure is easily adaptable to the mode with a lateral shift. Triboelectric system consists of two electrodes, bonded to the solid material. Cu leads are connected to the surface of electrode terminals designed for easy connection of the measuring device.

Triboelectric system with a free triboelectric layer is further designed and realized structure. The structure enables the contact the load on the stationary layer. This solution easily eliminates danger of mechanical damage to the measuring terminals. The backing layer was made of cardboard paper. Measuring leads were realized from Al foil. The free triboelectric layer was made from the teflon foil.

Triboelectric structure with a shift of one electrode was the last design and realization of the generator. The structure enables the load location between the ground terminal and triboelectric generator. Electrical connection enables elimination of undesirable electrostatic voltage.

3. Achieved results

Different structures have been designed, realized and measured, the example of structure with the free triboelectric layer was realized by mechanical machining of the cardboard paper, aluminum foil and teflon (PTFE) - Fig. 1. The oscilloscope was used to measure the output voltage. The output voltage reached amplitude of 40V at the load 1MΩ - Fig. 2.

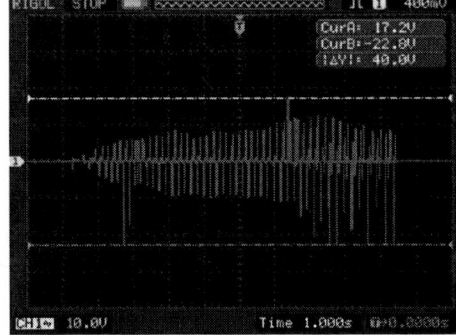

Fig. 1 The generator structure with a series of the electrodes, the principle of the free triboelectric layer

Fig. 2 The output voltage of the structure

The structure output current resp. voltage was measured depending on the load impedance - Fig. 3. The maximum output power of the generator according to the optimal load was created using the measured parameters - Fig. 4.

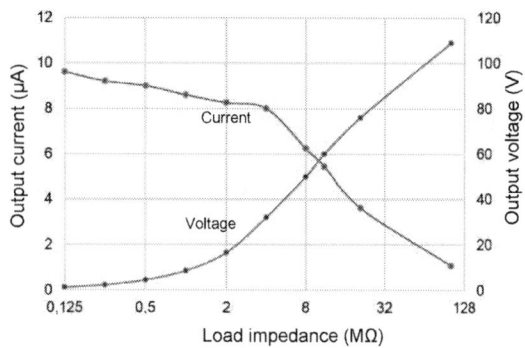 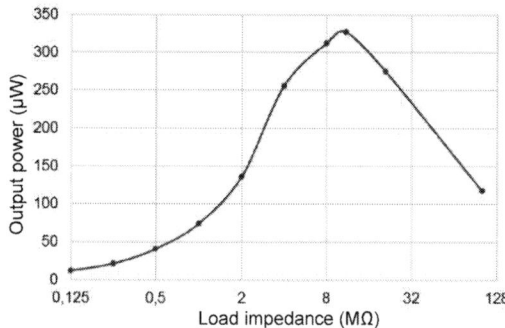

Fig. 3 Output voltage resp. current versus load impedance

Fig. 4 Optimal output power versus load impedance

4. Conclusions

Simple triboelectric structures have been designed and realized in the work. Structures were used to verify the functionality of the triboelectric effect. The basic parameters of the structures have been measured. The achieved results confirmed the possibility of using the structures as energy sources obtained from the environment. Designed and measured structures have been realized in the form of macromodels, achieved knowledge will enable further development of these structures and the improvement of parameters.

Acknowledgement

This research has been supported partially by the project of Ministry of Interior No. VG20102015015 „Miniature Intelligent System for Analyzing Concentrations of Gases and Pollutants, particularly Toxic", partially by CTU project No. SGS14/195/OHK3/3T/13 „Development of Smart Devices and Systems in the Field of Microelectronics, Nanoelectronics and Optoelectronics Micro-nano-optoelectronic structures, elements and systems".

References

[1]   Z. L. Wang, The Royal Society of Chemistry, 2014
[2]   Z. L. Wang, Sci. Am., 2008, 298, 82.
[3]   Z. L. Wang and W. Z. Wu, Angew. Chem., 2012, 51, 11700.
[4]   F. R. Fan, Z. Q. Tian and Z. L. Wang, Nano Energy, 2012, 1, 328.
[5]   *Z. L. Wang, ACS Nano, 2013, 7, 9533.*
[6]   S.M. Niu, S. H. Wang, L. Lin, Y. Liu, Y. S. Zhou, Y. F. Hu and Z. L. Wang, Energy Environ. Sci., 2013, 6, 3576.
[7]   S. H. Wang, L. Lin, Y. N. Xie, Q. S. Jing, S. M. Niu and Z. L. Wang, Nano Lett., 2013, 13, 2226.
[8]   17 S. H. Wang, Y. N. Xie, S. M. Niu, L. Lin and Z. L. Wang, Adv. Mater., 2014, 26, 2818.
[9]   Bily, A.: Verification of the triboelectric effect for nanogenerators, CTU in Prague, 1996

ASDAM 2016, The 11th International Conference on Advanced Semiconductor
Devices And Microsystems, November 13-16, 2016, Smolenice, Slovakia

Gas Analyzer for Quick Indicative Measurements

M. Husak, A. Laposa and J. Kroutil

Department of Microelectronics, Faculty of Electrical Engineering, CTU in Prague
Technicka 2, 166 27 Prague 6, Czech Republic
e-mail: husak@fel.cvut.cz

The article describes the design and realization of the gas analyzer using 4 types of gas sensors – catalytic, electrochemical, semiconductor and infrared sensor. The analyzer is intended for quick orientation measuring the presence of gas. The analyzer includes basic parts - the sensor part, the evaluation and control part, the display and the output actuator part. The analyzer is designed for easy modification of its connection to different types of chemical sensors. The sensor part includes a chemical sensor, an EEPROM and a temperature sensor. The type of the used sensor sets the operation mode of the both part - evaluation and control part. Actuators on the output are ready to control the security elements (ventilation, alarm, etc.). The display shows the basic information - gas concentration, temperature, and more. Evaluation and control part includes microprocessor, memory for long-term data, signaling LEDs, push buttons.

1. Introduction

Research of gas sensors is directed to a multisensor systems in recent years. Sensor systems integrate currently available and used detection technologies with different physical principles. The research is directed to use of identification capabilities gases from multiple detectors arranged in arrays (multisensor systems). Complex sensor systems are not necessary to use for quick orientation measurements of the concentration, the simple systems with quick response are often sufficient. This work has therefore focused on the design and implementation of a simple sensor system for gas analysis with computer control. The work is divided into three parts: Sensor part, evaluation and control part, the part with the display and output relays. The catalytic sensor has been used to verify system performance gas analyzer.

2. Design and realization of the gas analyzer

The gas analyzer consists of three main modules, possibility of a simple modification for all kinds of sensors has been monitored during the design. Block diagram of a gas analyzer is shown in Fig. 1 [1]. Sensor part contains suitable type of sensor, EEPROM and temperature sensor. Standardization of the output signal from various kinds of sensors is realized by the analog signal conditioning. The indicator on the board indicates the type of sensors, evaluation and control part chosen evaluation procedure for the used sensor.

The system has been designed and realized with the outlet part containing the display and the actuator. The output part is not necessary for the activity of the gas analyzer, but in some cases it may be useful, for example the output relay can handle safety features such as ventilation or alarm. The display shows basic information (gas concentration, temperature, date of next calibration etc.). This block is used in the cases where the gas analyzer is not connected to the control panel and partly takes over its function. An evaluation and control unit is the heart of the device, one contains the power supply of the analyzer, microprocessor, memory for storing long-term data, status LEDs, buttons and more.

978-1-5090-3084-2/16 $31.00 © 2016 IEEE

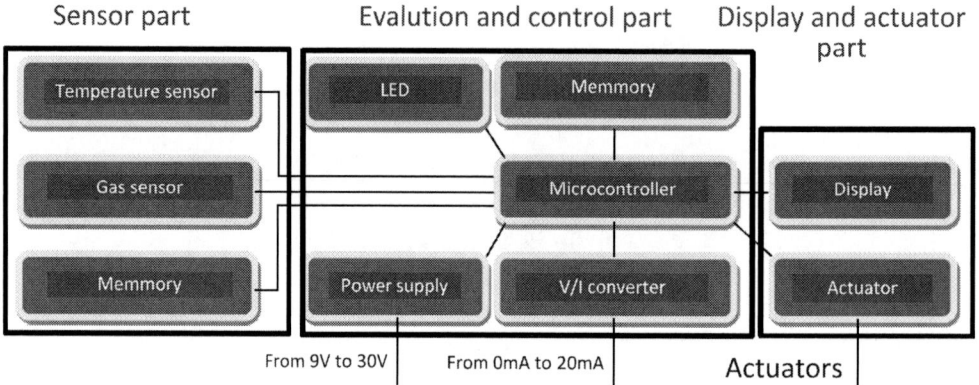

Fig. 1 Block diagram of the gas analyzer [1]

2.1 Sensor part

Sensor part was realized in four versions depending on the chemical sensor. Chemical sensors are of various kinds - catalytic, semiconductor, electrochemical and infrared. EEPROM (24AA04H from Microchip) communicates via I²C bus. Calibration values and sensor calibration data are stored in memory. Temperature sensor (Microchip MCP9701) has an analog output. The electrical connection of sensor analog part differs depending on the type of sensor. The *catalytic sensor* is connected into a Wheatstone bridge, electronic circuit is simple, other electronic circuits are not required for signal evaluation. The *semiconductor sensor* must be heated. The electronic circuit controls the current flowing through the heating coil of sensor. Change in resistance of the semiconductor sensor is used for signal evaluation. The output signal from an *electrochemical sensor* is evaluated by LMP91000 circuit, the output signal (Information about the concentration) is led by I²C bus to the microprocessor. Signal processing of the sensor together with the temperature sensor are in one integrated circuit. Resistance together with a transistor are used for an external heat sensor. *Infrared sensor* operates in an alternating mode, square-wave signal with a frequency of 2MHz. The sensor contains a thermistor and therefore the temperature can be measured directly by the sensor. Two circuits process the signals applied to the microprocessor. The circuits are two, because the sensor comprises a reference for comparison with the measured signal. Voltage to adjust the signal from the sensor is generated in the DC/DC converter.

2.2 Control and evaluation part

The analyzer contains three outputs - through RS458 (digital communication between sensor and control panel), USB (PC communication) and current loop (gas analyzer connection with the control panel). RS485 is used for digital communication between detectors and control panel. SN65HVD07 circuit ensures communication. USB is ready to communicate with a computer, USB makes it easier to calibrate, reading measured values from memory and the update date and time. These operations can also be performed via RS485, but you need a converter to the serial link. Current output is used to connect the gas analyzer with the control panel. The amount of current corresponds to the concentration. Current output is handled by an integrated circuit XTR111. The output was selected from 0mA to 20mA, 4mA to 20mA output is range for measuring the concentration, 2mA indicate an error and 0mA indicate sensor unconnected. Signaling LEDs indicate power supply, error, RS485 communication and signaling calibration of zero concentration and the prescribed concentration of the gas. The system contains buttons for manual calibration. EEPROM memory 24AA512 from Microchip. Memory has a large enough capacity (64K) for storing the history of the measured data. I²C bus provides communication with the memory.

Microcontroller (Microchip PIC24FJ128GC010) controls the operation of the analyzer. The microprocessor is a 16-bit, contains a number of peripherals. The microcontroller has a 12-bit A/D converters utilizing the principle of successive approximation, that can switch up to 50 channels. The microcontroller has two precision 16-bit sigma-delta converters. Converters are used for accurate voltage measurements from the sensor. The microcontroller has two resistive D/A converters. The converters are used as a voltage converter for current/voltage, as well as adjustable voltage reference for the sensor. The microcontroller has two I²C bus. In this type of bus transfer takes place over two wires (SDA and SCL). The microcontroller contains an asynchronous UART communication, which is transferred by the circuit SN65HVD07 to the communication according to standard RS485. The microcontroller enables communication via the USB specification with the computer. This feature is used primarily for calibrating, updating data and load measurement history.

2.3   Control software

The control program (written in C) is implemented in the development environment MPLAB X IDE from Microchip. The program starts initializing inputs and outputs, EEPROM, respectively displays. After initialization, control and evaluation part program found that is connected sensor. If there is no sensor connected, the program sets the error and waits for the sensor connection. The microprocessor selects the part of program after connecting the sensor, which corresponds to the sensor. Program loops are for all types of sensors identical, loops differ only by calculating or displaying on the display. The receiving values from the sensor part is followed, in the next step follows the calculation of the values of gas concentration. The next step in the program itself includes a manual calibration using buttons or calibration via USB. Manual calibration of the analyzer is calibrated with two levels of concentration. In the first step is saved value with zero gas concentration and in a further step the exact value at a non-zero concentration. Similarly, the calibration is carried out using a computer. After calibration, is followed by sending the output value, on converter voltage/current, when it is used as RS485. After sending the values, the loop repeats.

3.   Achieved results

Catalytically sensor MC119 has been used to verify the function of the proposed gas analyzer. Time response for the gas propane-butane was measured and one is presented in Fig. 2. The curve shows that the response time is 9.4s and the recovery time is 8.4s.

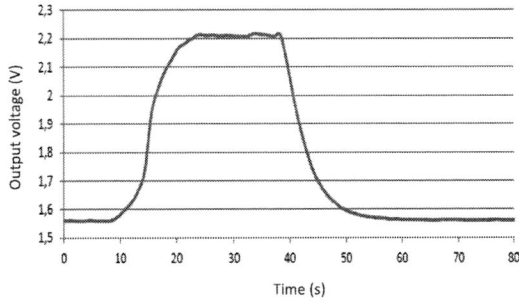

Fig. 2 Time response for propane-butane

The same measurements have been performed for benzene - Fig. 3 and the methanol - Fig. 4. The sensor used has been 2 years old, the sensor had longer response and recovery time compared with the datasheet. Methanol and benzene have been tested for two concentrations, time response of both gases corresponds to the catalog data. The output voltage of the sensor has been measured at a gas concentration of 1% for benzene voltage has been 1.704V, the voltage for the methanol has been 1.625V. The measured data shows that the sensor is more sensitive to benzene.

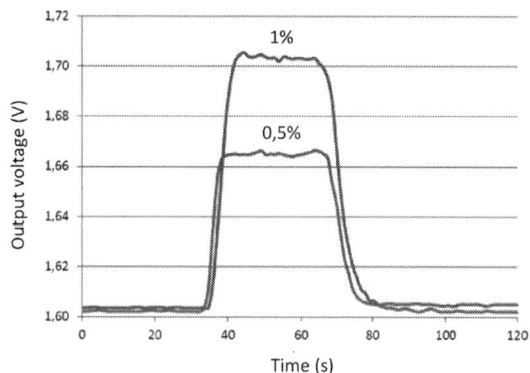 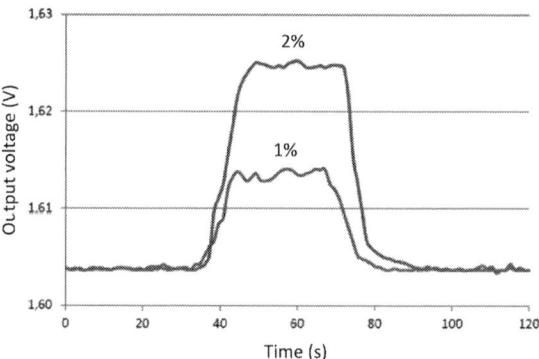

Fig. 3 Time response for benzene   Fig. 4 Time response for methanol

4. Conclusions

The gas analyzer has been designed as a modular compact detector with the ability to connect various types of sensors. Outputs are designed both analog and digital. The proposed analyzer contains most of the features offered by the competition, eg. company Dega [2]. The main part of the realized analyzer was tested - A/D converter, calculating the concentration of the catalytic sensor and converter voltage/current. The sensor was tested for presence of gas (benzene, methanol and propane-butane). Achieved results were in agreement with catalog values.

Acknowledgement

This research has been supported partially by the project of Ministry of Interior No. VG20102015015 „Miniature Intelligent System for Analyzing Concentrations of Gases and Pollutants, particularly Toxic", partially by CTU project No. SGS14/195/OHK3/3T/13 „Development of Smart Devices and Systems in the Field of Microelectronics, Nanoelectronics and Optoelectronics Micro-nano-optoelectronic structures, elements and systems".

References
[1] Mistr, L.: Gas Analyzer, CTU in Prague, 2014.
[2] *Dega* [online], http://www.dega.cz, 2014

Acoustic Method for Respiratory Monitoring

Jiří Kroutil, Alexandr Laposa, Miroslav Husák, Ratanak Sio

Department of Microelectronics, Faculty of Electrical Engineering,
Czech Technical University in Prague, Technicka 2, 16627 Prague, Czech Republic
e-mail: kroutj1@fel.cvut.cz

This paper describes a method of the respiration monitoring based on the sensing of acoustic signals in trachea. Further, the article describes method for signal processing of the acoustic signals. The respiration belongs among basic vital functions and the knowledge of their parameters and quality is necessary in medicine. The research is leading to the methods, which are inconvenienced patients. Monitoring of respiration is important for monitoring respiration towards observation quality sleeping or The Sudden Infant Death Syndrome (SIDS).

1. Introduction

Breathing is one of the essential functions. In medicine this function is needed to sense and to detect its parameters. The organism needs energy for the ensuring of basic vital functions, which is released by the oxidation of the energy matter (saccharides, lipids and proteins). This article describes a solution of respiration diagnostics. Concretely it is thought the external respiration.

Organism, which can be imagined as a biological system, is a source of biological signals. Bio signals are possible to divide according to origin into electrical, magnetic, acoustic, chemical, mechanical, optical, impedance, thermal, radiological and ultrasonic. This allows monitoring of respiration in various ways (Fleish pneumotachometer, extracting from ECG [1], using of pressure sensors [2], monitoring of humidity in the exhaled breath [3], non-contact respiration monitoring using slit light pattern projection [4], non-contact respiratory monitoring system using a ceiling-attached microwave antenna [5], monitoring by textile electrode [6] etc.). The method based on the monitoring of the acoustic signals appears as very attractive. One of the ways of study regulation respiration cycle is monitoring of breathing paradigm – depending among basic quantities of ventilation: respiration frequency, minute ventilation, inspiration and expiration time, apnoea pause, the respiration volume [7]. One of the ways of monitoring respiration cycle is monitoring of breathing paradigm [8].

2. Method

Monitoring of the respiration is based on detection acoustic signals from trachea. These system is composed of four parts: microphone part, signal preprocessing, data acquisition (DAQ – National Instruments NI USB-6009) and PC (Fig. 1). These signals are picked up by microphones. Two microphones (Wolfson Microelektronics WM7120) are used for sensing signal from trachea. Microphone 1 senses respiratory acoustic signal and additional signals added to the desired signal (external signals). Microphone 2 is used to eliminate external disturbance signals. Signals from both microphones are added to cancellation of external signals. These signals from microphones are amplified by signal preprocessing block and converted to digital form by interface. Processing of signals is provided in PC by software LabView.

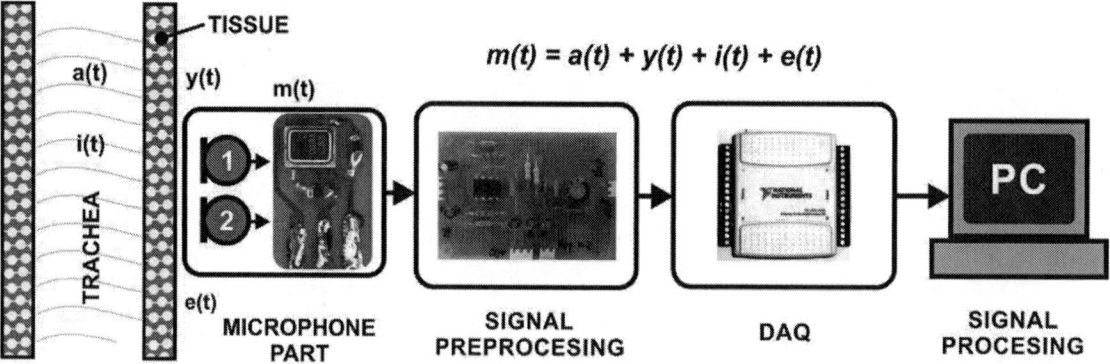

$$m(t) = a(t) + y(t) + i(t) + e(t)$$

m(t)-signal observed by the microphone, a(t)-vibration airflow in the trachea,
y(t)-disturbances from the interface between the microphone and skin,
i(t) is internal disturbances, e(t) is external disturbances.

Fig. 1 Block diagram of respiratory detection.

Fig. 2 shows breathing pattern acquired by microphone. The repeated breathing pattern and differences between inspiration and expiration are possible to read in the time behaviour of the signal. The smooth beginning and the abrupt ending are often obvious during the inspiration. On the other hand, the behaviour of the expiration shows the abrupt beginning and the smooth ending. Both phases are separated with the inspiratory and expiratory pauses. [7]

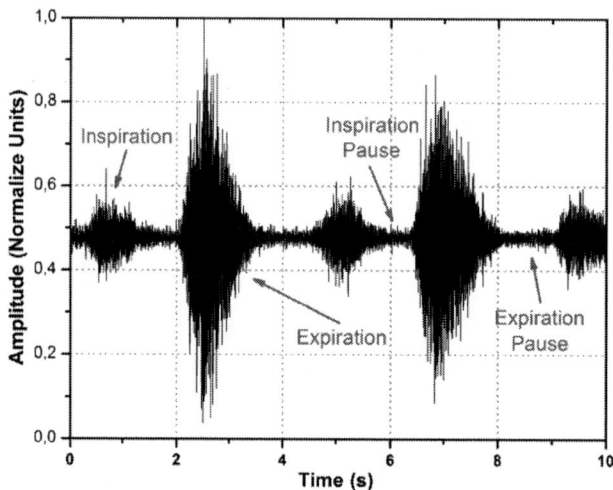

Fig. 2 Breathing pattern acquired by microphone.

Distinguish the different phases of breathing is important for monitoring of respiration. Fig. 3a) shows spectrum of inspiration phase and Fig. 3b) shows expiration phase. Analyse in frequency domain was used for recognise of inspiration and expiration [9]. According to the frequency spectrum (power of spectrum), we can detect, whether it is the inspiration or expiration. Fig. 4 shows block diagram of processing in frequency domain.

Sample frequency was fs=1800Hz. First, segmentation was performed, whose aim was to divide the signal into short sections. Frame length was chosen N=25 samples to correspond roughly N/fs=25/1800=14ms. Consequently, the power spectrum is calculated using FFT (Fast Fourier Transformation) for separation of active speech phase (inspiration and expiration) from silence phase (pauses). The last step is to determine whether it is the inspiration or expiration. First we have to define how many segments are used. As the normal breathing rate is 12-20 breaths per minute, we need a time frame around 1,4s which corresponds to the duration of inspiration and expiration. In our case it is approximately 2500 samples per respiratory phase (100 previous segments). To distinguish of inspiration from expiration was calculate by using the formula

$$Sum_k = \sum_{i=1}^{Nframk}(peak_i)$$

(1)

where N_{framk} is the number of frames per respiratory phase, $Peak_i$ is the maximum amplitude.

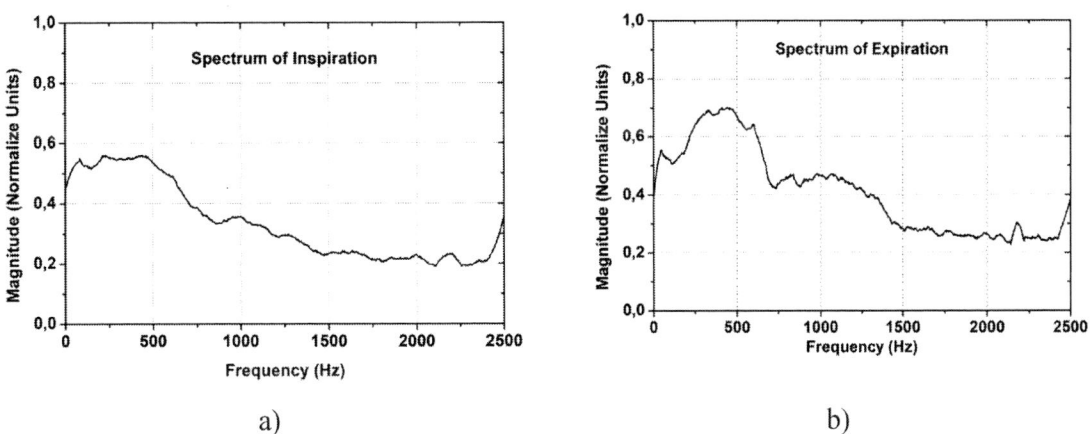

a)   b)

Fig. 3 Spectrum of inspiration phase a) and expiration phase b).

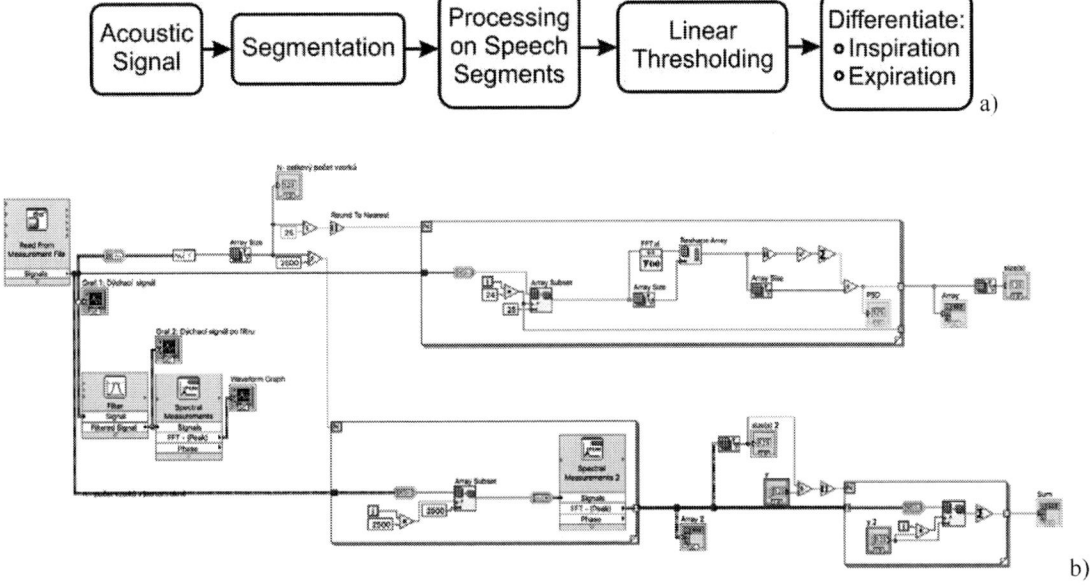

Fig. 4 Block diagram of processing in frequency domain a) and LabView implementation b).

3. Conclusion

In this paper was presented method based on monitoring acoustic signals from trachea. Acoustic method is simple, portable, easy to use and very interesting for the non-invasive measurement of the breathing. Distinguish the different phases of breathing is important for monitoring of respiration. This was provided in frequency domain. First, segmentation was performed, consequently calculated power spectrum using FFT for separation of active speech phase (inspiration and expiration) from silence phase (pauses). The last step was to determine inspiration and expiration phases. Using this method allows to determinate apnoea pause.

Acknowledgement

This research has been supported by the Grant Agency of the Czech Technical University in Prague, grant No. SGS14/195/OHK3/3T/13 and partially grant No. SGS16/089/OHK3/1T/13.

References

[1] Dobrev D., Daskalov I., *Two-electrode telemetric instrument for infant heart rate and apnea monitoring*, Medical Engeneering & Physics 20, 1998, pp. 729-734

[2] Brady, S., et al., *Garment-Based Monitoring of Respiration Rate Using a Foam Pressure Sensor*, G.M.P. Wearable Computers, 2005. Proceedings. Ninth IEEE International Symposium on Wearable Computers (ISWC'05), pp. 214 – 215

[3] Ma Y., Ma S., Wang T., Fang W., *Air-flow sensor and humidity sensor application to neonatal infant respiration monitoring*, Sensors and Actuators A 49, 1995, pp. 47-50

[4] Aoki H., Koshiji K., *Non-contact Respiration Monitoring Method for Screening Sleep Respiratory Disturbance Using Slit Light Pattern Projection*, World Congress on Medical Physics and Biomedical Engineering 2006 IFMBE Proceedings, 2007, Volume 14, Part 7, pp. 680-683

[5] Uenoyama M., Matsui T., Yamada K., et al., *Non-contact respiratory monitoring system using a ceiling-attached microwave antena*, Medical and Biological Engineering and Computing Volume 44, Number 9, 2006, pp. 835-840

[6] Ishijima M., *Cardiopulmonary monitoring by textile electrodes without subject-awareness of being monitored*, Med. Biol. Eng. Comput., 1997, 35, pp. 685-69

[7] Kroutil J., Husak M., *Monitoring of Breathing*, in Proceedings of the ASDAM'08 Conference, Smolenice, Slovakia, 2008, pp. 167 - 170.

[8] Bronzino J. D., *The Biomedical Engineering HandBook*, Second Edition, CRC Press LLC, ISBN 0-8493-0461-X.

[9] Abushakra, A., *Acoustic Signal Classification of Breathing Movements to Virtually Aid Breath Regulation*, IEEE Journal of Biomedical and Health Informatics [online]. 2013, 17(2), 493-500 DOI: 10.1109/JBHI.2013.2244901. ISSN 21682208.

ASDAM 2016, The 11th International Conference on Advanced Semiconductor
Devices And Microsystems, November 13-16, 2016, Smolenice, Slovakia

Gas sensor based on sputtered NiO thin films

M. Predanocy, I. Hotový, V. Řehaček

Institute of Electronics and Photonics, Slovak University of Technology, Ilkovičova 3, 81219
Bratislava, Slovakia
e-mail: martin.predanocy@stuba.sk

The nickel oxide (NiO) thin films were deposited by dc reactive magnetron sputtering from Ni target. Investigation of surface NiO was carried out by AFM measurements. NiO films are smooth and have an average roughness of 1.04 nm and 0.69 nm for 100 nm and 50 nm thickness of NiO film, respectively. Measured NiO films demonstrate to detect low concentration H_2 at temperature 250°C. It was observed the improvement detection characteristics for thinner 50 nm NiO films in all investigated concentration H_2 and operational temperature.

1. Introduction

Semiconducting metal oxides are the most attractive materials for using as active gas sensing layer in gas sensors for detection many different gasses. Nickel oxide is well-known and mostly expanded metal oxide material for its very good electrical, optical, structural and gas sensing properties. In over past years, there was an increased interest of scientific institutes in field of detection gasses in fabrication process of chemical company, fire protection, environmental monitoring and in use in healthcare. Nickel oxide is very important semiconducting material in electronic devices with the band gap energy in the range of 3.6 – 4 eV [1-2]. The selection of deposition method of NiO thin films has influence mostly on the microstucture and consequently on the structural, electrical and gas sensing properties of NiO material. This material is very often used in micro and nano-electronic applications, such as electrode materials in lithium ion batteries, catalyst in fuel cells, electrochromic devices, hole transporting layer in solar cells and gas sensing material for detection of hazardous and flammable gasses [1-4]. Gas sensors based on nickel oxide as sensitive layer are highly sensitive and reliable having a good performance/price ratio. There are many reasons for using NiO as gas sensitive layer in metal oxide gas sensors. Generally, gas sensors based on metal oxide material have fast response on the tested gasses, high sensitivity, high reliability, low price, low power consumption and easy implementation into the electronic devices. These attributes designate gas sensors for additional and more detailed research with aim to improve their properties.

In the present work there will be shown and discussed morphological properties of NiO thin films prepared by dc reactive magnetron sputtering for gas detection. There will be presented morphological and gas sensing properties NiO films on H_2 at lower operating temperature.

2. Experimental details

Nanocrystalline NiO films were deposited by dc reactive magnetron sputtering from Ni target in a mixture of oxygen and argon. For the deposition process of thin films have been used working gasses of high purity (99.999%), the flow rate prior to transport to the vacuum chamber were independently controlled by mass flowmeter. Material for the preparation NiO thin films was used Ni target minimum purity of 99.99% with a diameter of 4". The distance of prepared samples was approximately 75 mm from the sputtering target. A sputtering power of 600 W was used. The relative partial pressure of oxygen in the reactive mixture O_2-Ar was

978-1-5090-3084-2/16 $31.00 © 2016 IEEE   121

30%. Pressure at deposition apparatus prior to the deposition process was evacuated below 10^{-4} Pa. The working pressure in deposition process of NiO was kept at 0.5 Pa. The NiO films were prepared on unheated alumina substrates and glass - corning 1737 for morphological investigation and testing gas sensing properties. The film thickness measured by Talystep was in the range of $50 - 100$ nm for all examined samples. The different thicknesses of the deposited NiO films were achieved only time duration of sputtering process. In order to stabilize the properties of NiO thin films all samples have been annealed in a furnace at 500°C in N_2 atmosphere for 2 h.

The surface morphology of NiO thin films prepared on glass – corning substrate was studied by AFM. It was used special gas platform with SiO_2/Si_3N_4 membrane on micromachined Si substrate for detailed gas sensing investigation. The gas sensor structure contain Pt hotplate for heating active layer in the temperature range of 100 - 350°C. The electrical response of the gas sensors with a gas sensitive layer of NiO was measured under laboratory conditions for hydrogen. The gas measurement of NiO sensors mounted in TO-39 package was provided in gas chamber with small volume approximately 0.1 l. The small volume of the tested chamber ensures fast electrical response of the gas sensor to minimize delays desired concentration in the measuring chamber. Sensing the electrical response of the gas sensor to test gas was performed by measuring the electrical resistance of the gas sensitive layer between the Pt interdigitated electrodes using a SMU Agilent B2902B.

3. Results and discussion

3.1 Morphological analysis

The HRTEM and structural study of NiO thin films with 50 and 100 nm thickness was provided in previous work for the samples deposited at the same parameters [4]. These results indicate that best sensing properties are for thinner NiO film with bigger crystallites. Average size of grains calculated by XRD method was determined on the value 22 nm.

The fig. 1 shows surface morphology of the deposited and annealed NiO thin films. Surface of annealed NiO films is smooth and it has an average roughness of 1.04 nm and 0.69 nm for 100 nm and 50 nm thickness of NiO film, respectively. It can be seen on the surface of the annealed NiO films particle size up to 12 nm and the average surface roughness is low due to the small number of the largest grains. Differences of average surface roughness values for samples annealed NiO is almost twice thicker for layer with 100 nm thickness. However, this value is very small and it is close to the AFM measurement accuracy. Therefore, the values of R_a and RMS have only for information and these values indicate low magnification average surface roughness due to annealing of NiO layer thickness. The calculated R_a and RMS are shown in table 1 for annealed and deposited layer of NiO.

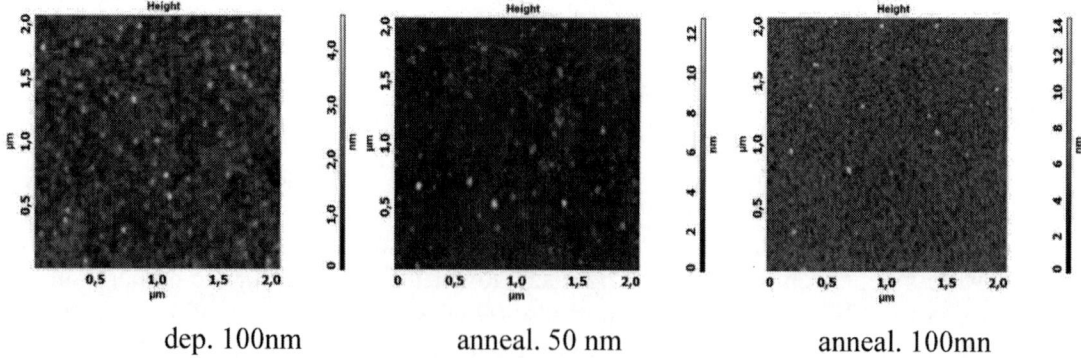

dep. 100nm   anneal. 50 nm   anneal. 100mn

Fig. 1. Surface morphology of NiO layer measured by AFM.

Table 1. Calculated values of roughness surface and *RMS* for deposited and annealed NiO layer.

	50nm	100nm	50nm 500°C	100nm 500°C
R_a [nm]	0.221	0.33	0.69	1.04
RMS [nm]	0.42	0.64	1.54	2.16

3.2 Gas sensing properties of NiO

It is necessary to determine type of contact between the Pt electrodes and NiO sensing layer before gas measurements of NiO gas sensor. Generally, it can be ohmic or Schottky contact. Measured I-U characteristics are linear in the range of -10V – 10 V and in the operating temperature from 100 to 400°C. Linear dependence of I-U characteristics determines the ohmic contact for all investigated parameters and thickness of NiO thin layers. Dynamic responses of resistivity NiO thin layers were not well identified for the temperature range from 100 to 150°C. The cause of this problem is insufficient stabilization of electrical resistance of NiO thin films exposed to reduction gas. Therefore, the detailed measurements of gas sensors were performed at the temperature larger than 150°C. Our attention was focused on determining the properties of NiO gas sensors at lower operating temperatures in order to minimize the input power supplied to the heater and achieve repeatable and reproducible values of electrical response to the tested gas.

The gas sensing responses of NiO thin films were investigated upon to exposure to hydrogen with different concentrations in the range from 10 to 1000 ppm. The gas sensing responses for different operational characteristics shown in fig. 2a revealed the behaviour of NiO gas sensors for temperature range from 150 to 450°C for 50 ppm concentration of hydrogen. The temperature of 350°C was correspondingly identified as the optimal operating temperature for NiO thin films with 50 and 100 nm applied in all investigations hereinafter. The highest dynamic electrical response (S_r=2.47, 50 ppm) of NiO thin film for 50 nm thickness was observed at the operating temperature 400°C and at 425°C in the case of the 100 nm thick NiO film. Fig. 2b shows the dynamic gas response characteristics of 50 nm NiO thin film towards hydrogen at 250°C. It can be seen that the electrical resistance of NiO gas sensors sharply increased due to the injection of tested and carrier gasses, until saturation of the NiO film. Saturation wasn't observed in NiO films with thickness 50 nm for concentration H_2 higher than 400 ppm. The saturation time is longer than 1 hour for hydrogen concentration above 600 ppm. This phenomenon is significantly stronger for thicker 100 nm NiO films. After many measuring cycles between tested and carrier gasses, the resistance of NiO thin films could recover their initial state, which indicates that NiO thin films have good reversibility. It can be easily found that the response increased rapidly with increasing hydrogen concentration for all investigated thicknesses of NiO thin films. The highest relative sensitivity 1.6 (200 ppm) of NiO films was achieved for 50 nm thickness NiO, and 1.36 for 100 nm thickness of NiO at 250°C operation temperature. Decrease of gas responses was observed upon exposure to hydrogen for the thicker NiO films in comparison to the thinner NiO film. The similar phenomenon was observed in other studies deal with the influence of the thickness for gas sensing responses [1,2,4]. Enhanced of gas sensing response for 50 nm NiO in comparison to thicker 100 nm film was approximately 20% at 250°C and it has increasing tendency up to 400°C (Fig. 2a). Fig. 3 shows NiO relative sensitivity of the NiO gas sensors depending on the hydrogen concentration in the range of 10 – 400 ppm for the temperature 250 and 400°C. It was observed the improvement detection characteristics for thinner 50 nm NiO films in all investigated parameters.

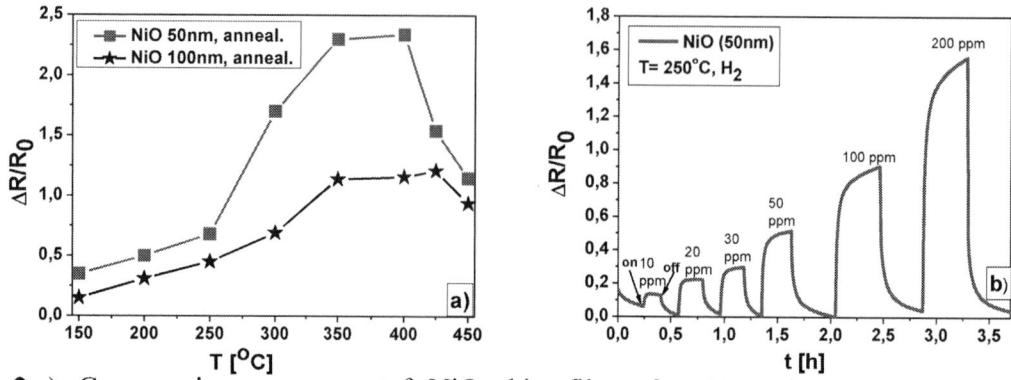

Fig. 2 a) Gas sensing responses of NiO thin films for 50 and 100 nm thicknesses, concentration of hydrogen was 50 ppm at the various operating temperature. b) Gas sensing transient response to hydrogen for 50 nm thick NiO at operating temperature 250°C.

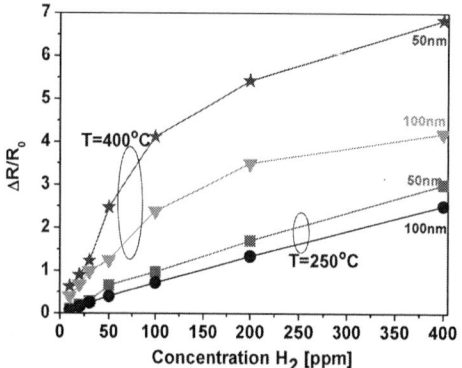

Fig. 3 The relative sensitivity of NiO thin films for different concentrations of hydrogen at the operating temperature 250 and 400°C.

4. Conclusion

NiO thin films with 50 and 100 nm thickness were successfully fabricated by dc reactive magnetron sputtering. Surface of annealed NiO films is smooth and it has the average roughness of 1.04 nm and 0.69 nm for 100 nm and 50 nm thickness of NiO film, respectively. Special attention was devoted to sensing properties towards hydrogen at low concentrations and the operation temperatures. Our results demonstrate the potential of NiO thin films with 50 nm thickness to detect 50 ppm of hydrogen at 150°C.

Acknowledgement

The work was supported by the Scientific Grant Agency of the Ministry of Education of the Slovak Republic and of the Slovak Academy of Sciences, No. 1/0828/16 and by project SAFESENS (agreement No. 621272-1) co-funded by the Slovak Republic and the ENIAC-JU.

References

[1] I. Castro-Hurtado, C. Malagù, S. Morandi, N. Pérez, G.G Mandayo and E. Castaño, *Acta Mater.* **61**, 4, 2013.

[2] H. Steinebach, S. Kannan, L. Rieth and F. Solzbacher, *Sensors and Act. B* **151,** 1, 2010.

[3] R. Kumar, C. Baratto, G. Faglia, G. Sberveglieri, E. Bontempi, and L. Borgese, *Thin Solid Films*, **583**, 2015.

[4] M. Predanocy, I. Hotový, M. Čaplovičová, V. Řeháček, I. Košč and L. Spiess, in *Proceedings of the ASDAM'2012 Conference*, Smolenice, Slovakia, 2012, p. 4.

ASDAM 2016, The 11th International Conference on Advanced Semiconductor
Devices And Microsystems, November 13-16, 2016, Smolenice, Slovakia

Gold nanoisland arrays prepared by sputtering and bio-functionalization of their surfaces

O. Szabó, V. Tvarožek, S. Kováčová, I. Novotný

Institute of Electronics and Photonics, Slovak Technical University in Bratislava
Ilkovičova 3, 81219 Bratislava, Slovakia
e-mail: *ondrej.szabo@stuba.sk*

Gold nanoisland arrays (Au NIA) were prepared by sputtering without masking and lithography and were investigated by local surface plasmonic resonance (LSPR) effect. Self-assembly of 11 mercaptoundecanoic acid (11-MUA) monolayers were used for bio-functionalization of Au surfaces because MUA improves an attachment of biomolecules (proteins, antibodies) to the surface of Au. Coverage of 11-MUA induced the blue or the red shift of LSPR wavelength depending on the Au NIA size (nominal thickness t). In accordance with the universal size-scaling model, smaller NIA structures with nominal thicknesses of 5 nm showed little blue shifts $\Delta\lambda = -11$ nm (Au NIA) and $\Delta\lambda = -13$ nm (Au/Ti NIA). In the case of the thicker Au NIA (thickness of 16 nm) the refractive index-based sensing was dominant and the red shift of LSPR wavelength was observed $\Delta\lambda = +24$ nm.

1. Introduction

Gold-based nanostructures have proven to be useful in the development of novel devices for medicine [1] - in particular due to the localized surface plasmon resonance (LSPR) – it is a coherent oscillation of the free electron gas in metal nanoparticles excited by electromagnetic radiation primarily in the Vis/NIR range. Thin film technologies significantly contribute to preparation of Au nanostructures, e.g. by: evaporation [2], sputtering and annealing [3, 4], combinations of solid-state dewetting, nanolithography, and dealloying for the fabrication of ordered nanoparticle arrays [5]. The nanostructure extinction and scattering spectra, especially the plasma resonance wavelength peak, depends on nanostructure composition/shape/size and local surrounding environment [6, 7]. Biochemical nanosensors often used 11 mercaptoundecanoic acid (11-MUA, Fig. 1) for bio-functionalization of Au surfaces because MUA shows good binding to Au through the Sulphur (S) constituent; MUA is not water soluble and its hydrophobic nature creates a major challenge to the surface modification of gold nanostructures [8]. The relative small MUA molecule (218 atomic mass unit, size < 1.5 nm) has a carboxylic

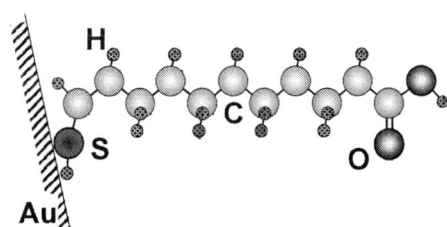

Fig. 1 An attachment of the 11-MUA molecule on the Au nanostructure surface.

group which is desirable for bioconjugation with amine-terminated biomolecules, such as proteins and antibodies.

Our aim was to form plasmonic Au nanoisland arrays (NIA) directly by sputtering without masking and lithography. Self-assembly of monolayers of 11-MUA on surfaces of Au NIA has been investigated by LSPR effect.

978-1-5090-3084-2/16 $31.00 © 2016 IEEE   125

2. Technology and Experimental

We used the deposition concept of sequential (cyclic) sputtering [9] of Au NIA or Ti ultrathin non-continuous film (as supplemental intermediate "seeds") on Corning glass substrates. The morphology of nanostructures was analysed by scanning electron microscopy (SEM) JEOL 7500F and evaluated by the open source image processing program ImageJ [10]. The nominal thickness t of non-continuous island Au films was measured by the mechanical Talystep method giving the value that characterizes only an "envelope" of the NIA which implicitly contains data of the amount of deposited material. Optical transmission spectra were measured using Ocean Optics 4000 Spectrometer in the spectral range 300-900 nm. After the Au NIR deposition on Corning glass substrate, the samples were subsequently transferred to 11-MUA 2 µmol/ml solution for 30 minutes. In order to remove the physical absorbed MUA molecule, all the samples were rinsed with ethanol and water and dried in nitrogen stream.

3. Results and discussion

The sensitivity of plasmonic nanosensors is particularly influenced by the sensing based on both plasmon coupling and target-induced local refractive index changes. The first effectively improves the detection sensitivity by increasing the size of aggregates and decreasing their distance. According to the universal size-scaling model of LSPR shift [11] the fractional shift $\Delta\lambda/\lambda_0$ of the plasmon resonance wavelength varies with the ratio of interparticle separation s and particle dimension D. We used very rough approximations to determine the nanoisland dimension D (assuming a circular-shape of the nanoisland area, $D = 2.(\sqrt{A_M/\pi})$) and inter-island separations s from the statistical evaluation SEM images ($s = NN_M - D$). NN_M is modus of the near neighbour distance of NIA. Sputtered Ti intermediate ultrathin non-continuous film (the nominal thickness of 0.6 nm) acts as a "seed" on the Corning glass substrate for Au growth. It improved the adhesion of Au NIA as well as it supports the forming of more defined Au/Ti NIA structures of smaller dimensions (modus of island size $D = 10$ nm and $s = 2$ nm, Fig. 2 a) in comparison with the Au NIA oneself ($D = 16$ nm, $s = 8$ nm, Fig. 2 b).

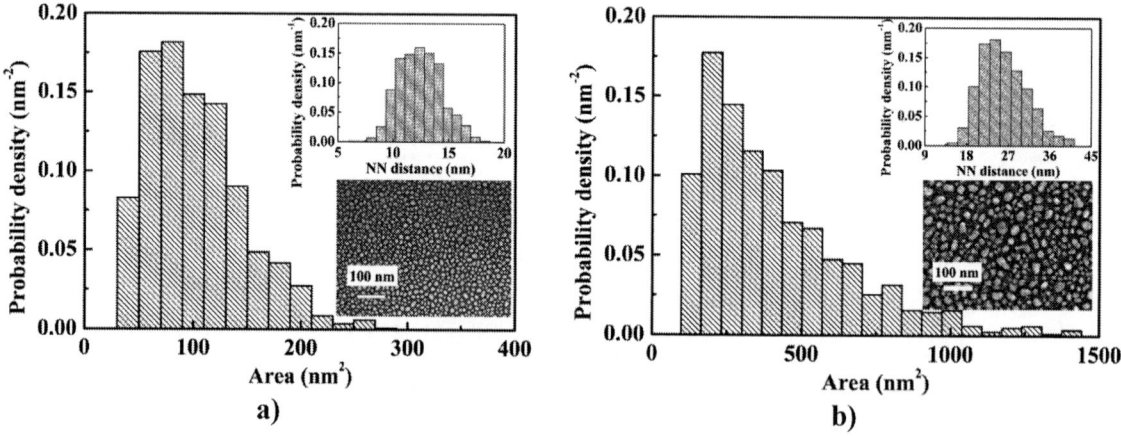

a)   b)

Fig. 2 SEM images of Au/Ti (a) and Au (b) NIA of nominal thicknesses $t \approx 5$ nm.

The LSPR wavelength blue shift $\Delta\lambda$ = - 21 nm corresponded to lower dimensions of the Au/Ti NIA and it is in agreement to the universal size-scaling model of LSPR [11], Fig. 3 a.

The refractive index-based (refractometric) sensing offers several advantages, as the possibility of miniaturization, the compatibility with microfluidics and the range of functionalization chemistry available is broader [12]. Specific properties of this sensing are: small size nanoparticles have high absorption ratios, while big size nanoparticles have high scattering ratios and anisotropic shape nanostructures exhibit higher refractive index sensitivity than spherical nanoparticles. The LSPR wavelength shift $\Delta\lambda$ in response to changes in refractive index is approximately described as $\Delta\lambda \approx m\ (n_{\text{adsorbate}} - n_{\text{medium}})(1 - e^{-2d/l_d})$ [12], where m is the sensitivity factor (in nm per refractive index unit (RIU)), $n_{\text{adsorbate}}$ and n_{medium} are the refractive indices (in RIU) of the adsorbate and medium surrounding the nanoparticle, respectively, d is the effective thickness of the adsorbate layer (nm), and l_d is the electromagnetic field decay length (nm).

Self-assembly of 11-MUA monolayer on surfaces of Au NIA induced a different LSPR wavelength shift depending on the size of Au nanoislands (their nominal thicknesses t).

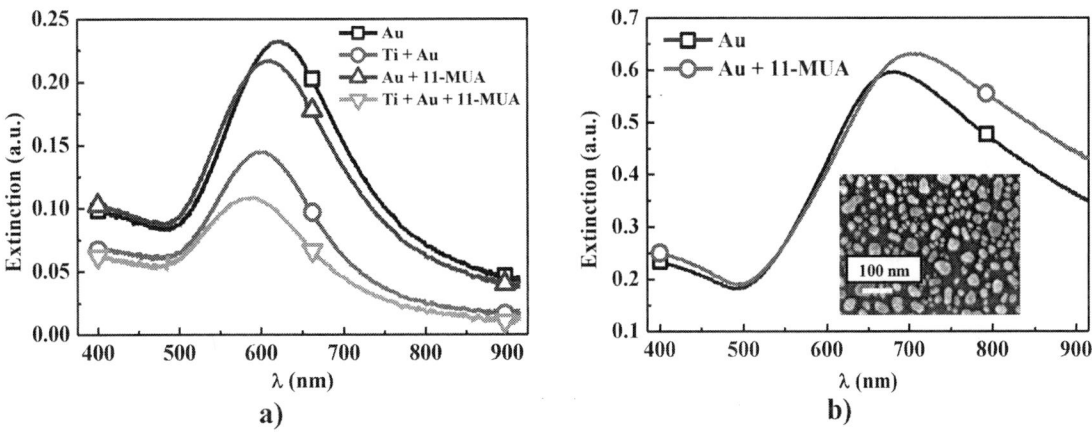

Fig. 3 LSPR wavelength blue shift corresponded to lower dimensions of the Au/Ti NIA ($\circ$) in comparison with Au NIA ($\square$). Influence of the 11-MUA coverage on extinction spectra (LSPR wavelength shift) of: (a) Au (Δ) or Au/Ti NIA (∇) of nominal thicknesses t = 5 nm; (b) Au NIA ($\circ$) of t = 16 nm (SEM image is inserted).

Smaller NIA structures with nominal thicknesses of 5 nm (Fig. 2) showed little LSPR blue shifts $\Delta\lambda$ = - 11 nm (Au NIA) and $\Delta\lambda$ = - 13 nm (Au/Ti NIA), after their modification by 11-MUA, Fig. 3 a. It was probably caused by the formation and local polarization of 11-MUA monolayer (of thickness about 1 nm) on Au individual islands only. This reduce the plasmon coupling strength among nanoislands what induced the blue shift of LSPR wavelength according to the universal size-scaling model of LSPR shift [11]. When the Au NIA was thicker (thickness of 16 nm), nanoislands were bigger (D = 21 nm) and their separations lower (s = 6 nm), Fig. 3 b. Then the coverage of 11-MUA was approximately continuous - therefore we can expect that the refractive index-based sensing was dominant. All optical measurements were performed in the air (n_{medium} = 1). The red shift of LSPR wavelength was observed $\Delta\lambda$ = + 24 nm in agreement to empirical and theoretical expectations (refractive index of MUA is $n_{\text{adsorbate}} \approx 1.5$) [13, 14], Fig. 3 b. Sputtered bigger NIAs of thicknesses > 10 nm can exploit specific properties of the refractive index-based sensing [15]: small size nanoparticles have high absorption ratios, while big size nanoparticles have high scattering ratios and anisotropic shape nanostructures exhibit higher refractive index sensitivity than spherical nanoparticles.

4. Conclusions

The technological procedure for preparation of plasmonic Au nanoisland arrays was simplified by using sputtering without masking and lithography. It was proven that Ti ultrathin non-continuous film, sputtered between Au and glass substrate, supported the formation of more defined Au/Ti NIA structures of smaller dimensions. Moreover, Au/Ti NIA were more stable and relatively non-sensitive to temperature treatment. Self-assembly of monolayers of 11-mercaptoundecanoic acid induced the blue or the red shift of LSPR wavelength depending on the Au NIA size (nominal thickness). Presented results paved the way of an exploitation of sputtered Au NIA in the development of refractometric LSPR biosensors.

Acknowledgement

The presented work was supported by SK VEGA Project VEGA 1/0651/16 and VEGA 1/0776/15. This work occurred in frame of the LNSM infrastructure, project number LM2015087 (MŠMT ČR). We are thankful to Dr. T. Ignat from Laboratory of Nanobiotechnology, IMT-Bucharest, for preparation of 11-MUA monolayers.

References

[1] D. Cornejo-Monroy, L. S. Acosta-Torres, A. I. Moreno-Vega, C. Saldana, V.Morales-Tlalpan, V. M. Castano, *J. of Nanosc. Lett.* **3:25**, 9 p, 2013.

[2] A. Vaskevich, I. Rubinstein, 333, *Nanoplasmonic Sensors*, A.Dimitriev (Ed.), Springer N.Y., 2012.

[3] J. Siegel, O. Lyutakov, V. Rybka, Z. Kolska, V. Svorcik, *Nanoscale Research Lett.* **6:96**, 9 p, 2011.

[4] Xin Sun, Hao Li, *Appl. Surf. Sci.* j. apsusc.2015.08.190, 2015.

[5] D. Wang, P. Schaaf, *Phys. Status Solidi* **A 210**, No. 8, 1544, 2013.

[6] L. De Sioa, T. Placido, R. Comparellid, M. L. Currid, M. Striccolid, N.Tabiryana, T.J.Bunninge, *Progress in Quantum Electronics* **41**, 23, 2015.

[7] L. Guoa, J. A.Jackmanb, H.-H. Yang, P. Chen, N.-J. Choa, D.-H. Kim, *Nano Today* **10**, 213, 2015.

[8] J. Cao, T. Sun, K. T. V. Grattan, *Sensors and Actuators* **B 195,** 332, 2014.

[9] O. Szabo, S. Flickyngerova, V. Tvarozek, I. Novotny, in *Proc. MIEL Conf.*, Nis, Serbia, 2014, p. 245.

[10] C.A. Schneider, W.S. Rasband, K.W. Eliceiri, *Nat. Methods* **9,** 671, 2012.

[11] P. K. Jain, W.Huang, M. A. El-Sayed, *Nano Lett.***7**, 7, 2080, 2007.

[12] M. A. Otte, B. Sepulveda, 317, *Nanoplasmonic Sensors*, A. Dimitriev (Ed.), Springer N.Y., 2012.

[13] H. Dadafarin, E. Konkov, S. Omanovic, *Int. J. Electrochem. Sc.*, **8**, 369, 2013.

[14] G. Barbillon, J.-L. Bijeon, J. Plain, M. L. de la Chapelle, P.-M. Adam, P. Royer, *Gold Bulletin* **40/3**, 240, 2007.

[15] S. S. Acimovic, M. P. Kreuzer, R. Quidant, 267, *Nanoplasmonic Sensors*, A. Dimitriev (Ed.), Springer N.Y., 2012.

ASDAM 2016, The 11th International Conference on Advanced Semiconductor
Devices And Microsystems, November 13-16, 2016, Smolenice, Slovakia

Study of Repetitive Avalanche Stress Invoked Degradation of Electrical Properties of DMOS and TrenchMOS Transistors

Juraj Marek[1], Ľubica Stuchlíková[1], Martin Jagelka[1,2], Aleš Chvála[1], Patrik Príbytný[1], Martin Donoval[1,2] and Daniel Donoval[1]

[1]Institute of Electronics and Photonics, Slovak University of Technology in Bratislava
812 19 Bratislava, Slovakia,
[2]NanoDesign, s.r.o, Drotárska 19a, 811 04 Bratislava, Slovakia
e-mail: juraj.marek@stuba.sk

An electrical ageing of three power MOS transistor types has been performed in order to investigate the gradual degradation in time (number of stress pulses) of the static electrical parameters. It is attributed to hot carrier injection in the space charge region of drain – P well blocking PN junction and involves different complex mechanisms mainly defects generation/activation in the drain side region of the gate oxide and near the oxide/silicon interface. Charge of defects leads to shift of electrical characteristics and parameters. The devices used in this experiment were low voltage vertical power DMOS transistor rated to 24 V and TrenchMOS transistors rated to 30 V.

1.   Introduction

MOSFETs in automotive systems can be subjected to events of unclamped inductive switching (UIS) over the lifetime of their application [1], [2]. UIS occurs when the MOSFET is connected to some kind of inductance (a lumped element or parasitic), and there is a rapid change in current [3], [4]. When a power MOSFET is used in circuit application an unclamped inductive load or parasitic elements present an extremely stressful switching condition for the power MOSFET since all energy stored in the inductor during the on state is dumped directly into the device during its turn off, causing the impact ionization within which avalanche conduction is enabled. Repetitive avalanching to which device is subjected for several millions of pulses generates high concentrations of electron–hole pairs that become

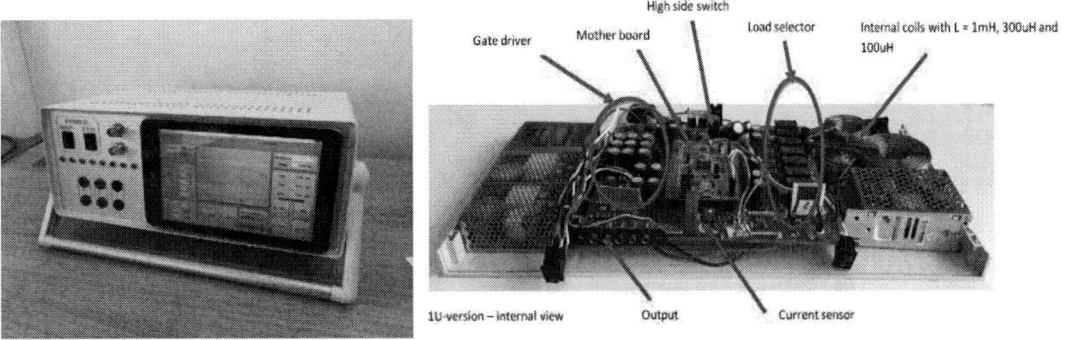

Figure 1. Custom-built repetitive avalanche test equipment.

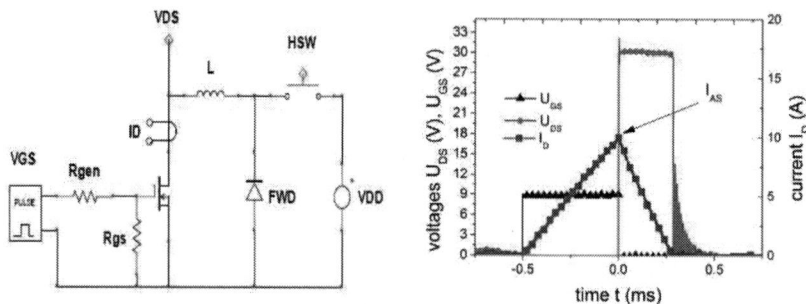

Figure 2. Basic UIS test circuit and typical current and voltage waveforms of the DUT under UIS test conditions.

hot carriers (HC) and can be injected into the gate dielectric [5]. Main effects of hot carrier injection (HCI) are change of threshold voltage V_{TH}, drain-source leakage current I_{Dleak}, breakdown voltage V_{BR} and ON-resistance R_{ON}. These changes during the lifetime operation of the device pose considerably risks against the requirement of a long-term reliability of automotive power MOSFETs.

2. UIS Test

A custom-built avalanche test equipment (Fig. 1) was used for repetitive ruggedness testing able to perform more than 10^8 of stress pulses. Commercial UIS tester ITC55100 was used for reference measurements and calibration. Commercial tester is not suitable for long term measurements because each set of 100 pulses is followed by time period needed for communication. This represents a considerable delay for larger number of pulses. In the Figure 2 the circuit diagram and typical voltage and current waveforms during the test are shown. The test circuit conditions were defined by following parameters: $V_{DD} = 20$ V, $I_{AS} = 5$ A, $L = 1$ mH, $V_G = 10$ V and $R_G = 25$ Ω. Duty cycle was set to 10% to maintain DUT temperature under 130°C. The device under test (DUT) is connected to the power supply through the inductor and high side switch HSW. When both DUT and HSW are turned ON, current, limited by inductor, starts to linearly rise. When current reach required value, DUT and HSW are turned off. The magnetic field induces a counter electromagnetic force (EMF) that can build up surprisingly high potentials across the switch (device under test) [5]. If no protective circuits are added to switch, all build in energy of inductor is dissipated directly in to device. [6] Different types of automotive grade MOSFETS were used for investigations: vertical DMOS rated to 24 V and two Trench MOS transistors rated to 24 V and 90V. To

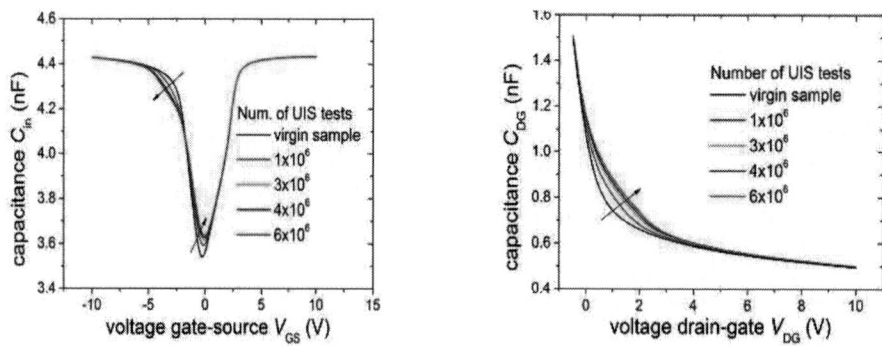

Figure 3. DMOSFET – Impact of repetitive UIS on input C_{in} and drain-gate C_{DG}

978-1-5090-3084-2/16 $31.00 © 2016 IEEE   130

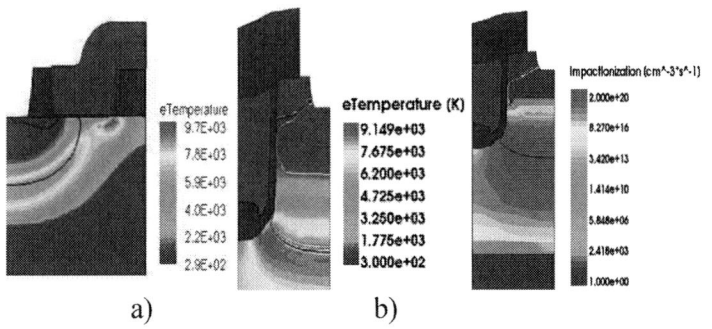

Figure 4. Distribution of impact ionization and hot electrons during avalanching period in a) DMOS and b) Trench MOS.

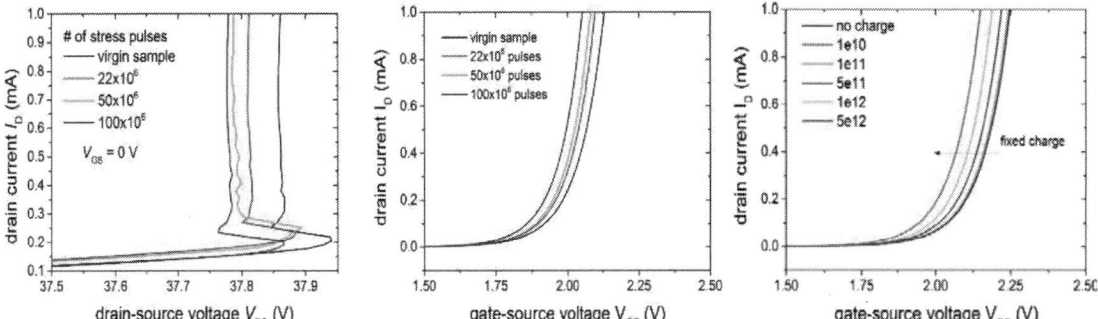

Figure 5. 30V TrenchMOSFET - Measured breakdown and transfer characteristics and Simulated impact of interface fixed charge on transfer characteristics

analyse parameter degradation I-V, C-V measurements were performed first on virgin samples and then after each set of stress pulses. To support the parameters shift analysis due to HCI, TCAD simulations and modelling were performed as well as DLTS measurements were performed. Measured spectra clearly confirm large number of defects generated or activated on drain side of gate oxide due to hot carrier injection.

3. Results and discussion

First DMOS transistors were subjected to repetitive UIS stress for 6×10^6 of stress pulses. Measured C-V curves are shown in the Fig. 3. The most influenced capacitances due to HCI are C_{GS} and C_{GD} whilst shift of drain to source capacitance C_{DS} and I-V characteristics was not observed. This is in good correlation with results from literature. TCAD simulations were used to visualize location and distribution of hot electrons in the volume of device (Fig.4). Generated hot electrons at the bottom of P-well are sufficiently far from transistor channel (Fig. 4), therefore their influence is limited. Only small amount of HC is close to gate dielectric. In case of 24 V TrenchMOSFET shift of transfer and breakdown characteristics was observed (Fig. 4b). The drain side of gate dielectric is more exposed to generated hot electrons (Fig. 4b). Therefore, shift of characteristics and parameters is more significant than in case of DMOS. To simulate impact of trapped charge on SiO_2/Si interface simulations were performed. Positive fixed charge was added on the SiO_2/Si interface where high concentrations of hot holes are present during avalanching period. Results are shown in the Fig. 5c. DLTS measurements were performed to analyse type and origin of charge causing shift of parameters. Obtained DLTS spectra for different sets of parameters and virgin - not stressed sample are shown in the Figure 6a. One can clearly see high increase of DLTS signal

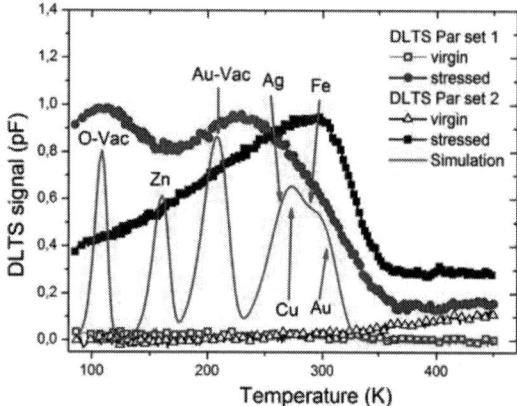

Trap #	origin	E_t-E_V (eV)	sigma (cm^2)
T1	O-Vac	0.180	3E-14
T2	Zn	0.293	9.48 E-14
T3	Au-Vac	0.320	2 E-15
T4	Ag	0.365	1.73 E-14
T5	Cu	0.489	1.68 E-14
T6	Fe	0.389	1.46 E-14
T7	Au	0.526	1.12E-14

Figure 6. Obtained DLTS spectra of virgin and stressed sample with simulated positions of dominant trap peaks. Table with parameters of dominant traps.

for stressed samples on comparison with virgin sample signalizing presence of large number of traps and defects. Using direct evaluation method we were able to determine seven dominant deep level traps originating in metal impurities and contributed vacancies. We think that presence of metals originate in technology process and that these traps were electrically activated by stress.

3. Conclusions

Influence of repetitive avalanche stress on electrical properties of DMOS and TrenchMOS transistors has been studied using custom built UIS test equipment. Degradation of capacitances C_{DG} and C_{in} was observed in all three types of structures. However degradation (shift) of I-V curves was observed only in Trench MOS devices. From analysis it is clear that DMOS transistors are less vulnerable to HCI in case of repetitive avalanching then Trench MOS devices and that breakdown voltage in modern high voltage Trench MOS devices is more affected by repetitive avalanching than in standard low voltage TrenchMOS devices.

Acknowledgement

This work was supported in part by the ENIAC JU Project no. 621270/2013 eRamp and in part by grant APVV-14-0749 and VEGA 1/0491/15 supported by Ministry of Education, Science, Research and Sport of Slovakia.

References

[1] Huard V., et.al, Mic.el. Reliability, 46, 1, 2006
[2] Rutter P., et.al, ISPSD, 2009,
[3] Dierberger K., Application Notes APT9402, 1994.
[4] Koh A., Application note AN10273, 2003.
[5] Donoval D.,et. Al, Solid State Electron., 52, 6, 2008.
[6] Doyle B., et al, IEEE Trans. El. Devices, 37, 8, 1990.
[7] Alatise, O., IEEE Trans. El. Devices, 57, 7, 2010

ASDAM 2016, The 11th International Conference on Advanced Semiconductor
Devices And Microsystems, November 13-16, 2016, Smolenice, Slovakia

Study of negative electron beam nanoresist HSQ on GaAs substrate

R. Andok[1], A. Bencurova[1], I. Kostic[1], A. Ritomsky[1], J. Skriniarova[2], K. Vutova[3]

[1] Institute of Informatics, Slovak Academy of Sciences, Dúbravská 9, 84507 Bratislava
[2] Institute of Electronics and Photonics, Faculty of Electrical Engineering and Information
Technology, Slovak University of Technology, Ilkovičova 3, 81219 Bratislava, Slovakia
[3] Institute of Electronics, Bulgarian Academy of Sciences, 72 Tzarigradsko shosse,
1784 Sofia, Bulgaria

e-mail: ivan.kostic@savba.sk

The aim of this paper is to characterize Hydrogen Silsesquioxane (HSQ) inorganic negative electron resist on GaAs substrate at 40 keV electron energy. The influence of process parameters on resist profiles was investigated with the aim of obtaining vertical sidewalls. The values of the proximity effect function parameters (β_f, β_b and η_E) were calculated using Monte Carlo approach. Observation of HSQ resist profiles were done for a set of linewidth dimensions.

1. Introduction

In the resist-based electron beam lithography (EBL), the performance of the structures is set by various factors such as beam size, resist material, exposure dose, development process as well as the writing strategy. Our aim in this paper is to characterize Hydrogen Silsesquioxane (HSQ) inorganic negative electron resist on GaAs substrate. HSQ is a high-resolution electron beam resist which exhibits high resolution (down below 10 nm half-pitch), small line edge roughness and high etching resistance (in comparison to polymer resist materials) [1]. Currently, commercially available HSQ XR-1541 (*Dow Corning*) resist is used for thicknesses of 50-150 nm.

2. Experimental

Direct writing e-beam lithography at 40 keV e-beam energy was used for the exposures utilizing the variable-shaped e-beam exposure system ZBA23 (*Raith, GmbH*) [2]. The beam current density was 0.8 A/cm^2, the minimal beam spot size was 50x50 nm^2 and the maximal beam spot size was 3000x3000 nm^2.

Negative resist HSQ XR-1541 (*Dow Corning*) was spin coated on GaAs substrate in order to obtain 150 nm thickness. After spinning, the HSQ layer was prebaked on a hotplate for 2 min at 150°C. Resist development was done in standard 2.38 wt% aqueous solution of tetra-methyl-ammonium hydroxide (TMAH) (AZ 326 MIF, *Merck*) for 120 sec and rinse 10 sec in deionised water. In addition, 25 wt% aqueous solution of TMAH referred as high-contrast developer [3] was applied. Profile measurements were done in scanning electron microscopy with field emission gun at magnification 300 000x.

3. Results and Discussion

Our motivation was the precise control of structure dimensions and resist 3D profile in order to get appropriate masking layer for dry etching and simulation of proximity parameters

978-1-5090-3084-2/16 $31.00 © 2016 IEEE

for preparation of photonic structures. Precision concerning the dimensions of structures in e-beam lithography depends on many factors, which is necessary to optimise.

A proximity effect problem can well be seen in Fig. 1. In the case of 40 keV electron energy, the range of the backscattered electrons is over a distance of around 15 micrometers, so the linewidth is changing from the border of the line grating towards the centre (from right to left in Fig. 1). So the simulation of proximity parameters is necessary to obtain precise linewidth over the whole grating.

Besides that, it is also essential to get vertical sidewalls of the resist profiles that are important for dry etching.

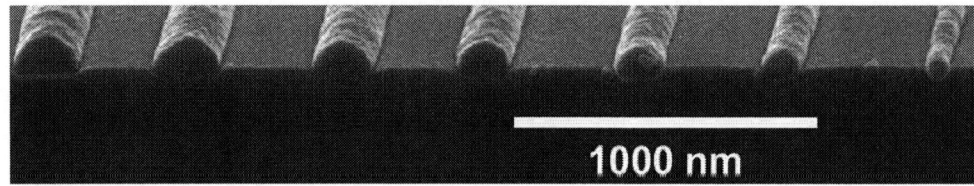

Fig. 1. Detail of a line grating in 150 nm thin HSQ resist on GaAs substrate; period of patterns 500 nm, beam shot 100x3000 nm^2.

3.1 Results for development rate of HSQ resist on GaAs substrate

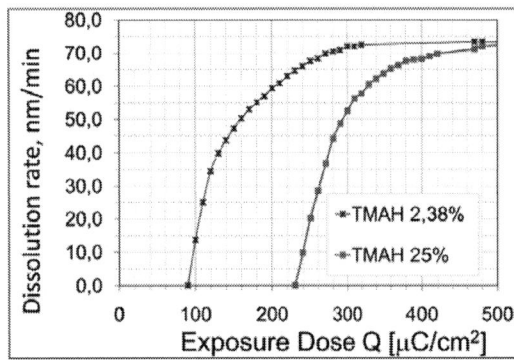

Evaluated dependencies of the development rate (solubility rate) on the exposure dose for thin HSQ layer (150 nm) on GaAs substrate and accelerating energy of the electrons of 40 keV using two developers (the first one containing 2.38 wt% aqueous solution of TMAH and the second one with 25 wt% aqueous solution of TMAH) are presented in Fig. 2.

Fig. 2. Solubility rate *vs* exposure dose in the case of 150 nm HSQ on GaAs substrate for aqueous solution of TMAH: (a) standard concentration 2.38 wt. % in H_2O developer; (b) 25 wt. % in H_2O developer, respectively.

3.2 Investigation of resist profiles for HSQ resist on GaAs substrate

The influence of the exposure dose on the dimensions and on resist profile of the structures was experimentally investigated for 40 keV electron energy on GaAs substrate in respect to the fabrication of periodic structures (lines, pillars or holes) (Figs. 3-4). In Fig. 3 resist profiles at various exposure doses for 500 nm linewidth are shown. At low dose 900 $\mu C/cm^2$ (Fig. 3a), sidewalls are rounded, nearly vertical sidewalls are attained at exposure dose 2320 $\mu C/cm^2$ (Fig. 3b), and sidewalls are sloped at large dose 3000 $\mu C/cm^2$ (Fig. 3c).

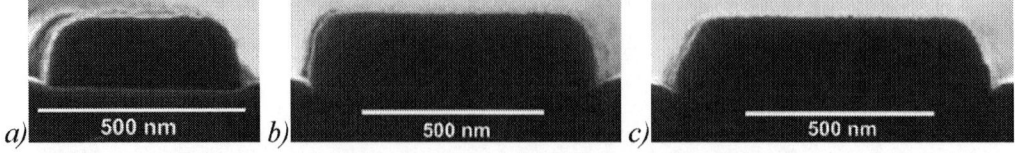

Fig. 3. Resist profiles for linewidth exposed with 500x3000 nm^2 beam spot. Used exposure doses: 900 $\mu C/cm^2$ (a), 2320 $\mu C/cm^2$ (b), 3000 $\mu C/cm^2$ (c), respectively.

In Fig. 4 is shown the resist profile for linewidth exposed with beam spot 200x3000 nm², exposure dose 1760 μC/cm² and 25 wt% aqueous solution of TMAH developer. Vertical sidewalls are achieved at around 2/3 of the resist thickness.

500 nm

Fig. 4. Resist profiles for linewidth exposed with 200x3000 nm² beam spot, exposure dose of 1760 μC/cm² and 25 wt% aqueous solution of TMAH developer.

3.3 Simulation results for e-beam exposure of HSQ resist on GaAs substrate

To simulate the exposure of HSQ film, simulation of electron scattering in the investigated HSQ resist on GaAs substrate was performed, applying our Monte Carlo simulation tool [4] for 10 000 incident electrons with an incident energy of 40 keV. Thus, using cylindrical coordination system (assuming that the scattering to azimuth direction is in symmetry) and dividing the resist layer along the resist depth (z-axis) into several thin sub-layers while dividing the resist layer into many concentric rings, discrete data for the electron energy deposition function in every unit ring (EDF(r,z), r representing the distance from the e-beam axis) were calculated in HSQ negative resist for a point source (with a negligible small beam diameter) (Fig. 5).

These distributions (stored as numerical data arrays) were approximated by an analytical function ("proximity effect function") using a sum of two Gaussians [4,5]. The values of the proximity effect parameters (β_f, β_b, η_E) were calculated using Monte Carlo approach [4,6]. Fig. 5 shows the calculated radial distribution of the energy deposition function EDF (discrete numerical data, represented by the symbol of a square) and its analytical fit (the curve) in the case of 150 nm HSQ/GaAs with electron beam energy of 40 keV. The calculated values of the parameters of the proximity effect function (β_f, β_b and η_E are the characteristic widths of the forward and the backward scattering particles and the ratio of the energy depth dissipation of the backward scattering particles to that of the forward scattering particles, respectively) at the HSQ resist/GaAs substrate interface (point source) are: β_f=0.051952 μm, β_b=2.422593 μm and η_E=0.447667. The spatial distribution of the deposited energy density in the resist (the values of parameters β_f, β_b and η_E) is the basic fact that determines the characteristics of the obtained latent image created during the e-beam exposure process.

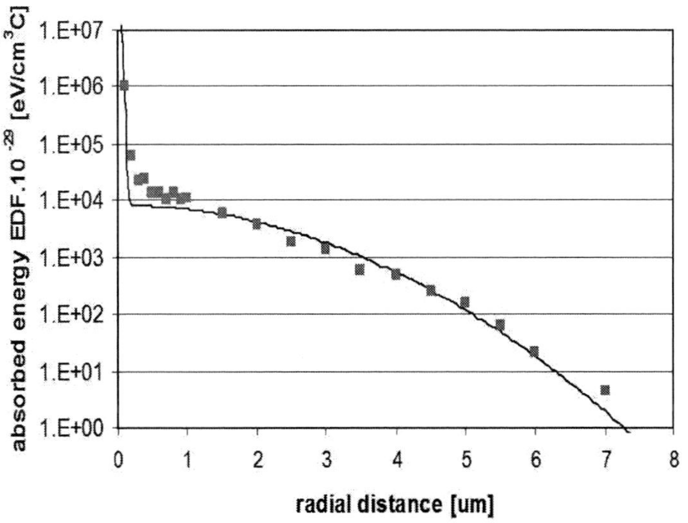

Fig. 5. Comparison of energy deposition function EDF (r) at the HSQ resist/GaAs substrate interface between the discrete data (squares) and the corresponding analytical fit (the curve). HSQ film thickness was 150 nm, e-beam energy was 40 keV (point source).

3.4 Verification of the process parameters

Proximity parameters from simulation and process parameters optimized for vertical sidewalls in 150 nm thin HSQ resist on GaAs substrate were verified in the patterning of large area gratings of periodic lines and circles. Typical periods achieved for the exposed patterns ranged from 600 to 100 nm. Examples of exposures are shown in Fig. 6.

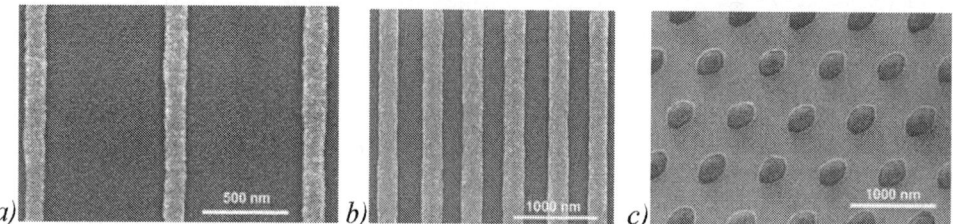

Fig. 6. Detail of gratings in 150 nm thin resist HSQ on GaAs substrate. (a) Beam spot 200x3000 nm^2 with 120/680 nm line/space, (b) 235/265 nm line/space; (c) beam spot 100x100 nm^2 with 300/700 nm diameter/period.

4. Conclusion

E-beam exposure of 150 nm thin HSQ negative resist on GaAs substrate was studied. Influence of process parameters on resist profiles was investigated with the aim to obtain vertical sidewalls. The values of the proximity effect function parameters (β_f, β_b and η_E) were calculated using Monte Carlo approach. Measurements on resist profiles were done for 200 nm, 500 nm and 1000 nm linewidth for various exposure doses at 40 keV electron energy. The significant reduction of sensitivity was observed using development in high-contrast 25 wt. % TMAH developer. Verification of process parameters was done on periodic gratings in resist.

Acknowledgement

This work was supported by the Slovak Research and Development Agency (SRDA) under the contracts APVV-14-0076 and APVV-14-0297, by the Bulgarian National Fund for Scientific Research under the contract DNTS/Slovakia01/1 (project NTS/Slovakia 01/25-13, in Slovakia by the SRDA under the contract APVV-SK-BG-2013-0030), by the Joint Research Project SAS-BAS 2015-2017 MAD, and by the Ministry of Education of the Slovak Republic and the Slovak Academy of Sciences under the Contract No. VEGA- 2/0134/15.

References

[1] H. Namatsu, Y. Takahashi, K. Yamazaki, M. Nagase, K. Kurihara, *Microelectronic Enginnering* **41/42**, 331–334, 1998.

[2] R. Andok, L. Matay, I. Kostic. A. Bencurova, P. Nemec, A. Konecnikova, A. Ritomsky. *Proceedings of the ASDAM'12 Conference*, Smolenice, Slovakia, 287-290, 2012.

[3] K.W. Joel, H. Yang, Bryan Cord, D. Huigao Duan, K. K. Berggrena, J. Klingfus, Sung-Wook Nam, Ki-Bum Kim, M. J. Rooks., *J. Vac. Sci. Technol.* **B 27**, 2622-2627, 2009.

[4] K. Vutova, G. Mladenov. *Lithography*, Chapter 17, Intech, Croatia (2010) p. 319-350.

[5] K. Vutova, G. Mladenov, I. Raptis, A. Olziersky, *Journal of Materials Processing Technology* **184** (1-3), 305-311, 2007.

[6] K. Vutova, E. Koleva, G. Mladenov, I. Kostic, *Journal Vacuum Science and Technology B -Microelectronics and Nanometer Structures* **27** (1), 52-57, 2009.

ASDAM 2016, The 11th International Conference on Advanced Semiconductor
Devices And Microsystems, November 13-16, 2016, Smolenice, Slovakia

Simple patterning method of sub-micro- and nanometer structures for gas sensor

P. Durina[1], A. Bencurova[2], M. Truchly[1], R. Andok[2], I. Kostic[2], B. Grancic[1],
A. Plecenik[1], P. Kus[1], K. Vutova[3], E. Koleva[3]

[1]Department of Experimental Physics, FMFI UK, Mlynska dolina F2, 84248 Bratislava
[2]Institute of Informatics, Slovak Academy of Sciences, Dubravska 9, 84507 Bratislava
[3]Institute of Electronics, Bulgarian Academy of Sciences, 72 Tzarigradsko shosse,
1784 Sofia, Bulgaria
e-mail: ivan.kostic@savba.sk

In this study, a simple patterning method of submicrometer structures is proposed for gas sensor development. Comb-like electrodes patterned in thin Pt layer were proposed to measure gas sensor electrical conductivity. Negative resist SU-8 was used as a masking layer for ion etching of electrodes in 35 nm Pt layer on sapphire substrate. This method was applied for the patterning of the comb-like structures with submicrometre dimensions for gas sensor conductivity measurements.

1. Introduction

In recent years, gas sensors have attracted focus in various applications - such as environmental monitoring, indoor-air-quality, health and safety at the workplace, and homeland security. There have been numerous attempts to develop sensing devices with high sensitivity, stability, and rapid response [1]. Metal-oxide-semiconductors are extensively used as gas sensors due to their property of varying their conductivity under gas exposure. TiO_2 seems to be one of the most interesting candidates for H_2, H_2S, CO and LPG gas detection [2]. Improved performance of a gas sensor is expected with submicrometer dimensions of electrodes.
The SU-8 negative resist is a relatively cheap chemically amplified epoxy based negative UV-photoresist [3]. Capabilities of SU-8 as a negative e-beam resist have been demonstrated [4]. In this study, proximity parameters of negative resist SU-8 were obtained, and a comb-like electrode was successfully fabricated by utilizing those parameters.

2. Materials and methods

Comb-like electrodes patterned in a thin Pt layer were proposed to measure resistivity resp. conductivity. We suggest a simple patterning method of submicrometer structures for that purpose. SU-8 negative resist was used as a masking layer for ion etching as it exhibits a good resistance.
A layer of SU-8 2002 resist was diluted with cyclopentanone in the ratio of 1:4 and spin coated for a thickness of 70 nm at 5000 r.p.m. on a 35 nm thin platinum layer on c-cut (0001) sapphire substrate with TiO_2 thin layer (100 nm). Subsequently, the resist was baked on a hot plate at 95° C for 5 min. The exposure was carried out at an accelerating voltage of 30 kV, beam current of 40 pA and the working distance of 4 mm. After the exposure, the resist was baked on a hot plate at 95°C for 2 min, and developed in a SU-8 developer at 23°C for 65 s.

978-1-5090-3084-2/16 $31.00 © 2016 IEEE

Lithography was performed using the scanning electron microscopy VEGA II SBH (*Tescan, Brno, Czech republic*) equipped with a control system for nanolithography (*Tescan*).

3. Results and Discussion

3.1 Lithography

For the high precision patterning of submicrometer structures, it was necessary to optimize the process of e-beam exposure, development process and baking time and temperature. The test for determining sensitivity and contrast consists of rectangles exposed with different exposure doses. Exposure doses were varied in the range of 1 to 30 $\mu C/cm^2$. The development was done in a SU-8 developer for 5 min at 23°C followed by a rinse in a solution of isopropanol and distilled water. The post-exposure bake was performed at 90°C for 2 min on a hot plate. The SU-8 resist exhibits a very high sensitivity of 0.5 $\mu C/cm^2$ (dose-to-clear) and contrast $\gamma \sim 0.99$.

The influence of the exposure dose on dimensions of the structures and resolution limitations were investigated from line tests. The resolution of the negative resist SU-8 at 30 keV electron energy and 18 nm probe size is demonstrated in Fig. 1. The minimal linewidth of 77 nm for single lines was achieved (Fig. 1a). In the case of periodical line grating, the resolution depends on the grating period and linewidth/space ratio due to the influence of electron scattering (better known as proximity effects). The minimal linewidth was 113 nm for a period of 600 nm (Fig. 1b) and 190 nm for 400 nm (Fig. 1c).

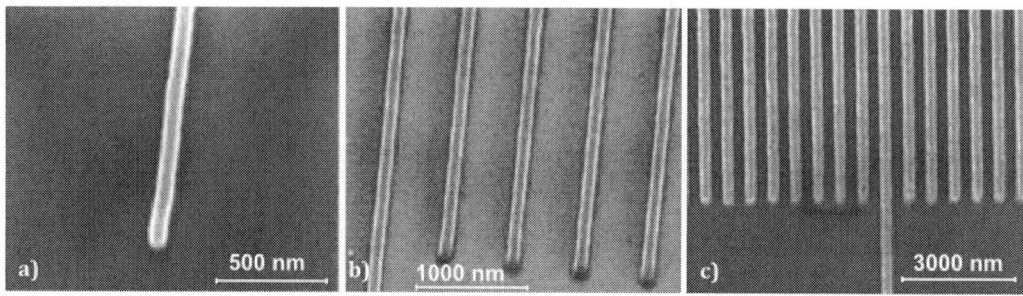

Fig. 1 Stripes exposed in SU-8 negative resist at 30 keV. (a) Single line of 77 nm linewidth. (b) Periodical line grating with the period 600 nm. (c) Periodical line grating with the period 400 nm.

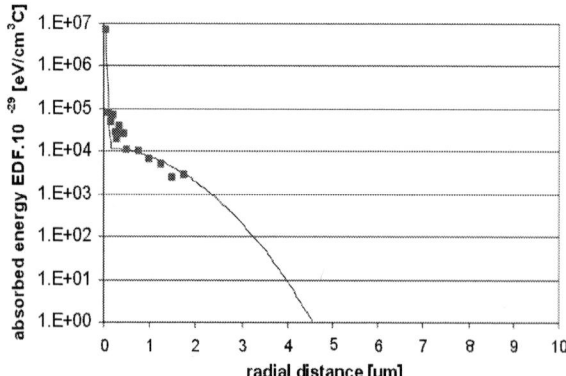

Proximity effect parameters have been simulated for the case of gas sensor fabrication [5], the thickness of the PMMA A2 resist and of the thin Pt layer on the Al_2O_3 substrate were 70 nm and 35 nm, respectively. All exposures were performed at 30 keV electron energy (point or Gaussian source). Discrete data for the energy deposition function (EDF) at the interface (at the depth of 70 nm) were obtained by Monte Carlo simulation [6,7] using a point source for 10 000 particles.

Fig. 2 Comparison between the obtained discrete data (▪) and its analytical fit (the curve).

Analytical approximation of the obtained discrete data for EDF was made using a sum of two Gaussians. Then, the values of the proximity effect parameters (β_f, β_b, η_E) were determined at the interface for the point e-beam source and for the Gaussian e-beam with δ_b=18 nm by Monte Carlo methodology [6]. Results from the Monte Carlo simulation are presented in Fig. 2. EDF discrete data obtained (•) and its analytical fit (the curve) are compared. The values of the proximity effect parameters for the point e-beam source are: β_f=0.4039827000E-01 [μm], β_b=0.1488017000E+01 [μm], η_E=0.5210086000E+00, and for the Gaussian e-beam source, δ_b=18 nm are: β_f=0.492000E-01 [μm], β_b=0.1488126000E+01 [μm], η_E= 0.5816625000E+00.

3.2 Etching

The resist SU-8 2002 was applied as a masking layer for ion etching of Pt comb-like sub-micrometres structures. Ion etching was made with Argon ions from PLATAR (KLAN-53M) ion source with ion energy E_{Etch}=500 eV, j_{Etch}≈1 mA/cm^2, pressure 10-2 Pa and 5 sccm, etch time was 6.5 min for 35 nm thin Pt layer. The etch rate of the SU-8 resist for ion etching of the Pt layer was 2 times slower than in case of PMMA.

The etch time for ion etching of 35 nm thin Pt layer was 12 min and the minimal resist thickness of 70 nm was necessary to achieve submicrometer Pt comb-like structures.

Patterned structures after ion etching of the Pt layer were examined from the viewpoint of critical dimensions (CD) control and line edge roughness (Fig. 3). The minimal space for 600 nm grating period was found to be 150 nm. The higher resist thickness leads to higher linewidth values due to the electron scattering. The period of 300 nm with resolvable lines of L/S=160/140 nm was demonstrated using exposure at 30 keV in a 70 nm SU-8 layer (Fig. 3c).

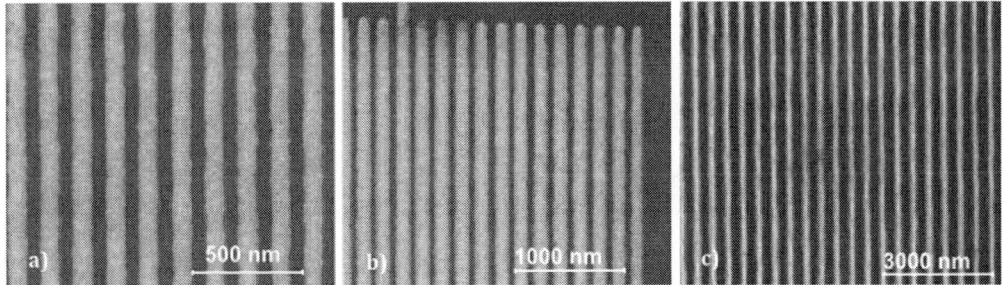

Fig. 3. The line gratings in a 35 nm Pt layer after ion etching (masking layer was a 70 nm thin SU-8 negative resist, exposure at 30 keV). a) Period 600 nm, Line/Space 360/240 nm, b) Period 600 nm, Line/Space 450/150 nm, c) Period 300 nm, Line/Space 160/140 nm

Finally, the optimized processes of e-beam lithography and ion etching were applied for the patterning of the comb-like structures for gas sensor conductivity measurements. Comb-like structures of various dimensions were etched into a 35 nm thin Pt layer on a sapphire substrate. Large structures as contacts and bridges were prepared using photolithography and the lift-off method. The 400 nm linewidth of structures of comb-like electrodes after ion etching of a 35 nm Pt layer is demonstrated in Figure 4.

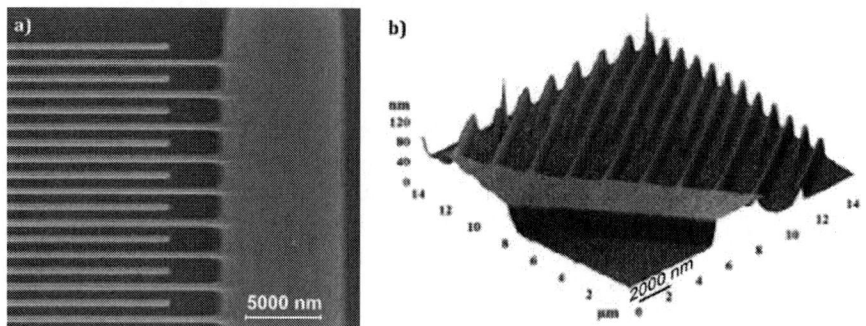

Fig. 4 Detail of comb-like electrode with the linewidth of 430 nm prepared by e-beam lithography and ion etching in a 35 nm thin Pt layer (a), AFM detail (b).

4. Conclusion

Comb-like electrodes patterned in thin Pt layer were proposed to measure electrical conductivity in gas sensor development. For this purpose, a simple patterning method of submicrometer structures was proposed and optimised. SU-8 negative resist was used as a masking layer for ion etching of electrodes in a 35 nm Pt layer on a sapphire substrate. Very high sensitivity of SU-8 resist was measured (0.5 $\mu C/cm^2$ dose-to-clear). Proximity effect parameters (β_f, β_b, η_E) have been simulated for the case of gas sensor fabrication. The minimal linewidth of 77 nm for single lines and the minimal period of 300 nm with resolvable lines of L/S=160/140 nm were demonstrated using exposure at 30 keV in a 70 nm the SU-8 layer. Optimized processes of e-beam lithography and ion etching were applied for the patterning of the comb-like structures with submicrometer dimensions for gas sensor conductivity measurements.

Acknowledgement

This publication was supported by the grants of the Ministry of Education, Science, Research and Sport of the Slovak Republic and the Slovak Academy of Sciences under contracts VEGA-1/0276/15 and VEGA-2/0134/15, by the Bulgarian National Fund for Scientific Research under the contract DNTS/Slovakia01/1 (project NTS/Slovakia 01/25-13, in Slovakia under the contract APVV-SK-BG-2013-0030), and by the Joint Research Project SAS-BAS 2015-2017 (supported by the Slovak and Bulgarian Academy of Sciences).

References

[1] Deen M.J., Kazemeini H.M., Holdcroft S. *J. Appl. Phys.* **103,** 124509:1–124509:7, 2008.

[2] Chaudhari G.N., Bambole D.R., Bodade A.B., Padole P.R., *Journal of Material Science* **41,** 4860–4864, 2006.

[3] LaBianca N. and Gelorme J., *Proc. SPIE* **2438,** 846-852, 1995.

[4] Bilenberg B., Jacobsen S., Schmidt M.S., Skjolding L.H.D., Shi P., Boggild P., Tegenfeldt J.O., Kristensen A. *Microelectronic Engineering* **83,** 1609–1612, 2006.

[5] Owen G., *Journal of Vacuum Science Technology B* **8,** 1889-1892, 1990.

[6] Vutova K. and Mladenov G., *Modelling and Simulation in Materials Science and Engineering* **2,**239-254, 1994.

[7] Vutova K., Mladenov G., *Lithography,* chapt **17,** Wang, M (ed) Intech, Croatia, 319-350, 2010.

ASDAM 2016, The 11th International Conference on Advanced Semiconductor
Devices And Microsystems, November 13-16, 2016, Smolenice, Slovakia

Particle detectors based on 4H-SiC epitaxial layer and their properties

B. Zaťko[1], L. Hrubčín[1,2], A. Šagátová[3,4], P. Boháček[1], F. Dubecký[1], K. Sedlačková[3],
M. Sekáčová[1], J. Arbet[1], V. Nečas[3], V. A. Skuratov[2]

[1]Institute of Electrical Engineering, Slovak Academy of Sciences,
Dúbravská cesta 9, SK-841 04 Bratislava, Slovakia
[2]Joint Institute for Nuclear Research, Joliot-Curie 6,
RUS-141980 Dubna, Moscow Region, Russia
[3]Slovak University of Technology in Bratislava, Faculty of Electrical Engineering and
Information Technology, Ilkovičova 3, SK-812 19 Bratislava, Slovak Republic
[4]University Centre of Electron Accelerators, Slovak Medical University,
Ku kyselke 497, SK-911 06 Trenčín, Slovak Republic
e-mail: bohumir.zatko@savba.sk

*We fabricated and characterized 4H-SiC Schottky diodes as a spectrometric
detector of α-particles. A Schottky contact of Ni/Au of 1.4 mm in diameter was
used. Current-voltage characteristics of the detector were measured and a current
density lower than 25 nAcm^{-2} was observed at room temperature. A ^{226}Ra used as
a source of α-particles within the energy range between 4.6 MeV and 7.7 MeV for
detector testing. The energy resolution below 48 keV was measured for 7.7 MeV
α-particles. The detector was irradiated by ^{132}Xe^{23+}ions with energy of 165 MeV
and fluencies of 5×10^9 cm^{-2} and 1.5×10^{10}cm^{-2}. Spectra of α-particles were
measured after irradiation and significant degradation of energy resolution was
observed.*

1. Introduction

Silicon carbide (SiC) is a suitable for fabrication of nuclear radiation detectors working
in harsh environments. 4H-SiC polytype is one of the mostly investigated. Detectors based on
epitaxial material show high spectroscopic performance for γ-rays (up to 60 keV) at room and
also elevated temperatures [1, 2]. Characterizations of the depletion region length of 4H-SiC
Schottky detector using a α, β and γ- radiation sources have also been realized [3-5]. Studies
using α-particles show also very promising results. Detectors achieve 100 % CCE, diffusion
length of holes up to 13 μm and energy resolution up to 0.25 % (Full Width at Half
Maximum, FWHM) for 5.48 MeV α–particles [6-8]. SiC detectors can be also utilized for
detection of neutrons. The conversion layer of ^{6}LiF and ^{10}B is necessary to use in case of
thermal neutron detection, while HDPE (High Density Polyethylene) for fast neutrons
detection. The thermal neutrons conversion layer transforms neutrons to heavy charged
particles (α, ^{3}H$^+$, etc.) which are easily detected [9, 10]. As silicon and carbide are light atoms
and therefore SiC detector can directly detect fast neutrons through elastic scattering. The
HDPE converter layer converts fast neutrons to protons, which impinge upon the detector and
increases the detection efficiency [11]. The effect of proton irradiation on 4H-SiC detector
properties, its influence on the quality of the rectification contact and elevated temperature
were also investigated [12-13]. Irradiation with 1 MeV neutrons shows a good spectrometric
performance of the SiC detector up to a fluency of 10^{14} cm^{-2} [14]. The γ-irradiation indicates
that detectors are able to operate up to doses about 5 MGy. The spectrometric performance of

978-1-5090-3084-2/16 $31.00 © 2016 IEEE   141

α-particles deteriorates minimal [15]. High energy electron irradiation and its influence on the diffusion length of minority carriers were studied and a non-negligible radiation damage recovery effect by low temperature annealing was observed [16].

In this work we present results of 4H-SiC detectors irradiated by ^{132}Xe^{23+} ions with energy of 165 MeV. We used two influences of ions, 5×10^9 cm^{-2} and 1.5×10^{10} cm^{-2}. Spectra of α-particles were measured before and after irradiation and significant degradation of energy resolution was observed.

2.   Detector fabrication

Several detector structures were prepared from a 70 μm thick nitrogen-doped 4H-SiC layer (with donor doping about 1×10^{14} cm^{-3} produced by ETC Catania) grown by LPE on a 3" 4H-SiC wafer with 0.5 μm thick n++-SiC buffer layer. The radiation detectors were prepared by evaporation of a double layer of Au/Ni with thicknesses 90 and 40 nm on the epitaxial layer using a high vacuum electron gun apparatus. A Schottky barrier contacts with a diameter of 1.4 mm was formed on the epitaxial layer through a contact metal mask while a full area ohmic contact of Ti/Pt/Au was evaporated on the opposite side (substrate). A typical reverse current-voltage characteristic of the fabricated detectors measured at room temperature is shown in Figure 1. The breakdown voltage exceeds 600 V and the reverse current is below 40 pA in the measured range which corresponds to current density below 25 nAcm^{-2}.

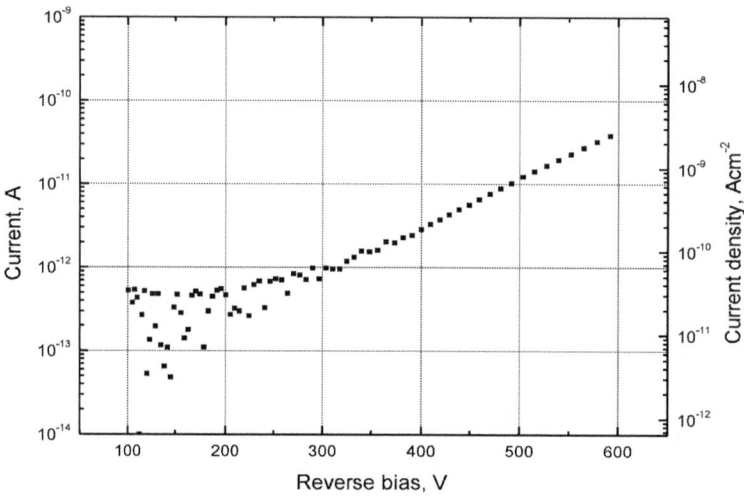

Figure 1. Typical reverse current-voltage characteristic of the 4H-SiC detector at room temperature.

2.   Results and discussion

The following study of 4H-SiC particle detector was performed using α-particles generated by ^{226}Ra. The α-particle gradually loses its' energy as it passes the detector material according to the Bragg curve. Fig. 2a shows SRIM simulated dependence of energy loss of α-particle vs. its depth in detector material. The most of energy losses are at the end of the particle track. Radioisotope ^{226}Ra generates monoenergetic α-particles with energies from 4.6 up to 7.7 MeV with ranges from 16 μm up to 27 μm in SiC detector material. The studied detector was placed in vacuum and connected to the spectroscopic chain based on CAEN. Fig.2b shows obtained calibrated spectrum of α-particle source at reverse bias of 120 V. The energy resolution in FWHM is 47.5 keV for α-particles with the highest energy of 7687 keV.

The detector was subsequently irradiated by heavy ions. The irradiation was performed at the Laboratory of Nuclear Reactions of the Joint Institute for Nuclear Research in IC-100 accelerator [17]. The energy of ions can be adjusted from 1 MeV/nucleon up to 1.25 MeV/nucleon. In our experiments we used $^{132}Xe^{23+}$ ions with energy of 165 MeV (corresponding to 1.25 MeV/nucleon) to study the influence of heavy ions irradiation which can displace atoms of in the semiconductor crystal. The range of Xe ions in SiC is similar like 5 MeV α-particles. The two fluencies of ions were used, 5×10^9 cm^{-2} and 1.5×10^{10} cm^{-2}. Fig. 3 shows significant deterioration of ^{226}Ra measured spectra. In case of 5×10^9 cm^{-2} fluency of Xe ions, the energy resolution for 7687 keV α-particles is more than 6 times worse. Also all peaks are shifted to lower energies despite of the reverse bias was increased up to 250 V. At fluency 1.5×10^{10} cm^{-2} the degradation of spectrometric performance is clearer. The two peaks with highest energy of α-particle are almost not present. This indicates that the thickness of space charge region is lower than range of α-particles due to new defects generated by heavy ion irradiation. These new defects prevent to spread the space charge region to larger depth of detector material.

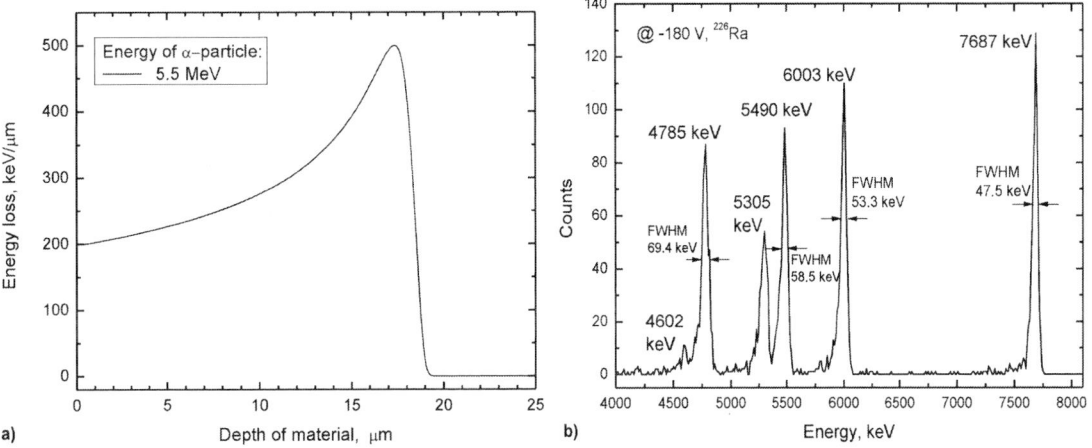

Figure 2. The range of 5.5 MeV α-particles in 4H-SiC detector (a); and the measured calibrated spectrum of ^{226}Ra α-particle source (b).

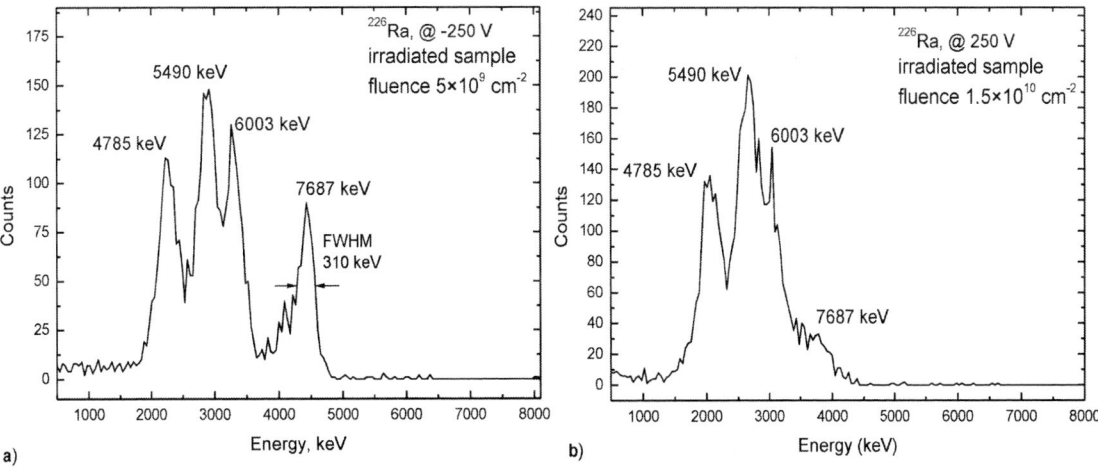

Figure 3. Measured α-particles spectra after 4H-SiC detector irradiation with fluency of heavy ions of 5×10^9 cm^{-2} (a); and 1×10^{10} cm (b).

3. Conclusions

We have prepared 4H-SiC Schottky contact detectors based on high-quality epitaxial layer. The breakdown voltage exceeds 600 V and the reverse current is below 40 pA in the measured range which corresponds to current density below 25 nAcm^{-2}. The α-particle source ^{226}Ra was used to study the spectrometric performance of the fabricated detector placed in vacuum. The energy resolution in FWHM of 47.5 keV for α-particles with energy of 7687 keV was obtained. The range of α-particles in detector material is up to 27 μm. The detector was irradiated by ^{132}Xe^{23+} ions with energy of 165 MeV to study the influence of heavy ions irradiation on detector material. The used fluencies of heavy ions were 5×10^9 cm^{-2} and 1.5×10^{10} cm^{-2}. Measured α-particle spectra of ^{226}Ra show significant deterioration of energy resolution. The 6 times worse energy resolution was observed at 5×10^9 cm^{-2} fluency of heavy ions. Results indicate that the thickness of space charge region is lower than range of α-particles due to new defects generated by heavy ions irradiation. New generated defects prevent to spread the space charge region to larger depth of detector material.

Acknowledgement

This work was partially supported by the Slovak Grant Agency for Science through grants 2/0152/16, and, by the Slovak Research and Development Agency under contract Nos. APVV-0321-11, APVV-0443-12, and by the Project Research and Development Centre for Advanced X-ray Technologies (ITMS code 26220220170) of the Research & Development Operational Program funded by the European Regional Development Fund (0.7).

References

[1] G. Bertuccio *et al*, *Nucl. Instrum. & Meth. in Phys. Res. A* **522** (2004) 413.
[2] G. Bertuccio *et al*, *Nucl. Instrum. & Meth. in Phys. Res. A* **518** (2004) 433.
[3] M. Bruzzi *et al*, *Diamond and Related Materials* **12**, (2003) 1205.
[4] F. Moscatelli *et al*, *Nucl. Instrum. & Meth. in Phys. Res. A* **546** (2005) 218.
[5] B. Zaťko *et al*, *JINST* **9**, (2014) C05041.
[6] F. Nava *et al*, *Nucl. Instrum. & Meth. in Phys. Res. A* **510** (2003) 273.
[7] K. J. Zavalla *et al*, *Proc. of SPIE Vol.* **8852** (2013) 88520D 1.
[8] B. Zaťko *et al*, *JINST* **10**, (2015) C04009.
[9] C. Manfredotti *et al*, *Nucl. Instrum. & Meth. in Phys. Res. A* **522** (2005) 131.
[10] A. Lo Giudice *et al*, *Nucl. Instrum. & Meth. in Phys. Res. A* **583** (2007) 177.
[11] R. W. Flamming *et al*, *Nucl. Instrum. & Meth. in Phys. Res. A* **579** (2007) 177.
[12] A. M. Ivanov *et al*, *Nucl. Instrum. & Meth. in Phys. Res. A* **597** (2008) 203.
[13] J. E. Lees *et al*, *JINST* **6**, (2011) C01032.
[14] F. Moscatelli *et al*, *Nucl. Instrum. & Meth. in Phys. Res. A* **583** (2007) 157.
[15] F. H. Ruddy and J. G. Seidel, *Nucl. Instrum. & Meth. in Phys. Res. B* **263** (2007) 163.
[16] A. Le Donne, *Diamond and Related Materials* **13** (2004) 414.
[17] B. N. Gikal, *Physics of Particles and Nuclei Letter*, Vol. **5**, No. 1 (2008) 33.

ASDAM 2016, The 11th International Conference on Advanced Semiconductor
Devices And Microsystems, November 13-16, 2016, Smolenice, Slovakia

Investigation of AlGaN/GaN Schottky Structures by Deep Level Fourier Transient Spectroscopy with Optical Excitation

L. Stuchlikova[1], A. Kosa[1], R. Szobolovszký[1], M. Petrus[1],
L. Harmatha[1], S. L. Delage[2], J. Kovac[1]

[1] Institute of Electronics and Photonics, Slovak University of Technology in Bratislava,
Ilkovičova 3, 812 19 Bratislava, Slovakia
[2] III-V Lab, 1 Avenue Augustin Fresnel, 91767 Palaiseau Cedex, France
e-mail: lubica.stuchlikova@stuba.sk

This paper highlights electrical characterization of Schottky structures prepared on AlGaN/GaN designed for HEMT's using Deep Level Transient Fourier Spectroscopy (DLTFS) method with electrical and optical excitation. In case of electrical excitation the density of minority and majority carrier traps had strong effect on the evaluation, whereas preliminary DLTFS with optical excitation (MCDLTFS) measurements made possible to identify minority carrier defects with higher precision. Parameters of two electron-like defects EL1 (0.47 eV), EL2 (0.54 eV) and three hole-like defects HL1 (1.13 eV), HL2 (0.93 eV) and HL3 (0.49 eV) were identified. HL1 and HL2 were confirmed by both methods: DLTFS and MCDLTFS. These are connected to a Carbon interstitial defect and well-know threading dislocations. Experimental results confirmed the benefits of different method utilizations in complex defect identifications.

1. Introduction

Improving cost effectiveness and performance of microwave devices using gallium nitride (GaN) high electron mobility transistors (HEMT) is an essential task of the current research. The performance of GaN based HEMTs is limited by the presence of electrical active deep defect states, and their charge carrier emission-capture processes [1, 2].

One of the most frequently used methods of electrically active defect investigation is Deep Level Transient Spectroscopy (DLTS) introduced by D.V. Lang [3]. It uses a simple idea to measure changes of capacitance transient signals as a function of temperature. DLTS is a high-frequency transient capacitance thermal scanning method, where the capacitance transient effect in a potential barrier of a semiconductor structure is induced by an electrical excitation pulse. The position of the DLTS spectrum peak on the temperature axis, outlined by changes in the transient signals, leads to the determination of fundamental defect parameters governing thermal emission and capture processes- activation energy and capture cross section. Since the introduction of the original method described in 1974 many other variations and upgrades were designed. Deep Level Transient Fourier Spectroscopy uses an automated computer measurement system, which is capable to measure whole capacitance transients and to save for further evaluations. Fourier [4] transformation is applied to process transient signals, thereby to achieve deep energy level parameters with a better emission time constant resolution.

DLTFS has a key role in defect characterization satisfying basic requirements of diagnosis, especially accuracy, non-destructivity and sensitivity. However in certain cases reliability of evaluated data is highly affected by the presence of mutually interacting complex impurities. It is difficult to effectively separate these responses therefore calculated parameters are obtained with reduced precision. Investigation of more complicated

978-1-5090-3084-2/16 $31.00 © 2016 IEEE

semiconductor structures as AlGaN/GaN HEMT also induces difficulties, since different materials and the presence of the two dimensional electron gas (2DEG) are involved. Various experimental and analytical evaluation methods are needed to be carried out and compared to establish a complex evaluation and defect investigation. Interaction between majority and minority carrier deep energy levels can be filtered out not only by theoretical evaluation approaches but also experimentally. Experimental solutions include basic DLTFS measurement parameter variations with electrical excitation pulses optimized to achieve separate signals, although this process is time consuming and not definite. A suitable more effective solution is the utilization of Minority Carrier Deep Level Transient Fourier Spectroscopy (MCDLTFS) with optical excitation impulses. In the conventional DLTFS method majority carrier excitation is observed by reverse pulse application, meaning an electrical excitation, while minority carrier traps can be observed when the excitation process is replaced by appropriate light pulses [5]. According to the theory a proper wavelength laser interacting with minority carriers and the valence band is able to filter out the majority response [6].

This work deals with identification of electrical active deep energy levels in AlGaN/GaN HEMT structures by DLTFS and MCDLTFS methods. Furthermore to utilise and confirm advantages and validity of MCDLTFS experiments in GaN based HEMT defect investigations.

2. Experimental

AlGaN/GaN Schottky diodes used in this experiment were prepared using conventional AlGaN/GaN HEMT processing steps on SiC substrates. Metallisation was prepared as follows: surface was prepared by cleansing and treated by desulfurization. Ohmic contacts were obtained by Ti/Al/Ni/Au deposition and annealing at 870°C for 30 s. Isolation was prepared by 160 keV Ar implantation (dosage 1×10^{13} cm^3). The Schottky contact with area 2×10^{-4} cm^2 was formed using Pt/Pd/Au (20/350 nm) metallization. The structure was then passivated with SiN. Quality of Schottky contacts were verified by I-V and character of HEMT structures by C-V measurements. Sudden decrease of capacitance observed in the reverse direction in the voltage range from -2 to -3 V was attributed to quantum well depletion on the border of AlGaN/GaN.

All DLTFS and MCDLTFS measurements were performed using the measurement system BIO-RAD DL 8000 DLTS in the laboratory Semitest at the Institute of Electronics and Photonics at Slovak University of Technology in Bratislava. DL8000 was equipped with a socket to connect the optical tool. The designed and fabricated illumination source was equipped with current adjustment, pulse modulator and an exchangeable cap for LED light sources. Optical excitation for MCDLTFS measurements were realized with four single emitter deep UV LEDs based on AlGaN with peak wavelengths of 250, 290 and 345 nm (optical output power of 30-70 μW). These wavelengths corresponded with energies 4.960, 4.275 and 3.594 eV, high enough to ensure minority charge carrier excitation.

3. Results and discussion

Fig. 1 left shows selected measured DLTFS spectra observed on the described Schottky AlGaN/GaN HEMT structures and Fig. 2 the corresponding MCDLTFS spectra (measurement conditions: reverse voltage V_R, filling pulse bias V_F, period width T_w and filling pulse time t_F are listed in the charts). All MCDLTFS spectra were determined as emission processes from minority carrier defect states where it can be noticed, that the spectrum peak amplitudes are significantly higher as the DLTFS result (Fig. 2 left).

978-1-5090-3084-2/16 $31.00 © 2016 IEEE

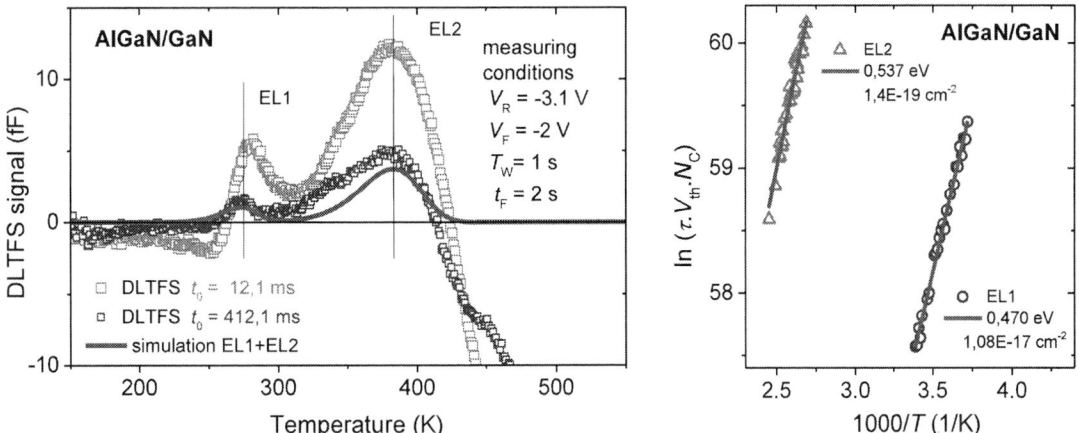

Fig.1 DLTFS experiment of AlGaN/GaN samples for two different time delay of capacitance transient signals t_0. Spectrum at electrical excitation for temperature set 100-550 K, two identified defect states EL1 and EL2 (left). Calculated Arrhenius plots of deep energy levels EL1 and EL2 with activation energies and capture cross sections (right).

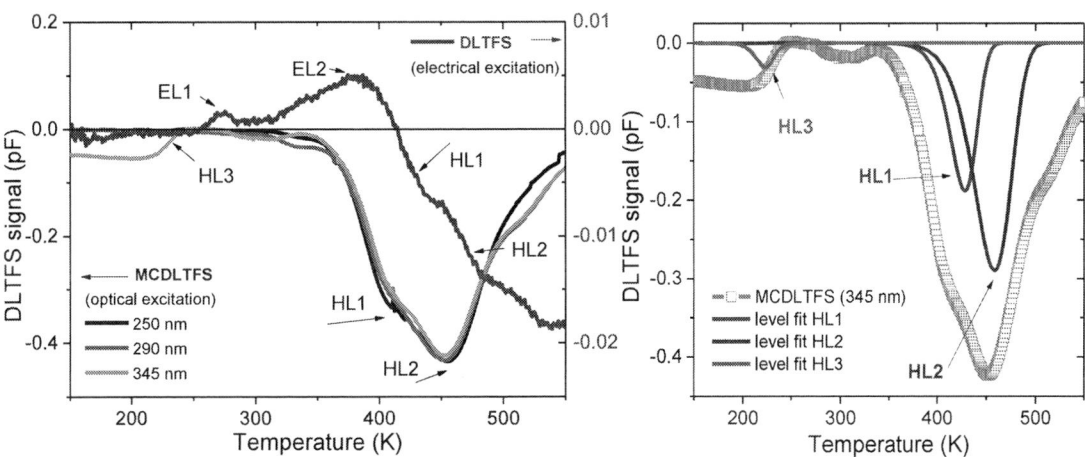

Fig. 2 MCDLTFS experiment of AlGaN/GaN samples at $V_R = - 3.1$ V, $V_F = -2$ V, $T_w = 1$ s and $t_F .= 2$ s. Spectral comparison of DLTFS and MCDLTFS at various wavelengths with identified deep energy levels (left), Simulated curves of MCDLTFS defect states (right).

Tab. 1 Parameters of identified deep energy levels with reference data.

Trap	Activation energy ΔE_T (eV)	Capture cross section σ_T (cm^2)	Comparison with parameters in references [7]
EL1	0.47	1.08×10^{-17}	C/O/H impurities, possibly in Ni substitutional position
EL2	0.54	1.40×10^{-19}	Nitrogen antisites
HL1	1.13	6.65×10^{-14}	Carbon interstitial defect
HL2	0.93	3.91×10^{-17}	Threading dislocations
HL3	0.49	1.36×10^{-15}	Unknown origin

Analysis of DLTFS measurements with electrical excitations confirmed two electron-like defects EL1 and EL2 and two hole-like traps HL1 and HL2. It was suggested that the defect EL1 (0.47 eV) probably corresponds with C/O/H impurities possibly in N substitutional positions [7], while the defect EL2 (0.54 eV) can be related to a nitrogen antisites [7]. HL1 (1.13 eV) probably corresponds with carbon interstitial defects and HL2 (0.93 eV) with well-know threading dislocations [7]. MCDLTFS spectra exhibited an extraordinary complex character (complex broad spectrum), where parameters of three hole like defect states were determined HL1, HL2 and HL3. HL1 and HL2 were confirmed by both methods, while HL3 (0.49 eV) was only visible at optical excitations (Fig. 2). The origin of HL3 is unknown. Tab. 1 shows all the calculated parameters of defects obtained by Arrhenius plots with reference data. These values were used for simulated DLTFS and MCDLTFS signals to confirm the validity and the spectral presence of observed defect states (Fig. 1 left and Fig. 2 right). Due to the complex nature of these spectral responses it was essential to utilise the MCDLTFS method, which is capable to verify and more precisely identify the presence of possible growth condition related defect states. Especially in this case, where many majority and minority carrier defect interactions were degrading the evaluation quality. Benefits of this method, hence the filtering out of majority carrier spectra (EL1, EL2) ensured a more precise calculation of hole like defect states (Fig 2 left). Moreover different light sources were examined revealing additional defects (HL3).

3.   Conclusion

This work deals with determination of deep levels in AlGaN/GaN HEMT structures by DLTFS method with electrical (DLTFS) and optical excitation (MCDLTFS). Parameters of 5 defect states were identified. The electron-like defects EL1 and EL2 were identified only by DLTFS. EL1 (0.47 eV) probably corresponds with C/O/H impurities, while EL2 (0.54 eV) was related to nitrogen antisites. Hole-like defects HL1 and HL2 were confirmed by both methods DLTFS and MCDLTFS, defect HL3 (0.49 eV) was identified only by MCDLTFS. With high probability HL1 (1.13 eV) is connected to carbon interstitial defects, and HL2 (1.13 eV) with threading dislocations. As expected, by utilisation of this proposed analysis we were able to evaluate complicated DLTFS spectra more precisely. Combination of results from MCDLTFS and DLTFS improved the defect evaluation process and ensured more possibilities of trap state interpretation.

Acknowledgement

The research leading to these results has received funding from the Electronic Component Systems for European Leadership Joint Undertaking under grant agreement No 662322, project OSIRIS. This Joint Undertaking receives support from the European Union's Horizon 2020 research and innovation programme and France, Norway, Slovakia, and Sweden. This publication reflects only the author's view and the JU is not responsible for any use that may be made of the information it contains.

References

[1]   A. Y. Polyakov, In-Hwan Lee. *Materials Science and Engineering* R. **94**, 2015.

[2]   M. Tapajna, O. Hilt, J. Würfl, J. Kuzmík, J., *Appl. Phys. Lett.*, **107**, 2015.

[3]   D. V. Lang, *J. Appl. Phys.* **45**, 3014, 1974

[4]   S. Weiss, R. Kassing, *Solid-State Electronics*, **31**, 12, 1733-1742, 1988.

[5]   M. Takikawa, T. Ikoma, *Japanese Journal of Applied Physics*, **19**, 7, L436, 1980.

[6]   J. H. Evans-Freeman, M. A. Gad, *Physica B: Condensed Matter*, **308**: 554-557, 2001.

[7]   D. Bisi, M. Meneghini, C. Santi, A. Chini, M. Dammann, P. Bruckner, M. Mikulla, G. Meneghesso, E. Zanoni. *IEEE Transactions on electron devices.* **60**, 10, 2013.

ASDAM 2016, The 11th International Conference on Advanced Semiconductor
Devices And Microsystems, November 13-16, 2016, Smolenice, Slovakia

DLTFS study of InGaAs/AlInAs heterostructures grown on n-InP:S substrates

A. Kosa[1], L. Stuchlikova[1], L. Harmatha[1], J. Kovac[1],
M. Badura[2], K. Bielak[2], B. Ściana[2], M. Tlaczala[2]

[1] Institute of Electronics and Photonics, Slovak University of Technology in Bratislava,
Ilkovičova 3, 812 19 Bratislava, Slovakia
[2] Wrocław University of Science and Technology, Faculty of Microsystem Electronics and
Photonics, Janiszewskiego 11/17, 50-372, Wrocław, Poland
e-mail: arpad.kosa@stuba.sk

In this paper authors are presenting the first results of electrically active defect investigations in the three $5 \times UD\text{-}In_{0.53}Ga_{0.47}As/Al_{0.42}In_{0.58}As$ MQW structures grown on n-InP:S substrates by low pressure metal organic vapour phase epitaxy at same growth conditions but at different growth rates. The defect distribution is different for each sample. As thinner are the quantum well layers as more significant minority carrier responses are measured. Parameters of nine deep energy levels were identified by Deep Level Transient Fourier Spectroscopy method. The three of them T3 (~0.38 eV), T5 (~0.17 eV) and T7 (~0.32 eV) were identified in all investigated structures. We assume that level T1 (~0.19 eV) and T5 (0.17 eV) are with high probability the result of carrier emissions from QWs.

1. Introduction

Quantum cascade lasers (QLC) are convenient, compact and effective sources of radiation in mid-infrared (wavelengths ~3-30 microns) and terahertz or far-infrared (wavelengths ~30-300 microns) regions of the electromagnetic spectrum. These structures offer unique applications in spectroscopy, sensing, and imaging devices [1]. However, these longer wavelengths are difficult to generate with solid-state devices due to lack of naturally occurring small bandgap materials. Various composition and different material based multi quantum well structures are continuously exploited in order to achieve cost and efficiency friendly solutions. At the same time electrically active defect states are formed during growth processes that can significantly affect electrical and optical properties of final structures. It is well known, that dislocations and many interface related defect states are formed between two materials acting as electrically active traps, that can be related to critical layer thicknesses, material compositions and growth conditions [2]. In order to improve the quality of such materials and to understand the generation of related impurities, various investigations and structure comparisons are needed to be carried out and investigated. One of the most frequently used techniques of electrically active defect state analysis is the Deep Level Transient Fourier Spectroscopy method (DLTFS) [3]. DLTFS stands out with its' positive features such as non-destructivity, sensitivity and reliability. This method is also adaptable for various cases and complex structures that can induce a really difficult analysis.

InGaAs/AlInAs is one of the perspective material systems, which is investigated from the QLC future application point of view. This paper reports the first results of electrical characterization carried out on Schottky, 5×UD-InGaAs/AlInAs MQW heterostructures grown on n-InP:S substrates by DLTFS methods.

978-1-5090-3084-2/16 $31.00 © 2016 IEEE

2. Experiment

The three investigated Schottky $5\times$UD-In$_{0.53}$Ga$_{0.47}$As/Al$_{0.42}$In$_{0.58}$As MQW heterostructures (labelled as A, B and C) were grown on n-InP:S substrates by low pressure metal organic vapour phase epitaxy (LPMOVPE) at same growth conditions but at different growth rates at the Wrocław University of Science and Technology in Poland. Composition of layers and growth rates for all three samples together with a calculated energy diagram for Sample A are displayed on Fig. 1.

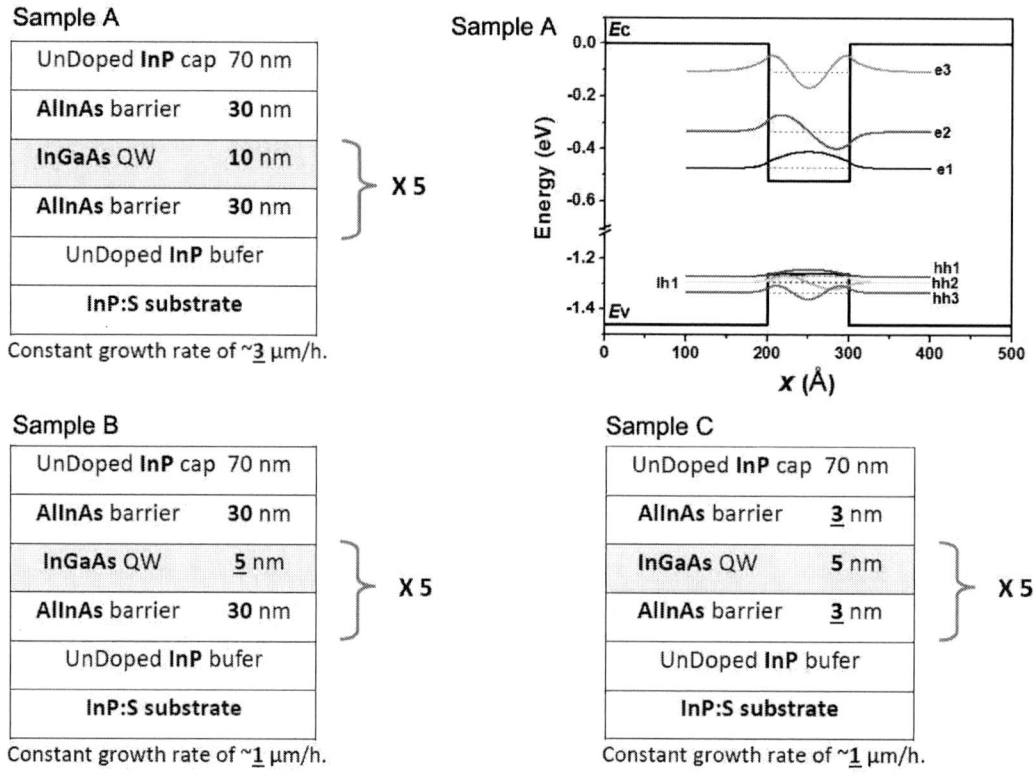

Fig. 1. Composition of layers and growth rates for all three investigated samples A, B and C and calculated energy diagram for Sample A

Experiments of deep energy level investigation were realized at the Slovak University of Technology in Bratislava by the DLTFS measurement unit BioRad DL 8000. Two modifications of DLTFS were used - DLTFS with electrical excitation (DLTFS) and DLTFS with optical excitation (Minority carrier DLTFS - MCDLTFS| with intention to achieve better separate emissions from minority carrier defect states [4]. In our case a GaAs laser (E_G = 1.42 eV) with P = 100 mW maximum power and near infrared wavelength λ = 850 nm was used fulfilling this task. Initial and particular results are discussed. Only values of activation energies of deep energy levels were possible to introduce from evaluation processes, because for correct values of capture cross sections electron and hole masses are needed.

3. Results and discussion

DLTFS spectra were obtained by applying same measurement parameters for all samples: time periods T_w = 20 ms, 500 ms, 1 s, filling time t_p = 20 ms, filling voltage V_P = -0.05 V and reverse voltage V_R = -0.1 V are show in Fig. 2 left. MCDLTFS spectra displayed in Fig. 2 right were measured at same measurement initial conditions, but filling voltage V_P = -0.05 V was substituted by an optical pulse. These measurements confirmed the existence of many electrically active defects. We have decided to focus above all on deep energy levels with similar parameters obtainable at all structures. These DLTS spectra were evaluated using the Fourier transform analysis with the "Direct auto Arrhenius single level evaluation mode.

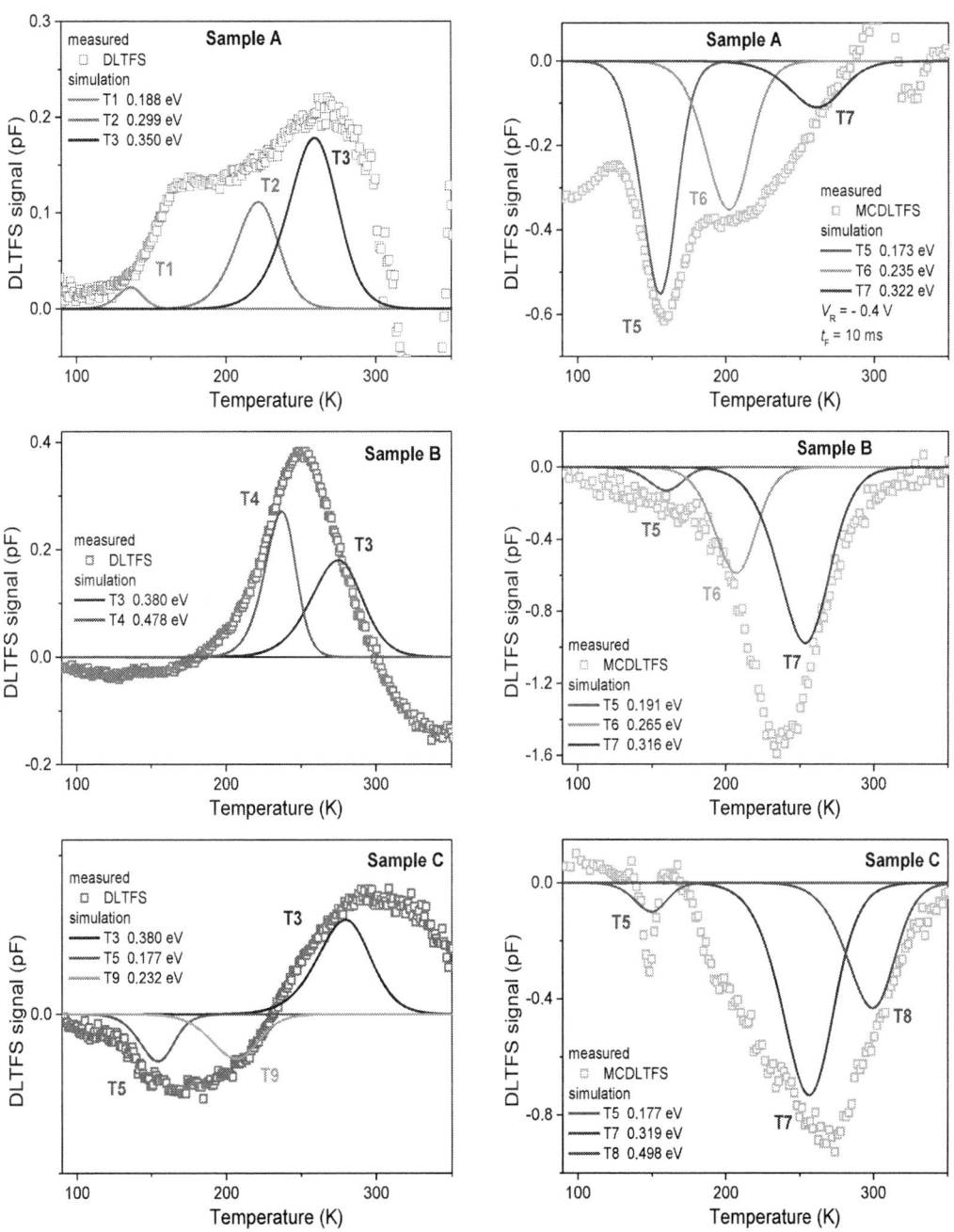

Fig. 2. Measured DLTFS and MCDLTFS spectra of the identified deep levels with simulated curves for level T1 - T9. Temperature scans with time period T_W = 20 ms are displayed.

Distribution of defects in measured DLTFS and MCDLTFS spectra are really distinct. For MCDLTFS results a minority (negative spectrum) response is typical as a result of minority carrier emission processes (Fig. 2 right). DLTFS signal indicates a connection between layer widths of each sample. By decreasing width of QW the spectrum changes from positive to negative (Fig. 2 left). After decreasing the width of barrier layers the positive signal is suppressed and shifted towards higher temperatures. In our interpretation as thinner are the quantum well layers as more significant minority carrier responses are measured. Shar transitions in spectra from positive to negative can be attributed to interacting majority and minority carrier defect complexes.

By Arrhenius curve calculation and various evaluation processes we were able to identify 9 deep energy levels labelled T1 - T8 with activation energies 0.19, 0.30, ~0.38, 0.48, ~0.17, ~0.25, ~0.32 eV, 0.5 and 0.23 eV. The levels T1 -T4 are results of majority carriers emisions. Due to the intrinsic layer involvement it was hard to definitely identify the conduction type of defects (hole like or electron like). According to the activation energy value it was assumed that lower energies (T1 (~0.19 eV) and T5 (0.17 eV)) were caused by QW emissions [5], while higher by possible material defect states (dislocations and many interface related defect states).

4. Conclusion

This paper deals with DLTFS investigation of electrically active defect states in $5\times$UD-$In_{0.53}Ga_{0.47}As/Al_{0.42}In_{0.58}As$ MQW heterostructures grown on n-InP:S substrates by LPMOVPE at same growth conditions but different growth rates and layer properties. This kind of material system, potential for future applications in quantum cascade lasers was examined in order to support the current research. To better understand related electrically active defect states Deep Level Transient Fourier Spectroscopy with electrical and optical excitations was utilized.

We confirmed the influence of different layer widths on defect distributions in investigated samples, where we identified activation energies of 9 deep levels. We have assumed that two of these were induced by the QW charge carrier emissions. All reported data and origin of identified defects are still under intensive investigations.

Acknowledgement

This work has been supported by the Scientific Grant Agency of the Ministry of Education of the Slovak Republic (VEGA 1/0651/16 and VEGA 1/0739/16). This work was co-financed by Wrocław University of Science and Technology statutory grants and the Polish National Centre for Research and Development under the project No. PBS2/A3/15/2013 "PROFIT".

References

[1] Ch. A. Wang, A. Goyal, R. Huang, J. Donnelly, D. Calawa, G. Turner, A. Sanchez-Rubio, A. Hsu, Q. Hu, B. Williams, *Journal of Crystal Growth*, **312**, 1157–1164, 2010.

[2] Z. Ohizumi, T. Tsuruoka, S. Ushioda, *Compound Semiconductors 2001*, **170**, 449, 2002.

[3] S. Weiss, R. Kassing, *Solid-State Electronics*, **31**, 12, 1733-1742, 1988.

[4] J. H. Evans-Freeman, M. A. Gad, *Physica B: Condensed Matter*, **308**: 554-557, 2001.

[5] J. F. Chen, J. S. Wang, *J. of Appl. Phys.*, **102**, 043705-1- 6, 2007.

ASDAM 2016, The 11th International Conference on Advanced Semiconductor
Devices And Microsystems, November 13-16, 2016, Smolenice, Slovakia

Impedance characterisation of NiO based gas sensor

M. Mikolášek*, M. Predanocy, P. Jom, and I. Hotový

*Institute of Electronics and Photonics, Slovak University of Technology,
Ilkovičova 3, 812 19 Bratislava, Slovakia
e-mail: miroslav.mikolasek@stuba.sk

*Nickel oxide, NiO is a perspective metal oxide with the ability to detect hazardous
and flammable gasses. The detection process is based on the change of the
conductivity of the NiO layer upon the presence of the gas. To tune the sensing
properties of NiO sensor, it is necessary to obtain more comprehensive
understanding of current transport processes in the structure. In this paper, we
present an impedance characterization of NiO based sensor, which allowed to
distinguish three conduction mechanism responsible for the current flow in the
structure.*

1. Introduction

In recent years, there has been strong research interest in miniaturization and
microfabrication of novel metal oxide gas sensors. Among the materials suitable for gas
sensing, nickel oxide (NiO) is relatively well-known and greatly expanded metal oxide for its
very good chemical, optical, electrical and gas sensing properties [1-4]. Such material
properties of NiO are strongly affected by deposition method as well as post-deposition
processing and can by tuned to obtain high sensitivity for detection of hazardous and
flammable gasses [1]. These gasses has ability to change the electro-physical properties of
NiO, which are reflected in the change of the conductivity. Simply measurements of the
conductivity response upon the presence of gasses provide direct information about the
sensing properties of NiO based sensor. More complex measurements, however, are required
to obtain deeper insight on the physical process responsible for gas sensing. Understanding of
such processes is required for further optimization of the NiO gas sensor.

The impedance spectroscopy is a powerful tool, which allows analysis of electric
properties and electronic transport within investigated structure [5-6]. In this paper, we carrier
out impedance characterization of 50 nm thin NiO gas sensor in the air ambient and provide
detail analysis and physical explanation of mechanisms responsible for current transport in the
structure.

2. Experimental

Nanocrystalline nickel oxide films with thickness 50 nm were deposited by DC reactive
magnetron sputtering from a Ni target (4" in diameter, 99.99% pure) in a mixture of oxygen
and argon at room temperature on a micromachined structure with SiO_2/Si_3N_4 membrane and
interdigitated Pt electrodes. A sputtering power of 600 W was used during the deposition. The
relative partial pressure of oxygen in the reactive mixture O_2-Ar was 30 %. In order to
stabilize the properties of NiO thin film, sample was annealed in a furnace at 500 °C in
nitrogen atmosphere for 2 hours.

The electrical behavior of the sensor was studied by an impedance spectroscopy at
various temperatures of the heater element and within wide range of AC frequencies from 100

978-1-5090-3084-2/16 $31.00 © 2016 IEEE   153

Hz to 1 MHz. The temperature of the build-in heater was set by adjusting its power according to the calibration curve by using Keithley 2612A. Before measurement, the sensor was first heated at temperature 300 °C for a few hours to stabilize properties of NiO layer. The LCR meter AGILENT 4284A was used to gather impedance data and EIS analysis software was used for calculation of parameters of proposed equivalent circuit. All measurements were carried out in the air.

3. Results and discussion

Figs. 1a, and b show impedance and phase, respectively, measured as a function of frequency (Bode representation) at temperatures in the range of 0 - 245 °C. The increase of the temperature results into the decrease of the impedance at low frequencies. This is in accordance with decrease of resistance upon the increase of temperature. At high frequencies, however, the impedance decreases with linear trend. Such decrease is due to the presence of temperature independent serial resistance in the structure. The phase vs. frequency curves exhibit two peaks, which suggests two parallel combination of R and C components in the equivalent circuit. In the case of studied NiO structure, the presence of two R-C elements indicates two kind of depletion layers at the NiO grain surface. Numerical simulation was carried out to determine the individual parameters of the AC equivalent circuit and detail assessment of the processes in the structure. The example of fit to the measured data drawn in Nyquist representation is shown in Fig. 2a. Very good correlation between the proposed AC circuit (inset of Fig. 2a) and the measured impedance data by numerical simulation was found. The equivalent circuit is formed by a parallel combination of R2 - C1 connected in series with parallel combination of R3-CPE1 and series combination of R1, which together represents three conduction paths in the structure. To explain the origin of such conduction paths, it is important to consider the structure of the NiO based sensor. NiO is formed by grains, which are prepared on Pt interdigitated electrode (Fig. 3). The ionosorption of oxygen at the grain surface results into the creation of a potential barrier and the corresponding depletion layers at the contacts between grains. Such potential barrier, presumably Schottky barrier is represented electronically by a resistor R and a capacitor C or constant phase element CPE1 in parallel. In our cases, the CPE element has capacitance character and can be replaced by capacitance with low error.

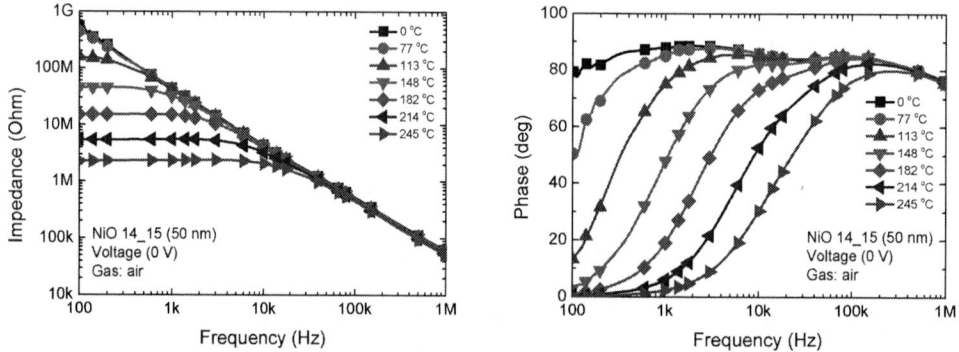

Fig.1: Bode representation of the temperature dependent a) impedance and b) phase as a function of frequency for 50 nm thin NiO based gas sensor.

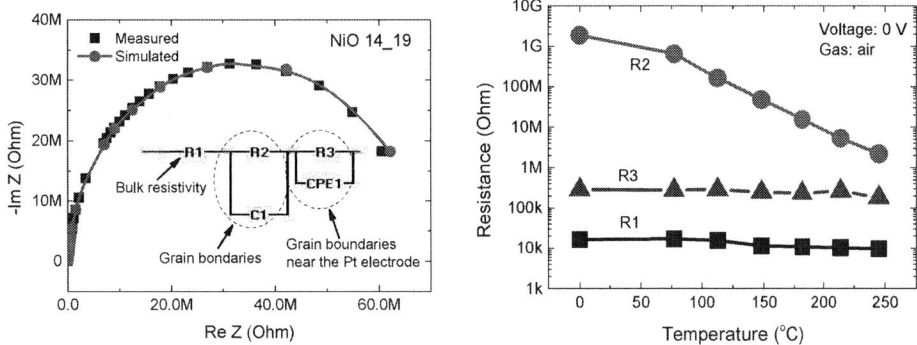

Fig. 2: a) Example of Nyquist diagram of measured and simulated curve. Inset represents an equivalent circuit used to fit measured data. b) Temperature dependence of R1, R2 and R3 resistors extracted by simulation.

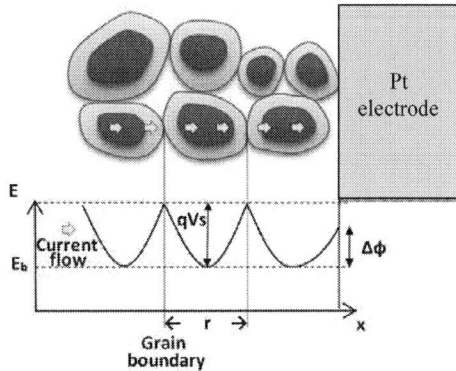

Fig. 3: Grain structure of the NiO layer prepared on Pt electrode. The bottom part represents change of the potential barrier in the structure showing two kind of barriers.

The presence of two R-C (or R-CPE) elements indicates two different depletion layers at the grains surface. One possible explanation of this is that two type of grains are presented in the structure, forming different Schottky barriers between them. Another more possible explanation is that one R-C (R-CPE) component is linked with the surface of NiO grains, which are close to the interdigitated electrode and might be populated due to the catalytic activity of the Pt with a different concentration of reactive species than those which are distant from the electrodes. In this way, two regions with NiO grains having different grain depletion layers are formed in the sensor. First one is in the region near the surface and in the middle of the sensor layer, and is in the equivalent circuit represented by the parallel combination of R2-C1 and in the Bode representation of phase vs. frequency is linked with the peaks at low frequencies (Figs. 1b-3b). The second one is in the region near the Pt electrode, which is in the equivalent circuit represented by the parallel combination of R3-CPE1 and linked with high frequencies in the bode diagrams (Figs. 1b-3b). The element R1 is linked with the grain bulk resistance in the NiO structure.

The analysis of the impedance data measured at different temperatures allowed us to extract individual elements of the equivalent circuit as a function of temperature, which can further provide insight on the carrier transport properties in the structure. Fig. 2b shows temperature dependence of R1, R2 and R3. The R1 has low temperature dependence. This element represents the resistivity of the grains in the structure as well as contact resistance of measurement system. We can expect negligible role of this conduction path in the gas sensing properties of the sensor. The R2 and R3 elements reflect the behavior of the barriers between grains in the region distant from the Pt electrode and close to the Pt electrode. Different temperature behavior of R2 and R3 indicates different conduction mechanism presented in these region. Temperature dependence of R2 exhibits an Arrhenius behavior with activation energy of 0.52 eV. We can expect thermionic emission through such barrier as a dominant conduction mechanism. In case of R3, however, negligible temperature dependence indicates tunnelling transport of carriers. Assumed is that electrons located at the periphery of one sphere are transported to another by tunneling mechanism through a small gap (around 0–0.1 nm) in between. Both types of conductance are proportional to the density of electrons at the periphery, which is determined by the surface barrier height [7] and strongly affected by gas. The impedance measurements in the gas ambient will be carried out in the next step to determine the impact of individual transport properties at the gas sensing behavior.

3. Conclusion

Impedance analysis of NiO gas sensor allowed to separate three conduction paths: 1) resistance of the grains, 2) resistance due to the Schottky barriers at the grain boundaries of NiO in the middle of the layer and near the surface and 3) resistance due to the NiO grain boundaries near the Pt contact. Two different transport mechanism, tunneling and thermionic emission, were suggested for current paths through grain boundaries. The impedance characterization in the gas environment is required to reveal the role of such different grain boundaries and current transports in the gas sensing properties.

Acknowledgement

This work was supported by the Scientific Grant Agency of the Ministry of Education of the Slovak Republic and of the Slovak Academy of Sciences no. 1/0651/16 and no. 1/0828/16 and by project SAFESENS (agreement No. 621272-1) co-funded by the Slovak Republic and the ENIAC-JU.

References

[1] H. Steinebach, S. Kannan, L. Rieth, F. Solzbacher, *Sensors and Act.B* **151,**162-168, 2010.

[2] C. Wang, J. Liu, Q. Yang, P. Sun, Y. Gao, F. Liu, J. Zheng, G. Lu, *Sensors and Actuators B: Chemical*, **220**, 59-67, 2015.

[3] R. Kumar, C. Baratto, G. Faglia, G. Sberveglieri, E. Bontempi, and L. Borgese, *Thin Solid Films*, **583**, 233-238, 2015.

[4] I. Kosc, I. Hotovy, V. Rehacek, R. Griesseler, M. Predanocy, M. Wilke, L. Spiess, *Applied Surface Science*, **269**, 110-115, 2013.

[5] J. Moon, J.A. Park, S.J. Lee, J.I. Lee, T. Zyung, E.C. Shin, J.S. Lee, *Physical Chemistry Chemical Physics*, **15** (23), 9361-9374, 2013.

[6] M. Perný, V. Šály, M. Váry, M. Mikolášek, J. Huran, J. Packa, *Journal of Electrical Engineering*, **65** (3), 174-8, 2014

[7] N. Yamazoe, and K. Shimanoe, *Journal of Sensors*, **2009**, ID 875704, 2009.

ASDAM 2016, The 11th International Conference on Advanced Semiconductor
Devices And Microsystems, November 13-16, 2016, Smolenice, Slovakia

Schottky contact metallization stability on AlGaN/GaN heterostructure during the diamond deposition process

O. Babchenko[a], G. Vanko[a], J. Dzuba[a], T. Ižák[b], M. Vojs[c], T. Lalinský[a] and A. Kromka[b]

[a]Institute of Electrical Engineering SAS, Dúbravská cesta 9, 841 04 Bratislava, Slovakia
[b]Institute of Physics of the AS CR, Cukrovarnícka 10/112, 162 00 Prague, Czech Republic
[c]Institute of Electronics and Photonics STU, Ilkovičova 3, 812 19 Bratislava, Slovakia
e-mail: oleg.babchenko@savba.sk

The issue of gate metallization stability on AlGaN/GaN heterostructure during the diamond deposition process has been studied. Among tested Ni, Ir, NiO and IrO$_2$ materials the iridium-based has the most promising characteristic to be used. The diamond growth in focused microwave plasma system on transistors with Ir and IrO$_2$ Schottky contact metallization has been demonstrated and discussed.

1. Introduction

The current mankind is oriented onto intense use of various electronic devices with their performance constantly forced upwards. Therefore, the numerous researchers devote their interest to advanced materials able to satisfy such challenging requirements. In this situation the AlGaN/GaN heterostructure based high electron mobility transistors (HEMTs) are among the most promising candidates for high power applications [1–5]. In consequence, the high power applications intend functional stability of material itself and contact metallization at high temperatures. While thermal stability of AlGaN/GaN heterostructure is attributed to its nature, the situation with contact metallization is following. The high temperature stable ohmic contact (source and drain) of HEMTs are mostly formed by Ti/Al based metal system [3, 4, 6, 7]. For high temperature Schottky contact (gate) with stable interface the use of Ni, Ir and their conductive oxides, has been proposed [8–12].

Unfortunately, as well as the most of other materials proposed for high power application, the AlGaN/GaN based transistors suffer from the "self-heating" effect. The high power dissipation during the device operation and, as the result, temperature increase, cause the degradation of electron mobility and channel current collapse [5, 13]. One of the proposed solutions in this case is the use of diamond as a heat sink for efficient thermal management [4, 5, 13–15]. On the other hand, the issue of diamond growth on the AGaN/GaN based HEMTs is not a trivial task. The gas chemistry used for diamond deposition at high temperatures is able to degrade GaN material properties as well as damage the metallic contacts [14–17]. Thus the appropriate protective coating, which must effectively passivate whole HEMT surface without short circuits and do not introduce disturbance in thermal management, is desirable. The other option, is use of diamond deposition process conditions that will not damage neither AlGaN/GaN heterostructure nor contact metallization [15, 17].

In this article we focused on the problematic of diamond growth on the AlGaN/GaN based HEMTs. Particularly, the stability of Schottky metallization during diamond growth performed using microwave plasma enhanced chemical vapor deposition was studied.

2. Experiment

For the experiments the commercially available (NTT-AT, Japan) low tensile stressed AlGaN/GaN heterostructure grown on a Si (111) substrate was used. The thickness

978-1-5090-3084-2/16 $31.00 © 2016 IEEE

of the AlGaN and GaN layers were 20 nm and 4.2 µm, respectively. The Al mole fraction in the AlGaN layer was nominally set to 0.25 without intentional doping. The surface defects (traps) of AlGaN top surface were passivated by 2 nm GaN cap layer [11].

In the first set of experiments the Ni, Ir, NiO and IrO_2 were tested as promising materials for high temperature stable Schottky interface formation on AlGaN/GaN heterostructure. The Ir and Ni were deposited on AlGaN by the UV lithography, metal evaporation and lift-off. The subsequent oxidation was realized by 1 min rapid thermal annealing (RTA) at 800°C in O_2 atmosphere. Next, samples were seeded by the diamond seeds using ultrasound nucleation in the ultra-dispersed water based diamond powder suspension [17–19]. The diamond deposition was realized in the low pressure low temperature linear antenna microwave plasma system [19] at: microwave power 1.7 kW, gas mixture of 2.5% CH_4 and 10% of CO_2 in H_2 at pressure 0.1 mbar and average temperature of 330°C for 30 h.

For the set of experiments related to the functionality of transistor structure the circular HEMTs with the source-drain spacing 180 µm and variable (from 40 µm to 160 µm) gate length were fabricated. Briefly, the ohmic contacts of circular source and drain surround were formed by Nb-Ti/Al/Ni/Au metallic system annealed at 850°C in N_2 atmosphere for 35 s. The different type of Schottky contacts (gate ring) were represented by Ir/Au, and IrO_2/Au (1 min RTA in oxygen at 800°C) metallic systems. The details of ohmic and Schottky contacts fabrication can be found elsewhere [7, 11, 12]. Next, the samples were processed by the selective area nucleation based on polymer/seeding layer/polymer arrangement [17, 18]. The diamond deposition was realized in the focused microwave plasma system [20] at: microwave power 2 kW, gas mixture of 5% CH_4 and 1.5% of CO_2 in H_2, process pressure 40 mbar and average temperature 600°C for 6 h.

The obtained layers and changes in samples morphology were investigated by scanning electron microscopy (SEM) and Raman spectroscopy.

3. Results and discussion

The SEM images from the first set of experiments are shown in Fig. 1. It should be noted, that in all cases a clear border between GaN surface and metallization was observed. Next, in all cases on GaN surface the fully closed diamond film with crystals size up to 200 nm and morphology typical for low pressure plasma system [21, 22] can be found.

Concerning the diamond growth on metallization the situation is following. In case of Ni and NiO metallization (Fig. 1a, 1b) no diamond film with clear facets was found. In case of Ir (Fig. 1c) the diamond film has morphology similar to that observed on GaN surface. Finally, the diamond film on IrO_2 (Fig. 1d) is not closed and consists of large clusters (up to 350 nm). The reason of this can be the oxygen present in IrO_2 layer and diffused during the diamond deposition affecting the early stage of diamond growth.

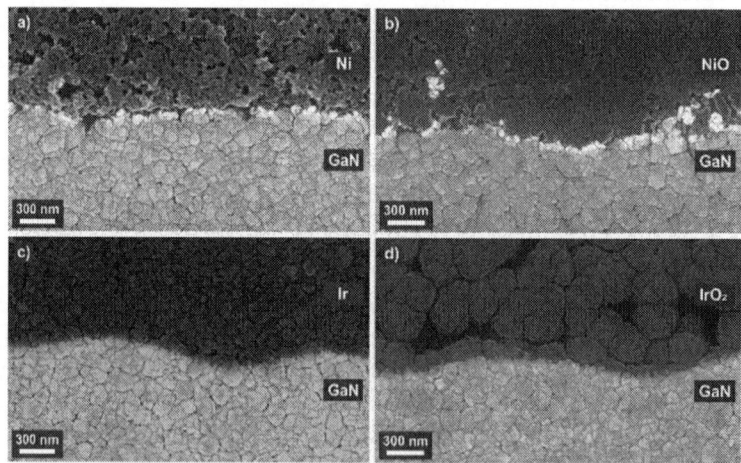

Fig. 1. SEM images of diamond film deposited on AlGaN/GaN heterostructure with: a) Ni, b) NiO, c) Ir, and d) IrO_2 metallization.

The Raman measurements (not presented) reveal on GaN and Ir (IrO_2) typical spectra for low pressure low temperature diamond with broaden diamond peak and high graphitic phases signal [19, 21, 22]. On the contrast, the measurement on Ni (NiO) reveals intensive graphitic phases signal with no detectable sign of diamond. The most probably due to Ni ability to absorb carbon atoms [23], the high carburization rate appears even at low temperature. Thus, in the case of Ni and NiO, the graphitic film instead of diamond was formed. Therefore, the Ir-based metallization seems to be more suitable for gate contact fabrication.

The SEM images and Raman spectra from the second set of experiments are shown in Fig. 2. As it can be seen, in the case of Ir/Au metallization (Fig. 2a) the diamond deposition process caused delamination (peeling-off and bubbles formation) of gate metallization. On the contrary, the IrO_2/Au metallization (Fig. 2b) seems to survive diamond deposition process without any damage. The fully closed diamond films of similar morphology with large (up to 400 nm) well facetted crystal were found in both cases. These results indicate that annealing in oxygen not only improves the electronic characteristic of the device [10, 12] but also improve its mechanical properties (gate metallization adhesion).

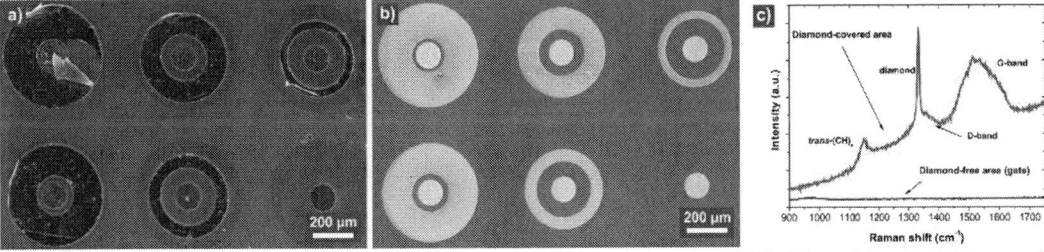

Fig. 2. The SEM images of diamond film grown on circular HEMTs with: a) Ir/Au and b) IrO_2/Au gates, and Raman spectra (c).

The Raman measurements (Fig. 2c) show excellent efficiency of strategy chosen for selective diamond growth. The spectrum typical for nanocrystalline diamond film (the sharp diamond peak together with signal from sp2 phases from grain boundaries) [16–18] was found on diamond-covered area. The signal from diamond-free area (in this case measured on gate) shows only background noise (i.e. possible signal is below the detection limit).

The I-V characteristics before and after the diamond deposition on circular HEMT for two different gate length are shown in Fig. 3. The output characteristics indicate that transistor with shorter gate (Fig. 3b) was more negatively affected by the diamond deposition than the long-gated transistor (Fig. 3a). Similar decrease of saturation current with increasing of R_{ON} resistance after the diamond deposition has been already reported [15]. The origin of this effect is probably in the larger

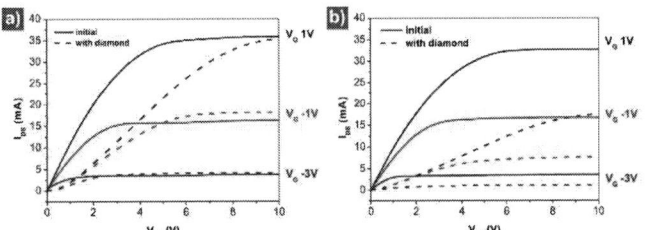

Fig. 3. The I-V characteristics (I_{DS} – drain current, V_{DS} – drain-source voltage, V_G – gate voltage) of HEMT with: a) 140 μm and b) 40 μm long gate.

opened (gate-free) area in the source-drain space of short-gated transistor that was exposed to plasma during the diamond deposition. It means that although there was no etching of heterostructure surface for such HEMTs the diamond growth process need to be optimized.

4. Conclusions

We can summarize that, according the first set of experiments, Ni and NiO under chosen conditions were not favorable for application as a gate contact for diamond covered devices. In this case the Ir metallization show the most promising results for diamond growth on it while employing of IrO_2 require optimization of diamond seeding technology. According the results of the device functionality test, the IrO_2/Au gates reveal higher stability, i.e. better adhesion, than Ir/Au during the diamond deposition. Nevertheless, both structures remain functional after processing what indicates that the Ir-based materials can be used for stable Schottky contact formation and further covered by diamond. Finally, in case of short-gated HEMTs the optimization of diamond deposition process or technology modification is required.

Acknowledgement

The authors acknowledge financial support of the SASPRO Programme project 0068/01/01, co-financed by the European Union and the Slovak Academy of Science, the Slovak Research and Development Agency, grant Nr. APVV-0455-12, the Grant Agency of the Czech Republic, grant Nr. 14-16549P and the framework of the LNSM infrastructure, and the bilateral project between Czech and Slovak Academy of Sciences Nr. SAV-16-02, 2016-2017. The authors appreciate help of K. Hruška (IoP AS CR) in scanning electron microscopy.

References

[1] O. Ambacher, *J. Phys. Appl. Phys.*, vol. 31, no. 20, pp. 2653–2710, Oct. 1998.
[2] S. T. Sheppard *et al.*, 2000, pp. 232–236.
[3] T. Kikkawa, *Jpn. J. Appl. Phys.*, vol. 44, no. 7A, pp. 4896–4901, Jul. 2005.
[4] G. D. Via *et al.*, *Phys. Status Solidi C*, vol. 11, no. 3–4, pp. 871–874, Feb. 2014.
[5] R. Zhang *et al.*, *Diam. Relat. Mater.*, vol. 52, pp. 25–31, Feb. 2015.
[6] N. Shigekawa *et al.*, *IEEE Electron Device Lett.*, vol. 28, no. 2, pp. 90–92, Feb. 2007.
[7] G. Vanko *et al.*, *Vacuum*, vol. 82, no. 2, pp. 193–196, Oct. 2007.
[8] C. M. Jeon *et al.*, *Appl. Phys. Lett.*, vol. 82, no. 3, p. 391, 2003.
[9] C. M. Jeon and J.-L. Lee, *Appl. Phys. Lett.*, vol. 82, no. 24, p. 4301, 2003.
[10] C. M. Jeon *et al.*, *J. Vac. Sci. Technol. B Microelectron. Nanometer Struct.*, vol. 24, no. 3, p. 1303, 2006.
[11] T. Lalinský *et al.*, *Appl. Phys. Lett.*, vol. 100, no. 9, p. 92105, 2012.
[12] T. Lalinský *et al.*, *Appl. Surf. Sci.*, vol. 283, pp. 160–167, Oct. 2013.
[13] A. Wang *et al.*, *Semicond. Sci. Technol.*, vol. 28, no. 5, p. 55010, May 2013.
[14] M. Alomari *et al.*, *Diam. Relat. Mater.*, vol. 20, no. 4, pp. 604–608, Apr. 2011.
[15] N. Govindaraju and R. N. Singh, *Mater. Sci. Eng. B*, vol. 176, no. 14, pp. 1058–1072, Aug. 2011.
[16] T. Izak *et al.*, *Mater. Sci. Forum*, vol. 821–823, pp. 982–985, Jun. 2015.
[17] T. Izak *et al.*, *Phys. Status Solidi B*, vol. 251, no. 12, pp. 2574–2580, Dec. 2014.
[18] O. Babchenko *et al.*, *Phys. Status Solidi B*, vol. 247, no. 11–12, pp. 3026–3029, Dec. 2010.
[19] A. Kromka *et al.*, *Vacuum*, vol. 86, no. 6, pp. 776–779, Jan. 2012.
[20] M. Füner *et al.*, *Appl. Phys. Lett.*, vol. 72, no. 10, p. 1149, 1998.
[21] T. Izak *et al.*, *Phys. Status Solidi B*, vol. 249, no. 12, pp. 2600–2603, Dec. 2012.
[22] O. Babchenko *et al.*, *Surf. Coat. Technol.*, vol. 271, pp. 74–79, Jun. 2015.
[23] A. V. Melechko *et al.*, *J. Appl. Phys.*, vol. 97, no. 4, p. 41301, 2005.

ASDAM 2016, The 11th International Conference on Advanced Semiconductor Devices And Microsystems, November 13-16, 2016, Smolenice, Slovakia

Characterization and gas sensor testing of single and mixed metal oxides based on NiO and TiO$_2$ thin films

I. Hotový, M. Predanocy, M. Mikolášek, P. Benko and V. Řeháček

[1]Institute of Electronics and Photonics, Slovak University of Technology, Ilkovicova 3, 812 19 Bratislava, Slovakia
ivan.hotovy@stuba.sk

We present results on single and mixed metal oxides based on NiO and TiO$_2$ thin films which are able to detect 1000 ppm of hydrogen and to operate at 200°C. These films with thickness in the range of 50 - 100 nm were prepared by DC reactive magnetron sputtering on alumina and silicon substrates. The XRD measurements revealed polycrystalline character of annealed thin films at 600°C, whereas as-deposited samples were confirmed to be amorphous. Sensor structures based on NiO/TiO$_2$/Si worked at very low operation temperature of 150°C though their responses to hydrogen appear lower relative sensitivity in comparison with those prepared from single oxides.

1. Introduction

Due to rising interests in human safety, automotive industry, regenerative and green sources of energy and also domestic and industrial use, concepts for new methods of detecting combustible gases such as H$_2$ and CH$_4$ are required. There is a high demand for new combination of gas sensitive materials improving standard gas sensing responses and properties [1-2]. Nanocrystalline materials are in focus of research as a consequence of the superior properties which are possible to achieve when materials are built up from structural units nanometer scale size.

Chemo-resistive semiconducting metal oxides are perspective candidates for modern gas sensors due to their fast response/recovery times, good sensitivity, low cost, easy maintenance and ability to detect various gases. Almost all semiconducting metal oxides are n-type because electrons are naturally produced via oxygen vacancies. From this group of metal oxides TiO$_2$ is very promising material with unique parameters (electrical and optical) and wide use: solar cells, photo-catalysis, bio-active materials and gas sensors [3]. Also several p-type gas sensors which majority carriers are holes were identified such as CuO, CoO, NiO. However, there was not enough research in the combination of these kinds of p-n materials and their mutual interactions and properties.

In this work we focused on the structural and gas sensing properties of titanium and nickel-based oxides for H$_2$ detection deposited on sapphire and silicon substrates.

2. Experimental procedure

Single and mixed NiO and TiO$_2$ films were deposited by DC reactive magnetron sputtering from a Ni and Ti target in a mixture of oxygen and argon. The relative partial pressure of oxygen in the reactive mixture O$_2$-Ar was 30%. Details of these sputtering deposition conditions had been described elsewhere [4]. The film thicknesses measured by Talystep were in the range of 50 and 100 nm for all examined samples. The films were prepared onto unheated alumina substrates for structural characterization. Investigated thin

978-1-5090-3084-2/16 $31.00 © 2016 IEEE

films based on NiO and TiO_2 were deposited on silicon chips with thermal oxide and with Ti-Pt-based interdigitated electrodes for electrical and gas measurement tests. The thickness of the thermal oxide was 2.5 μm, the electrode thickness was 190 nm. Reticles of 15 x 15 mm² with 25 sensor chips of 3 x 3 mm² were inserted into a two-side trenched silicon shadow mask of 150 mm diameter for structuring of the sensitive layers. In order to stabilize the properties, all films were annealed in a furnace at 600°C in dry air for 2 hours.

The crystal structure was identified with a Theta-Theta Diffractometer D5000 with a Goebel mirror in the grazing incidence geometry with CuKα radiation. The average crystalline grain size of TiO_2 and NiO nanoparticles was estimated from the integral breadths and the peak position of an XRD line broadened according the Scherrer formula.

The temperature variation of electrical resistance has been measured in the special vacuum chamber PLV-50 (fy Cascade Microtech) in the temperature range of 150÷300°C. The responses from prepared sensor structures to 1000 ppm of hydrogen were obtained by measuring their electrical resistance. Relative sensitivity was determined as a ratio of R_g/R_a where R_g is the MOX electrical resistance value at the end of the exposure time to hydrogen and R_a is the electrical resistance value of the base-line in nitrogen. The sensor resistance was measured every 0.5 s by a multimeter and recorded using a GPIB interface for communication with a computer by LabVIEW language. Mass flow meters and controllers with a nominal flow of 100 and 1000 sccm were used to set the flow of carrier and tested gases. The MOX sensor structures were placed in the chamber with a constant flow mixture carrier (N_2) and test (H_2) gases while the total pressure in the chamber was kept at a constant pressure of 5×10^3 Pa. All prepared MOX sensor structures were situated direct on the heating table and they were measured in sequence at each setting in the operation temperature (150, 200, 250 and 300°C).

3.   Results and discussions

3.1   Structural properties

The structural changes and identification of phases for mixed oxides $TiO_2/NiO/Al_2O_3$ (L1) and $NiO/TiO_2/Al_2O_3$ (L2) were studied with the help of XRD technique. The observed XRD patterns were compared with standard JCPDS (ICDD PDF-2) data files (anatase TiO_2: 12–1272, rutile: 21–1276, Ni:4–0850, NiO 44–1519) [30,31]. The XRD measurements revealed polycrystalline character of annealed thin films at 600◦C (Fig. 1b), whereas as-deposited samples (Fig. 1a) were confirmed to be amorphous. For both investigated layouts strong and sharp anatase TiO_2 phase (tetragonal) with main (101) diffraction peak can be observed. For the first layout L1 also weak and slightly shifted rutile (110) can be found. The XRD patterns of the both investigated layouts exhibit face-centered cubic Ni phase. For the first layout are remarkable broader and weaker Ni diffractions with a possibility of rhombohedral NiO phase inclusion. Slight NiO (202) diffraction at 79.37° can support this consideration.

3.2   Electrical characterization of MOX films in air

In the initial study, we carried out the measurements of the electrical resistance of the prepared samples as a function of temperature in air. For this characterization the chip No. 20 was selected whose geometry contains the biggest number of Pt intergitated electrodes. Figure 2a shows the temperature dependences of the studied samples measured in air. All samples exhibit a decrease of the electrical resistance with increasing operation temperature. In general, the results indicate that the samples prepared on the basis of TiO_2 films exhibit a higher resistance in air than NiO-based films. The measured curves allowed us using the Arrhenius equation to determine the activation energy E_a for the studied sensors. Figure 2b summarizes the values of the activation energy calculated in the temperature range $200 \div 300$ °C for NiO and TiO_2 based structures. It is observed that for NiO-based sensors, the values of the activation energy lie between 0.5 and 0.6 eV. On the other hand, the sample prepared by combination of NiO and TiO_2 films reached the value of $E_a \sim 0.6$ eV, which is caused by incorporation of TiO_2 into the common layer. All examined samples prepared on the basis of TiO_2 exhibited a higher value of the activation energy around 0.7 eV in comparison with NiO-based films. The smallest value of about 0.5 eV belonged to the sample prepared by combination of TiO_2 and NiO.

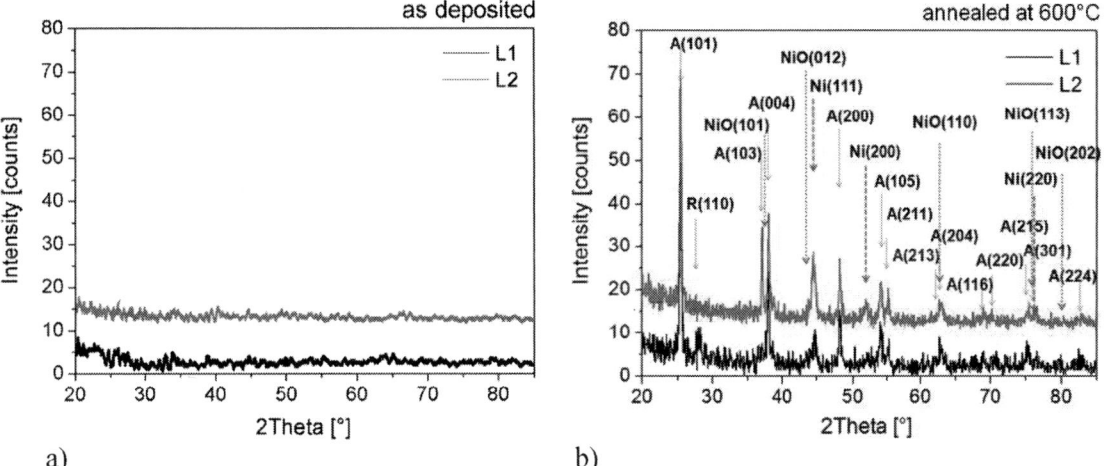

Fig. 1: XRD patterns of annealed thin films L1 (TiO_2/NiO/Al_2O_3) and L2 (NiO/TiO_2/Al_2O_3) at 600°C. Observed diffraction peaks belong to A (anatase), R (rutile), Ni (metallic nickel) and NiO.

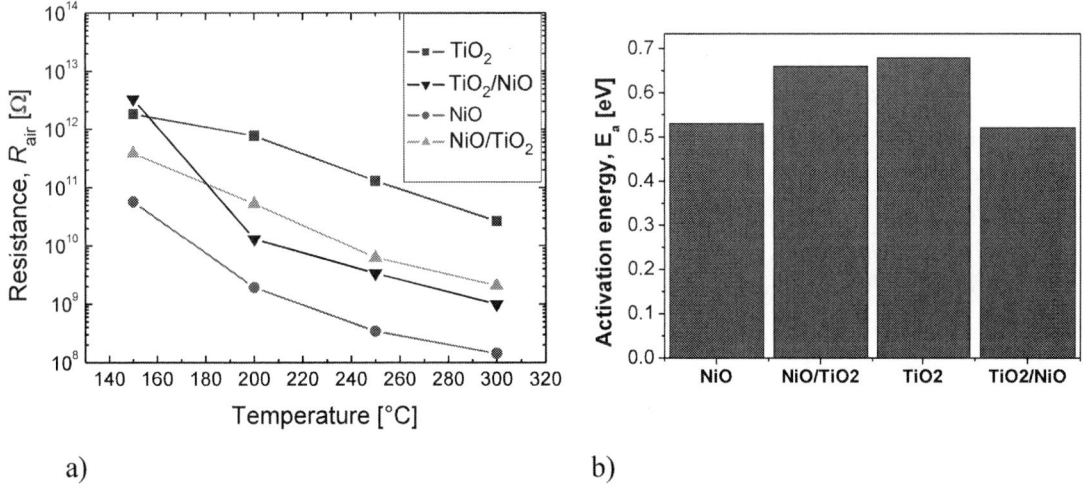

Fig. 2: Temperature dependencies of resistance of NiO-based films (a) and TiO_2-based thin films (a), effect of the temperature on resistance and the activation energies of all examined samples (b).

3.3 Gas sensing properties of MOX sensor structures towards hydrogen

Figure 3 represents the values of relative sensitivity in dependence on the operation temperature for all investigated samples. We can see that optimal operation temperatures are 200 and 250°C and they provide very high relative sensitivities for both material systems. The best results were reached at 200°C when the relative sensitivity is of the order of 10^4. It seems that this temperature is optimal for all examined samples. Another situation is in the case of the samples prepared as a combination of NiO and TiO_2. These samples show that the best operation temperature is 150°C though the responses to hydrogen appear very low in comparison with those prepared as NiO-based and TiO_2-based structures.

The last experiment was devoted to investigation of the effect of the size and the number of Pt interdigitated electrodes occurring under the gas sensing material. As we expected it was found that if the contact area was higher (the number of electrodes and their area), then we can record a higher value of relative sensitivity. This statement is valid for all measured samples.

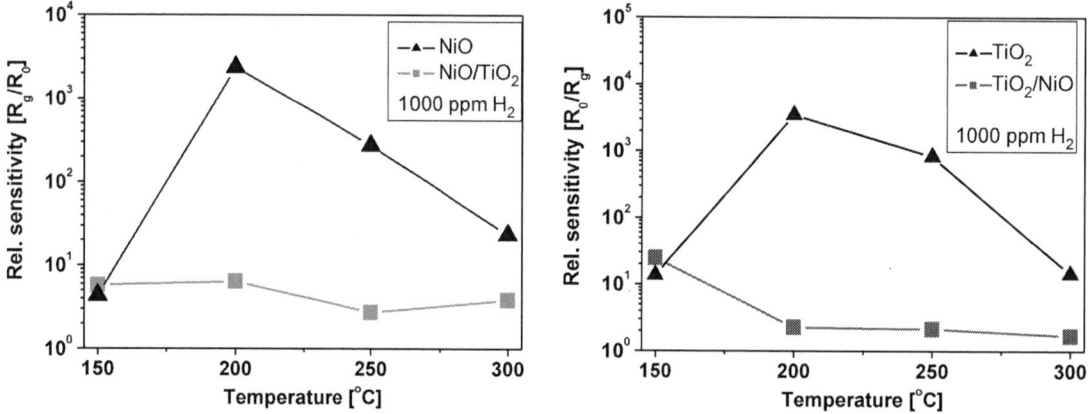

Fig. 3: Variations of relative sensitivity vs operation temperature for all investigated sensor structures.

4. Conclusion

In summary, our results demonstrate the potential of sputtered single and mixed metal oxides based on NiO and TiO_2 thin films to detect 1000 ppm of hydrogen and to operate at 200°C. The XRD measurements revealed polycrystalline character of annealed thin films at 600°C, whereas as-deposited samples were confirmed to be amorphous. $NiO/TiO_2/Si$ sensors were able to work at very low operation temperature of 150°C and their responses to hydrogen is higher in comparison with those prepared from the single TiO_2 films.

Acknowledgement

The work was supported by the Scientific Grant Agency of the Ministry of Education of the Slovak Republic and of the Slovak Academy of Sciences, No. 1/0828/16 and by project SAFESENS (agreement No. 621272-1) co-funded by the Slovak Republic and the ENIAC-JU.

References

[1]   N. Yamazoe, K. Shimanoe, New perspectives of gas sensor technology, *Sensors and Actuators B* **138** (2009) 100.

[2]   F. Yang, S.C. Kung, M. Cheng, J.C. Hemminger, R.M. Penner, *ACS Nano* **4** (2010) 5233.

[3]   N.M. Shaalan, T. Yamazaki, T. Kikuta, *Sensors and Actuators B* **156** (2011) 784.

[4]   I. Kosc, I. Hotovy, V. Rehacek, R. Grieseler, M. Predanocy, M. Wilke, L. Spiess, *Applied Surface Science* **269** (2013), 110.

ASDAM 2016, The 11th International Conference on Advanced Semiconductor
Devices And Microsystems, November 13-16, 2016, Smolenice, Slovakia

3-D Device and Circuit Electrothermal Simulations of Power Integrated Circuit Including Package

Aleš Chvála, Peter Benko, Patrik Príbytný, Juraj Marek and Daniel Donoval

Slovak University of Technology in Bratislava, Faculty of Electrical Engineering and
Information Technology, Institute of Electronics and Photonics,
Ilkovičova 3, 812 19 Bratislava, Slovakia
e-mail: ales.chvala@stuba.sk

In this paper we present an advanced methodology for effective 3-D device electrothermal simulation of power structures and power integrated circuits. The proposed electrothermal simulation is based on direct interconnection of a 3-D FEM thermal model and electrical circuit model of the device using a mixed-mode setup supported in Synopsys TCAD Sentaurus environment. This approach combines the speed and accuracy, and couples temperature nonuniformity to the active device electrothermal behaviour. The simulation results are compared with circuit electrothermal simulation which uses electrical RC network as an equivalent of the thermal system. The features and limitations of the methods are analyzed and presented.

1. Introduction

Today's device and circuit simulators are standard tools in the development, characterization, and optimization of electronic systems and devices. However, device finite element method (FEM) electrothermal simulations are very time consuming and require powerful hardware equipment, particularly for complicated 3-D structures. Circuit simulations have been limited to electronic models at a preselected temperature or the electrothermal feedback is provided only by an implemented simple single-pole *RC* equivalent network [1]. Introduction of an external Cauer or Foster *RC* network as thermal impedance model have been widely used in power electronic circuit simulations [2]. However, the *RC* elements need to be extracted from measurement and/or FEM thermal simulation. 3-D geometry based *RC* thermal mesh is described in [3]. Another widely used solution is based on coupling the separately solved thermal and electrical equations. The FEM is used for thermal simulation and a SPICE-like program is used for electrical simulation [4][5]. However, this complex solution requires a proper synchronization and data transfer between two different tools.

Our designed 3-D device electrothermal simulation is based on direct coupling between FEM thermal and circuit electrical simulation using mixed-mode setup supported by Synopsys TCAD Sentaurus environment [6][7]. The mixed-mode setup allows direct interconnection of a 3-D FEM thermal model of the whole system (semiconductor layers, package, and printed circuit board up to cooling assemblies) and an equivalent electrical temperature dependent circuit model of the device. The advantages of the proposed method are in the high speed of simulation and simplicity of implementation for complete, high complexity structure analysis.

2. Structure and electrothermal simulation methodology description

The device under investigation is a power IC realized in a BCD technology and packaged in ceramic DIL 24 package (Fig. 1a). The IC integrates two power LDMOSFETs

978-1-5090-3084-2/16 $31.00 © 2016 IEEE

(MOS$_{VER}$ and MOS$_{HOR}$), eight temperature sensors (TS) distributed along the die and four TS placed in the centre and edge of the transistor active areas (VC, VE, HC and HE) (Fig. 1b). The temperature sensors are created by the emitter-base junction of a bipolar transistor [8]. In our proposed methodology for effective 3-D electrothermal simulation, the 3-D FEM thermal model of the whole system and the electrical circuit model of the LDMOSFETs are directly connected using a mixed-mode setup in Sentaurus Device (Fig. 1c). This setup is built to allow heat flux from LDMOSFET circuit model to the thermal contacts of the 3-D thermal model via thermal nodes. The MOS$_{VER}$ and MOS$_{HOR}$ active areas are split into 8×8 segments. Each segment represents one equivalent temperature-dependent electrical circuit model. It allows taking into account distributed properties through active area. Heat generation and heat transfer are solved in the FEM thermal model. The temperatures on the thermal nodes are taken to drive the temperature dependent electrical parameters of the LDMOSFET circuit model. The short time of simulation is assured by solving only the heat equation in the 3-D FEM thermal model, reduction of the mesh optimized for thermal flow, and fast solving of the equivalent circuit electrical model.

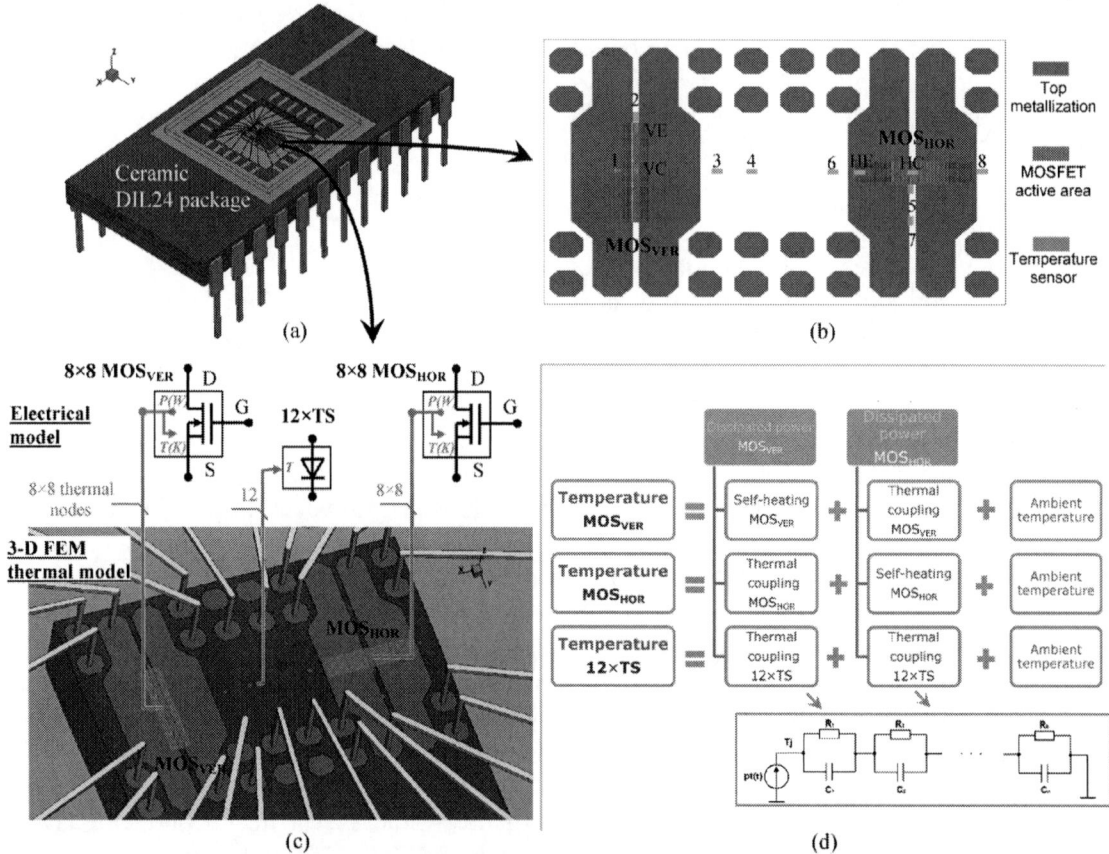

Fig. 1. (a) 3-D thermal model of power IC in ceramic DIL 24 package. (b) Layout of the analyzed power IC (c) Equivalent MOSFET circuit model connected to 3-D package thermal model by the thermal nodes in the mixed-mode simulation. (d) RC thermal network block schematic of the power IC.

The second proposed methodology is modified circuit electrothermal simulation which uses electrical *RC* network as an equivalent of the thermal system. The *RC* network for circuit electrothermal simulation is built to allow thermal coupling of all components in the power IC (Fig. 1d) [9]. Foster-type electrical equivalent networks represent the thermal impedance of self and mutual heating. The current sources proportional to the MOS_{VER} and MOS_{HOR} dissipated powers are fed into the thermal equivalent network. The T_j node voltages represent the junction temperatures of the components due to self-heating and thermal-coupling to the others devices. The resulting temperature of each components (MOS_{VER}, MOS_{HOR} and temperature sensors) is a sum of self-heating, thermal-coupling due to power dissipated in the others devices, and ambient temperature. The self-heating of the temperature sensors is neglected due to their very low power dissipation. The *RC* elements are extracted automatically by MATLAB from thermal impedance curves obtained by the 3-D FEM thermal simulations of the IC.

3. Simulation results

Fig. 2a shows a comparison of the measured and simulated transient evolution of MOS_{HOR} drain current and temperature of sensors integrated in analyzed power IC. Very good agreement between simulations and measurement confirms the validity of the proposed methodologies.

Our designed 3-D mixed-mode electrothermal simulation splits the structure along the active area into several parts, which allows analysis of the inhomogeneous distribution of temperature and electrical properties. The inhomogeneous distributions of temperature and current density distribution along MOS_{HOR} active area are shown in Fig. 2c. The lower temperature and lower current density at the edge segments are caused by more effective cooling of the structure edges compared with the central parts.

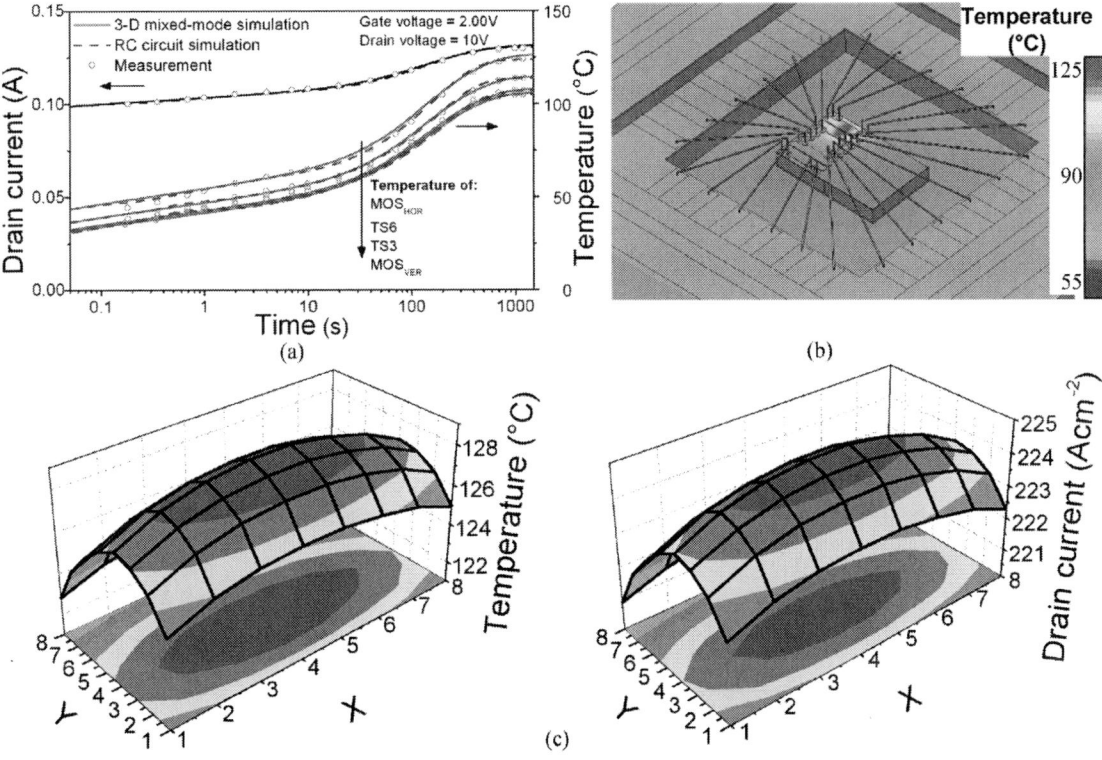

Fig. 2. (a) Transient characteristics of loaded MOS_{HOR}. (b) Temperature of the device. (c)) Temperature and current density distributions in MOS_{HOR} active area at time 1500 s obtained by 3-D mixed-mode simulation.

The circuit electrothermal simulation takes ~2 s. However, this simulation requires additional time for *RC* network extraction from thermal simulations of thermal impedances (~4 × 6 min) and *RC* elements extraction in MATLAB (~0.5 min). The 3-D mixed-mode electrothermal simulation based on direct coupling between the 3-D thermal FEM model and circuit electrical model takes ~6 min for the designed full structure model. The advantages of the proposed mixed-mode method are in the high speed of device simulation and simplicity of implementation for complete, high complexity structure analysis.

3. Conclusions

Effective 3-D electrothermal device simulation based on the direct coupling FEM thermal and circuit electrical simulation in the mixed-mode Sentaurus device setup was designed and verified. The designed methodology is developed for Synopsys TCAD Sentaurus environment and allows decreasing of the simulation time for complicated 3-D structures, power devices, and IC. The power IC was used to perform validation of the proposed electrothermal simulation. The simulation approach helps to assess the device properties by means of evaluating both temperature and current distributions in the structures operating under different conditions and topology. In comparison to *RC* circuit electrothermal simulations, the implemented 3-D thermal flow and distributed parameters of the MOSFET active area provide more realistic simulation results.

Acknowledgement

This work was supported in part by the ENIAC JU Project no. 621270/2013 eRamp, in part by Grant VEGA 1/0491/15, and in part by Grant APVV-14-0749 supported by Ministry of Education, Science, Research and Sport of Slovakia.

References

[1] H. Agarwal, S. Venugopalan, M.-A. Chalkiadaki, N. Paydavosi, J. P. Duarte, S. Agnihotri, et al., in *Proceedings of the SISPAD'13 Conference*, Glasgow, United Kingdom, 2013, p. 53.

[2] L. Nagy and V. Stopjaková, in *Proceedings of the TELFOR'13 Conference*, Belgrade, Serbia, 2013, p. 541.

[3] M. Pfost, C. Boianceanu, H. Lohmeyer, and M. Stecher, *IEEE Trans. on Electron Devices*, **60**, 2, p. 699, 2013.

[4] G. De Falco, M. Riccio, G. Breglio, and A. Irace, *Microelectron. Rel.* **54**, 9, p. 1833, 2014.

[5] A. Chvála, D. Donoval, J. Marek, P. Príbytný, M. Molnár, and M. Mikolášek, *IEEE Trans. Electron Devices*, **61**, 4, p. 1116, 2014.

[6] F. Nallet, L. Silvestri, C.-S. Yun, S. Holland, M. Rover, and T. Cilento, in *Proceedings of the ISPSD'14 Conference*, Waikoloa, HI, 2014, p. 334.

[7] A. Chvála, D. Donoval, A. Šatka, M. Molnár, J. Marek, and P. Príbytný, *IEEE Trans. Electron Devices*, **62**, 3, p. 828, 2015.

[8] M. Pfost, D. Costachescu, A. Podgaynaya, M. Stecher, S. Bychikhin, D. Pogany, and E. Gornik, *IEEE Trans. Semiconductor Manufacturing*, **25**, 3, p. 294, 2012.

[9] A. Raciti, D. Cristaldi, G. Greco, G. Vinci, and G. Bazzano, *IEEE Trans. Industrial Electronics*, **62**, 10, p. 6260, 2015.

ASDAM 2016, The 11th International Conference on Advanced Semiconductor
Devices And Microsystems, November 13-16, 2016, Smolenice, Slovakia

Analysis and modelling of the electric field
in the Gate oxide of 4H-SiC DMOSFET

Luigi Di Benedetto, Gian-Domenico Licciardo and Alfredo Rubino

Department of Industrial Engineering, University of Salerno,
Via Giovanni Paolo II, 132, 84084, Fisciano (SA), Italy
e-mail: ldibenedetto@unisa.it, gdlicciardo@unisa.it and arubino@unisa.it

For the first time, we report an analytical model of the electric field in the gate oxide of power DMOSFETs. The model describes the dependency of E_{OX} on doping concentrations and thicknesses of the drift-region as well as on the JFET-region geometry when reverse bias is applied to the device. There are not fitting parameters in the equations so that the model can be used as an a priori tool of analysis for reverse behaviour. Comparisons with numerical simulations are reported in order to show the goodness of the model results.

1. Introduction

Nowadays, 4H–SiC Double–implanted Metal-Oxide-Semiconductor Field-Effect-Transistors (DMOSFETs) are the very promising candidates for very–high voltage applications [1]. It has been shown that DMOSFETs can manage up to *15kV* achieving the physical limitations of SiC material [1]. In contrast to its silicon counterpart, in a standard 4H–SiC DMOSFET structure (see in Fig.1), the electric field at the p^+–base/n–drift junction can be *10* times higher due to the high critical electric field in 4H–SiC and the SiO_2 layer may break down or reduce its reliability, if the width of the JFET region, X_{JFET}, reaches a critical value. In fact, many efforts have been taken in order to reduce the electric field into the gate oxide, E_{OX}, for example introducing p^+–doped regions under the oxide [2]. Similar issues are also considered in planar IGBTs in which the size of the JFET–region affects both reverse and forward characteristics [3].

The aim of a DMOSFET design is to minimize both the series resistance in forward behaviour and the excursion of E_{OX} in reverse bias conditions [4]. Therefore, the design of such transistor requires an optimization of the doping concentration and geometries values of the drift–region, of the JFET–region size and of the oxide/semiconductor properties, i.e. thickness, maximum strength, interfaces defects and so on. The only available design tool useful to analyze the reverse behaviour is TCAD numerical simulations software [5]. Up to now, an analytical model able to describe thoroughly the dependency of E_{OX} on Drain–Source voltage, V_{DS}, which takes into account all geometrical and physical parameters and avoids the use of fitting parameters, has not been yet proposed. Such an analytical model can efficiently initiate the TCAD design allowing the calculation of an initial set of geometrical quantities by satisfying all relevant electrical specifications.

2. Analytical model

In our analysis we use an ordinary DMOSFET base structure of Fig.1.a with the main physical and geometrical parameters. Applying reverse bias conditions ($V_{GS}=0V$ and $V_{DS}>0V$), the oxide–semiconductor interface of the JFET–region is shielded by the p^+/n junctions [6]

978-1-5090-3084-2/16 $31.00 © 2016 IEEE   169

similar to Vertical–JFETs [7] or Junction Barrier Schottky diodes [8]. When a high V_{DS} is applied, p^+–well/n–DRIFT space charge regions overlap into the JFET–region, interact up to $(x,y)=(0,X_{JFET}+Y_W)$, as schematized in Fig.1.b [9], and reduce the electric field at the semiconductor–oxide interface, $E_{S/O}$, as well as E_{OX} due to Gauss law ($|E_{ox}|=\varepsilon_s\varepsilon_{ox}^{-1}|E_{S/O}|$).

From the oxide towards the substrate, the electric field distribution along the y-axis, E_Y, has a negative slope and covers an area of depth, $X_{JFET}+Y_W$, and of width, X_{JFET}, then, after this boundary ($y>X_{JFET}+Y_W$), it linearly extends in the DRIFT–region with a slope of qN_{DRIFT}/ε_S, starting with the minimum value at $(x,y)=(0,X_{JFET}+Y_W)$ equal to [7]:

$$E_{V_D} = \left| E_{pn} - \frac{qN_{DRIFT}X_{JFET}}{\varepsilon_s} \right|$$

(1)

where E_{pn} is the electric field at the junction edge of the vertical p^+–well/n–DRIFT junction and is equal to $-\sqrt{2qN_{DRIFT}\varepsilon_S^{-1}(V_{DS}+V_{bi})}$ or $-qN_{DRIFT}\varepsilon_S^{-1}Y_{DRIFT}-(V_{DS}-V_{PT})Y_{DRIFT}^{-1}$, respectively, for no-punch-through ($V_{DS}<V_{PT}$) or for punch-through ($V_{DS}\geq V_{PT}$) conditions, $V_{PT}=0.5qN_{DRIFT}\varepsilon_S^{-1}Y_{DRIFT}^2-V_{bi}$ the punch–through voltage, $V_{bi}=V_T\ln(N_{DRIFT}N_W n_{i,eff}^{-2})$ the p^+–n built–in voltage, and $n_{i,eff}$ the effective intrinsic concentration. Taking into account the previous results, in order to evaluate $E_{S/O}$ and E_{OX}, the following 2D Poisson's equation must be solved in the X_{JFET}–width and ($X_{JFET}+Y_W$)–depth regions:

$$\nabla^2\phi(x,y)=-\frac{qN_{DRIFT}}{\varepsilon_S}, \quad \phi(0,X_{JFET}+Y_W)=\phi^*, \quad \left.\frac{\partial\phi}{\partial y}\right|_{(0,X_{JFET}+Y_W)}=E_{V_D}$$

(2)

where $\phi^*=X_{JFET}0.5(E_{pn}+E_{V_D})-V_{bi}+V_{FB}+\phi_s$, $V_{FB}=WF-\chi-0.5E_Gq^{-1}+V_T\ln(N_{DRIFT}n_i^{-1})$ is the flat–band voltage, WF the Gate work function, χ 4H–SiC electron affinity, E_G 4H–SiC band-gap and Φ_S the potential voltage at the oxide/semiconductor interface. From the superposition principle, the potential can be written as $\Phi(x,y)=\Phi_L(x,y)+\Phi_P(y)$ where $\Phi_L(x,y)$ is solved as a Laplace problem, having as boundary condition $\phi_L(x,0)=\phi_L(x,X_{JFET}+Y_W)=0$ and $D_x\phi_L|_{x=0}=0$ [9]-[11], and $\Phi_P(y)$ is a 1–D Poisson problem solution that takes into account the charge in the JFET-region with $\phi_P(0)=\phi_s$ [12]. From their derivative the electric field distribution is as follows:

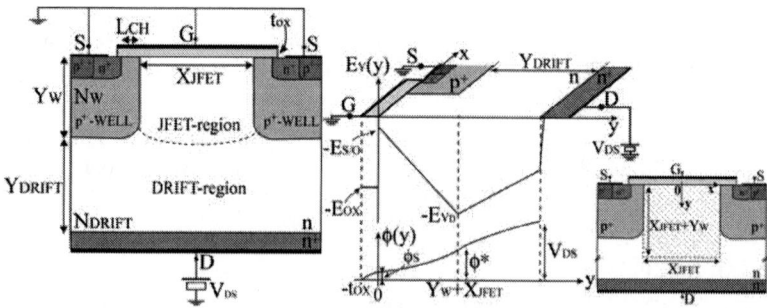

Fig. 1. a) Cross-section view of a DMOSFET structure with the main physical and geometrical parameters. b) Cutaway of DMOSFET structure and schematic of $E_Y(0,y)$ and $\Phi(0,y)$ highlighting the main model quantities. In the insert the domain region for 2D-Poisson problems is the dashed area.

$$E_Y(0,y) = -\frac{\pi k_I}{X_{JFET}+Y_W}\cos\left(\frac{\pi}{X_{JFET}+Y_W}y\right) + \frac{qN_{DRIFT}}{\varepsilon_S}y - k_2 \tag{3}$$

where $\qquad k_I = -(X_{JFET}+Y_W)\pi^{-1}\left[E_{V_D}+qN_{DRIFT}\varepsilon_S^{-1}(X_{JFET}+Y_W)-k_2\right]$ and

$k_2 = \left[E_{pn}X_{JFET}+0.5qN_{DRIFT}\varepsilon_S^{-1}\left(2Y_WX_{JFET}+Y_W^2\right)-V_{bi}+V_{FB}\right]\left(Y_W+X_{JFET}\right)^{-1}$. Finally, using (3) at the semiconductor–oxide interface, namely $(x,y)=(0,0)$, the electric field in the gate oxide can be expressed:

$$|E_{OX}| = \frac{\varepsilon_S}{\varepsilon_{OX}}\left|E_{pn}+\frac{qN_{DRIFT}}{\varepsilon_S}Y_W-2k_2\right| \tag{4}$$

3. Results

In Fig. 2 and 3 comparisons with numerical simulation results [5] are reported to validate the model. DMOSFET structure of Fig.1 has $Y_W=0.6\mu m$, $N_W=10^{18}cm^{-3}$, $t_{OX}=50nm$, $L_{CH}=0.5\mu m$, $Y_{SOURCE}=0.2\mu m$ and $qWF=4eV$, which are often used in 4H–SiC DMOSFETs [14]–[17]. X_{JFET}, N_{DRIFT} and Y_{DRIFT} values are changed. Furthermore, physical models include field and doping dependent carriers mobility, SRH recombination, incomplete ionization and bandgap narrowing with parameters taken from [7],[18]-[22]. Finally, all the following curves have $V_{GS}=0V$ and are swept from $V_{DS}=0V$ up to the blocking voltage, V_{BL}, which is defined as

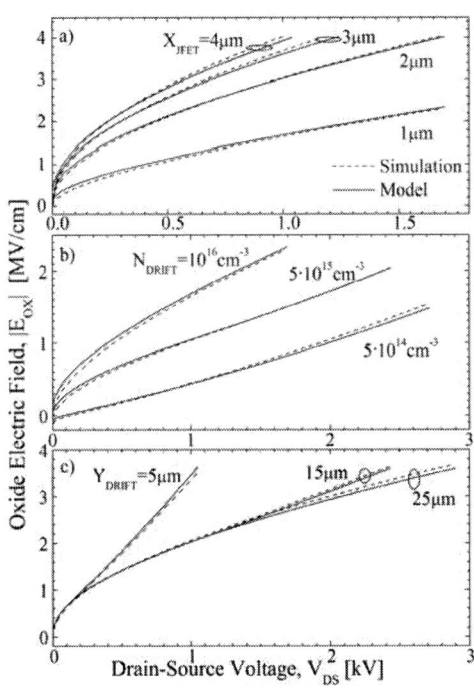

the V_{DS}-value for which either the avalanche breakdown of p^+–well/n–DRIFT junction appears, i.e. $E_{pn}=E_{crit}\simeq 2.49\cdot10^6/\left(5-0.25\log N_{DRIFT}\right)$ V/ cm, or the oxide electric field is $E_{OX}=E_{OX,lim}=4MV/cm$, which permits to satisfy the oxide reliability concerns [23].

Fig.2 shows $|E_{OX}|-V_{DS}$ curve for different values of a) X_{JFET}, b) N_{DRIFT} and c) Y_{DRIFT}. The shielding effect of p-n junction is clear observing Fig.2.a: for $X_{JFET}<2\mu m$, the blocking voltage is due to the breakdown of the p/n junction, whereas, for $X_{JFET}\geq2\mu m$, E_{OX} achieves the maximum value and, hence, the blocking voltage reduces. Similar effect can be obtained with N_{DRIFT}: the space charge region is wider for low N_{DRIFT} reducing the electric field under the oxide. The accuracy of the model is also described observing the slope of E_{OX} vs. V_{DS} of Fig.2.c: the DRIFT-region changes from punch-through to no-punch-through condition increasing Y_{DRIFT} from $5\mu m$ to $25\mu m$.

Fig. 2. Comparisons between model and numerical simulation of $|E_{OX}|-V_{DS}$ for different device structure of Fig.1 at $V_{GS}=0V$ and $T=300K$. a) $Y_{DRIFT}=15\mu m$, $N_{DRIFT}=10^{16}cm^{-3}$ and various values of X_{JFET}. b) $Y_{DRIFT}=15\mu m$, $X_{JFET}=1\mu m$ and various values of N_{DRIFT}. c) $X_{JFET}=2\mu m$, $N_{DRIFT}=5\cdot10^{15}cm^{-3}$ and various values of

In Fig.3 it is reported X_{JFET} value as function of the blocking voltage, V_{BL}, for a fixed N_{DRIFT}. To get these curves, (4) has been manipulated obtaining an expression for the JFET-region width:

$$X_{JFET} = Y_W \left(E_{pn} + \varepsilon_{ox}\varepsilon_s^{-1} \left| E_{OX} \right| \right) \left[E_{pn} - \varepsilon_{ox}\varepsilon_s^{-1} \left| E_{OX} \right| \right]^{-1}$$

. For a fixed N_{DRIFT}, the increase of the blocking voltage due to a decrease of X_{JFET} is accurately described as well as the trend of the curves.

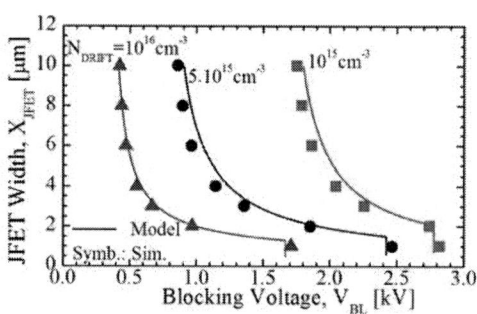

Fig. 3. Comparisons between model and simulations of V_{BL}–X_{JFET} at $V_{GS}=0V$, $T=300K$ and for $Y_{DRIFT}=15\mu m$ and different values of N_{DRIFT}.

4. Conclusions

The proposed analytical model predicts the electric field dependency on V_{DS} in reverse bias conditions and takes into account both geometrical and physical parameters without using any fitting parameters. Comparisons with TCAD numerical simulations show the accuracy of the model for different DMOSFETs device structures. Beside to give a further description of the device physics under reverse bias conditions, it can be useful at the first step of the DMOSFET power device design.

References

[1] S. Allen *et al.*, *Materials Science Forum*, **821–823**, 701–704, 2015.

[2] Q. Zhang *et al.*, *Proceedings of the 27th ISPSD*, 89–92, 2015.

[3] S. Chowdhury *et al.*, *IEEE Electron Dev. Lett.*, **37**, 3, 317–320, 2016.

[4] L. Di Benedetto *et al.*, *IEEE Electron. Dev. Lett.*, **37**, 2016. *Doi: 10.1109/LED.2016.2613821*

[5] SILVACO Int., Santa Clara, CA, ATLAS User's Manual, 2005, Ver. 5.10R.

[6] B. J. Baliga, Fundamentals of Power Semiconductor Devices, Springer–Verlag, 2008.

[7] S. Bellone *et al.*, *Solid State Electron.*, **109**, 17–24, 2015.

[8] S. Bellone *et al.*, Solid State Electron., *120, 6–12, 2016.*

[9] S. Bellone *et al.*, *Procedings of the International Semiconductor Conference (CAS)*, **2**, 405-408, 2010.

[10] S. Bellone *et al.*, *IEEE Trans. on Electron Dev.*, **56**, 12, 2902-2911, 2009.

[11] S. Bellone *et al.*, *IEEE Trans. on Instrum. and Meas.*, **57**, 6, 1112-1117, 2008.

[12] L. Di Benedetto *et al.*, *IEEE Electron. Dev. Lett.*, **63**, 6, 2474–2481, 2016.

[13] C. Bulucea et al., *Solid State Electron.*, **30**, 1227–1242, 1987.

[14] A. Saha *et al.*, *IEEE Trans. on Electron. Dev.*, **54**, 10, 2786–2791, 2007.

[15] G. D. Licciardo *et al.*, *IEEE Trans on Power Electron.*, **30**, 10, 5800–5809, 2015.

[16] G. D. Licciardo *et al.*, *IEEE Trans. on Electron. Devices*, **63**, 4, 1783-1787, 2016.

[17] K. Matocha *et al.*, *Materials Science Forum*, **778–780**, 903–906, 2014.

[18] S. Bellone *et al.*, *IEEE Trans on Power Electron.*, **29**, 5, 2174–2179, 2014.

[19] M. L. Magherbi *et al.*, *Solid State Electron.*, **109**, 12-16, 2015.

[20] S. Bellone *et al.*, *IEEE Trans on Power Electron.*, **29**, 1, 514–521, 2014.

[21] L. Di Benedetto *et al.*, *IEEE Electon. Dev. Lett.*, **35**, 2, 244-246, 2014.

[22] S. Bellone *et al.*, *IEEE Trans. on Electron Dev.*, **59**, 9, 2546-2549, 2012.

[23] L. Di Benedetto *et al.*, *IEEE Elect. Dev. Lett.*, **63**, 9, 3795–3799, 2016.

ASDAM 2016, The 11th International Conference on Advanced Semiconductor
Devices And Microsystems, November 13-16, 2016, Smolenice, Slovakia

Impact of repetitive UIS on modern GaN power devices

Juraj Marek[1], Ľubica Stuchlíková[1], Martin Jagelka[1,2], Aleš Chvála[1], Patrik Príbytný[1], Martin Donoval[1,2] and Daniel Donoval[1]

[1]Institute of Electronics and Photonics, Slovak University of Technology in Bratislava
812 19 Bratislava, Slovakia,
[2]NanoDesign, s.r.o, Drotárska 19a, 811 04 Bratislava, Slovakia
e-mail: juraj.marek@stuba.sk

In this paper we present the results from repetitive Unclmped Inductive Switching - UIS measurements on power GaN HEMTs. Experimental analysis was performed on two types of power devices - normally ON and OFF HEMTs. UIS test was used to simulate real switching conditions in which power device has to operate. Analysis has shown that charge trapping effects on interfaces and in bulk layer have strong impact on electrical performance of devices. DLTS study was performed on virgin and short and long stressed sample for better understanding of trapping effects.

Keywords high-electron mobility transistor (HEMT), UIS, inductive switching, DLTS,

1. Introduction

Recent progress in GaN-based high-electron mobility transistors (HEMTs) has confirmed them to be the leading transistor technology for future high-power devices at high-frequency operation utilizing their excellent electronic properties, high electron saturation velocity, and high breakdown voltage [1]. However, one of the critical disadvantages of pure GaN power devices compared with their Si counterparts is their lack of unclamped inductive-switching (UIS) capabilities, and this problem has not received the proper attention it deserves. It has been reported [2,3] that cascade GaN HEMT devices exhibit intrinsic UIS capability even it is low and mainly due to capacitive charging it might be sufficient in most switching applications. However, for normally-off power GaN devices, that becoming more and more commercially available situation is more complicated and protective elements are needed to prevent destruction of device. The lack of UIS capability of GaN based devices is due to absence of avalanche breakdown and defects/traps invoked breakdown mechanism. In this paper we would like to show that even in case when power GaN HEMT is protected by clamping diode, repetitive exposure to high voltage in off state and high current in on state together with significant self heating may cause degradation of electrical parameters like threshold voltage shift, leakage current and on state resistance increase.

2. Experimental samples

For our experimental analysis we used two types of commercially available power GaN devices. First, normally on device, with threshold voltage $V_{th} = -6V$. Second device was normally off HEMT with threshold voltage $V_{th} = 1$ V. Maximum rated drain current for both devices was $I_D = 20A$. Both devices were fabricated on Si wafers with AlGaN buffer, AlGaN barrier and GaN channel layer. Normally on device was MIS HEMT with SiN gate dielectric. For normally off device p-GaN technology was used to form gate contact.

978-1-5090-3084-2/16 $31.00 © 2016 IEEE

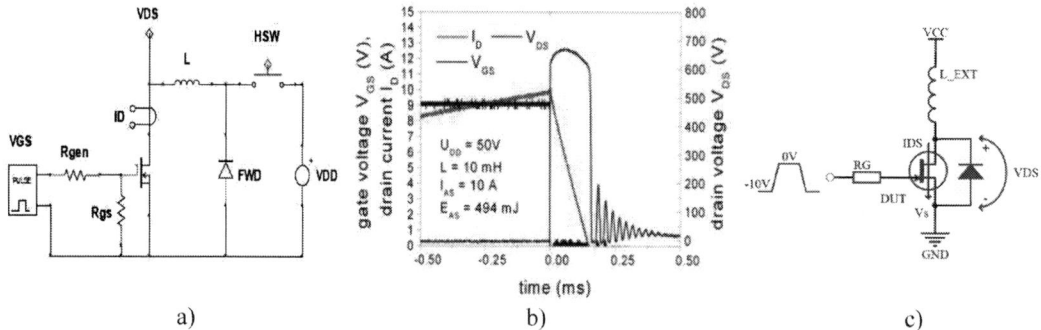

Figure 1 Basic UIS test circuit and typical current and voltage waveforms *of Si MOSFET* under UIS test conditions.

3. Inductive switching stress

When switching a resistive load, the *I–V* loci follow the Ohm's law and are relatively simple. However, when switching a reactive load, the switching becomes complicated. Voltage and current can have phase differences, resulting in high voltage and high current to occur at the same time [2]. Typical turn-off transients for Si-FET are shown in the Fig.1b. Therefore, when no additional clamping element is added to protect power transistor, switching of inductive loads represent very stressful condition for power switch that may leads to catastrophic failure and malfunction of whole system. Especially in the systems where significant parasitic inductances cannot be sufficiently restrain. In case when protective antiparallel diode is added to power element whole energy during discharging period of inductance is handled by this diode causing significant self heating. This heating then also causes increase of power HEMT operating temperature. High induced voltage on this protective element is also present on power HEMT due to antiparallel configuration (Fig. 1c). Both increased temperature and high voltage represent significant stress. Test circuit parameters were: supply voltage V_{DD} = 30V, max on state current I_{AS} = 5A, inductance of switch load L= 0.3mH, breakdown voltage of protective element V_{AV} = 125V (strongly temperature dependent), typical length of discharging (avalanching) period t_{av}= 35-40 µs. Duty cycle of stress pulses was set to 10% to keep DUT temperature under 130°C.

4. Results and discussion

First normally on transistors were subjected to repetitive high voltage pulsing stress. Figure 2 reports representative I_D–V_{DS} in pulsed conditions and *C–V* curves measured on the analyzed samples before and after stressing. One can clearly see that stress caused significant shift of threshold voltage, increase of on resistance and increase of drain-source leakage current. Surprisingly impact on gate-source capacitance was very low. This led us to the conclusions that in accordance whit literature shift of parameters was due to the injection of electrons towards traps located in the gate–drain access surface [3,4]. The existence of surface states, may lead to the so-called virtual gate effect, which determines a strong increase in the on-resistance. Change of leakage current can be explained by the trapping of electrons in the buffer; this may occur either in the semi-on state, due to the presence of hot electrons, [5] or in the off-state, due to the exposure to high drain–substrate bias [6]. We think that the injection

978-1-5090-3084-2/16 $31.00 © 2016 IEEE   174

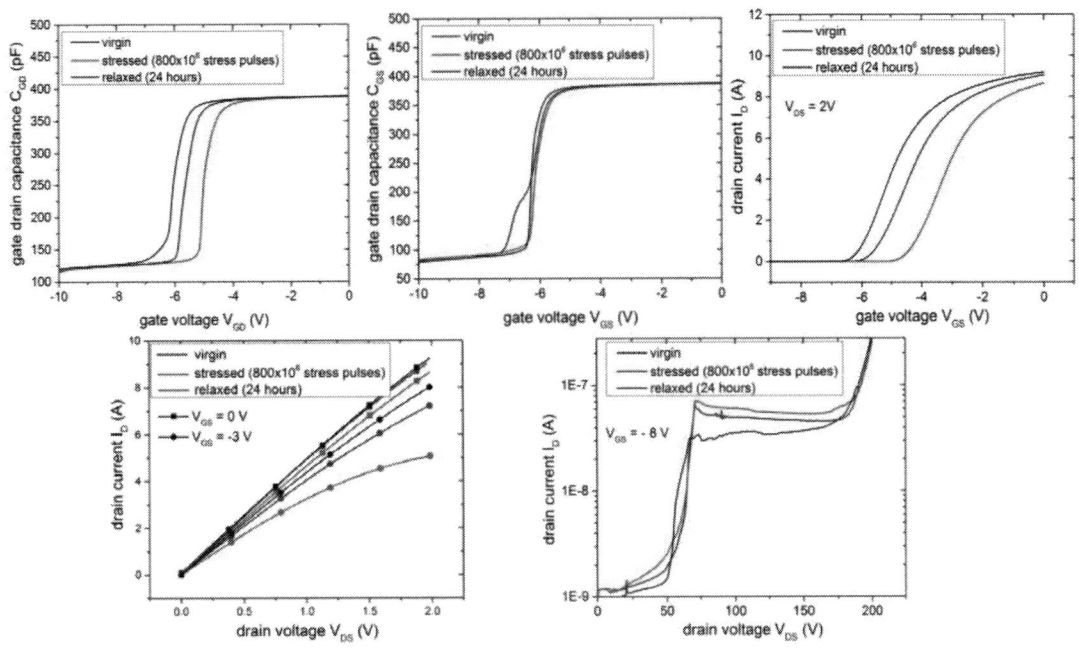

Figure 2 Measured gate-drain C-V, transfer and output characteristics of normally on GaN MIS-HEMT before and after multipulse UIS stress with protective antiparallel diode.

of electrons in the AlGaN region under the gate was minimal - explaining low impact on C_{GS}. Next set of measurements was performed on normally off devices. Representative I_D–V_{DS} and C–V curves measured on the analyzed samples before and after stressing are shown in the Fig. 4. Fact that no impact of stress on threshold voltage and only very small impact on capacitances C_{GS}, C_{GD}, and C_{DS} was observed indicates that trapping effects on barrier interfaces and close to gate electrode were low. This was most likely achieved by using proper pasivation layers. Increase of R_{DSon} good visible from output characteristics (Fig. 4a) is due to trapped charge in the buffer layer. To explain observed change in leakage current DLTS analysis was performed on virgin sample and after stress. From DLTS spectra we can determine three dominant peaks attributable to electron trap ET1, and hole traps HT1 and HT2. After stress, concentration of electron traps ET1 was slightly decreased while concentration of hole traps HT1 was increased and HT2 was decreased. From this

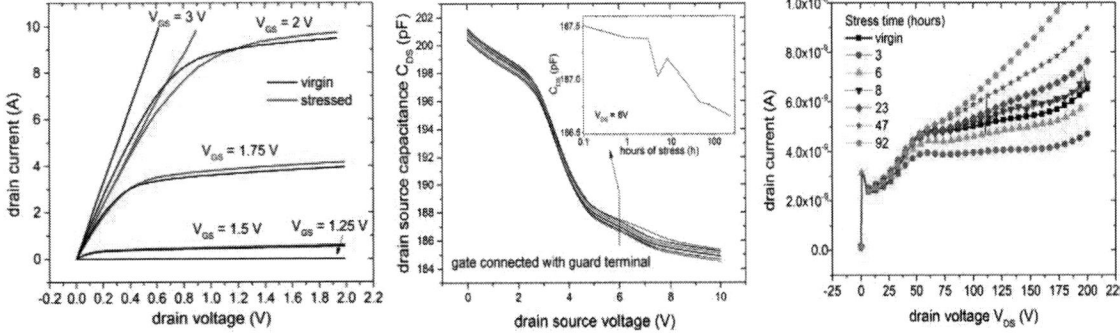

Figure 3 Measured gate-drain C-V, transfer and output characteristics of normally on GaN MIS-HEMT before and after multipulse UIS stress with protective antiparallel diode.

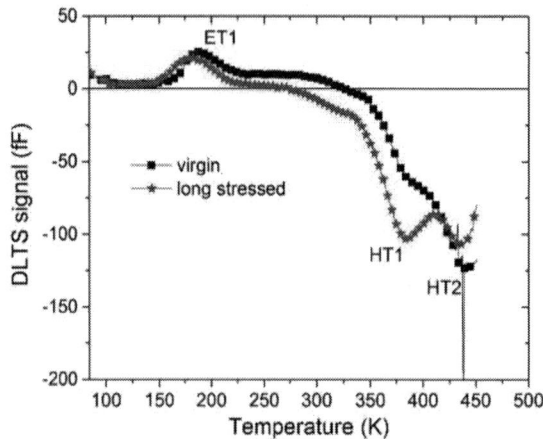

Figure 4 Measured DLTS spectra for virgin, and stressed sample

measurement it is clear that samples are degraded by stress, new traps are generated by stress and this leads to increase of leakage current. We suppose that first decrease of leakage current can be attributed to short trapping effects and that this effect is reversible.

Conclusion

Impact of repetitive high voltage switching on electrical performance of normally on and off power GaN HEMTs was studied. Observed shift of electrical parameters indicate that a strong trapping effects occurs in buffer layer or on barrier interfaces. DLTS analysis was performed to study traps distribution and concentration for virgin and stressed samples. It is clear that trapping effects have significant impact on switching performance of power GaN HEMTS and have to be considered during power switches usage.

Acknowledgement

This work was supported in part by Grant VEGA 1/0491/15, in part by Grant APVV-14-0749 supported by Ministry of Education, Science, Research and Sport of Slovakia and in part by project 'PowerBase'. This project has received funding from the Electronic Component Systems for European Leadership Joint Undertaking under grant agreement No 662133. This Joint Undertaking receives support from the European Union's Horizon 2020 research and innovation program and Austria, Belgium, Germany, Italy, Netherlands, Norway, Slovakia, Spain and United Kingdom.

References

[1]   N. Kaminski, Proc. IEEE PEAC, pp. 1-9, 2009.
[2]   Chen, Wen-Yi et al, IEEE Trans. on Device and Materials Reliability, Feb. 2012
[3]   M. Meneghini et al., Electronics, vol. 5, no. 14, pp. 1–8, 2016.
[4]   M. Ťapajna et al., Appl. Phys. Lett., vol. 107, p. 193506, 2015.
[5]   M. Meneghini et al., Appl. Phys. Lett. 104, 143505 (2014)
[6]   D. Jin et al, Proc. IEEE 24th ISPSD ICs, 2012, p. 333

ASDAM 2016, The 11th International Conference on Advanced Semiconductor
Devices And Microsystems, November 13-16, 2016, Smolenice, Slovakia

Post-deposition annealing and thermal stability of integrated self-aligned E/D-mode n[++]GaN/InAlN/AlN/GaN MOS HEMTs

M. Blaho[1], D. Gregušová[1], Š. Haščík[1], A. Seifertová[1], M. Ťapajna[1], J. Šoltýs[1], A. Šatka[2], L. Nagy[2], A. Chvála[2], J. Marek[2], J. Priesol[2] and J. Kuzmík[1]

[1]Institute of Electrical Engineering Slovak Academy of Sciences, Dúbravska cesta 9, 841 04 Bratislava, Slovakia

[2]Institute of Electronics and Photonics Slovak University of Technology, Ilkovičova 3, 812 19 Bratislava, Slovakia

E-mail: michal.blaho@savba.sk.

We describe technology and evaluate thermal performance of enhancement/depletion (E/D)-mode n[++]GaN/InAlN/AlN/GaN HEMTs with a self-aligned metal-oxide-semiconductor (MOS) gate processing, where n[++]GaN layer was etched away only under the gate for E-mode and for D-mode stay intact. Gate contacts were isolated using a dielectric layer deposited at low temperature through an e-beam resist to retain the self-aligned approach. Threshold voltage of the as deposited E- and D-mode HEMTs was +0.6 V and −2.4 V, respectively. After post-deposition annealing (PDA) at 300 °C in N_2 atmosphere the threshold voltage has been changed to +3 V and -4,4 V for E- and D-mode HEMTs, respectively. These effects were explained by decreasing density of deep interface states in the D-mode HEMTs and decreasing surface donors at the semiconductor-oxide interface in case of the E-mode HEMTs. After PDA, electrical performance of both types of transistors was evaluated from room temperature to 150 °C. At elevated temperatures, injection and trapping of electrons from the gate metal to the oxide was found in D-mode HEMTs, while emission of electrons from the oxide-semiconductor interface states was crucial for the E-mode ones.

1. Introduction

GaN-based electronics became very promising for high-performance digital and analog circuits applications. High operation temperature is one of the main advantages of these materials; therefore studies of thermal stability of these devices are desirable. Next challenge is design of epi-stricture for enhancement (E)-mode GaN high electron mobility transistors (HEMTs). In digital circuit it is necessary to design structures with both types of (E- a D-mode) transistors.

We recently suggested approach of E-mode InAlN/AlN/GaN GaN MOS HEMTs capped with n++GaN. Only a single epi-structure growth was needed and similarly only a single lithographic step for a gate recessing, oxide deposition and metallization was applied [1]. This self-aligned design was shown to eliminate the gate space charge extension towards the drain and consequently, in future fast devices the parasitic drain delay component can be minimized [2]. Later we published on logic invertor where both E and D-mode InAlN/GaN HEMTs were integrated on one wafer [3]. However, no posted-deposition annealing (PDA) was applied by

978-1-5090-3084-2/16 $31.00 © 2016 IEEE   177

processing the invertor. In the present study we investigate the effect of PDA on integrated E/D devices, as well as performance at elevated temperatures.

2. Sample preparation

A 6 nm n^{++}GaN (2×10^{20} cm^{-3})/ 1.5 nm In$_{17}$Al$_{0.83}$N/ 1 nm AlN/ 950 nm GaN heterostructure has been grown on sapphire by MOCVD at EPFL Lausanne [4]. Mesa etching was performed by Ar milling; Ti/Al/Ni/Au ohmic contacts were formed 10μm apart and annealed at 850 °C. By lithography we applied patterned gate with 2μm length and 60μm width. Gate recessing of the E-mode HEMT was performed using SiCl$_4$:SF$_6$ (25 %) gas mixture in the inductive-coupled plasma reactive-ion etching (ICP RIE) system. Etching was performed for 2.5 minutes at 2.7 Pa keeping about 75 V self-induced bias. By using an atomic-force microscopy (AFM) the etch initiation time, surface roughness inside and outside of the mesa region and the selectivity of the GaN cap etching against InAlN are studied directly in gate trenches [3]. Consequently we avoid possible ambiguity in determining the etching selectivity if dummy bare (oxidized) InAlN sample was used instead. Subsequently a 10-nm thick Al$_2$O$_3$ was grown in an atomic-layer deposition (ALD) system using trimethylaluminium and water at 100°C on the sample which was etched for 2.5 min.

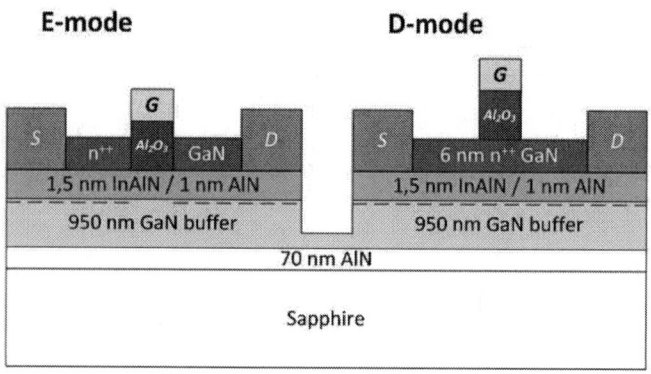

Figure 1 *Schematic drawing of the integrated self- aligned E/D-mode n^{++} GaN/InAlN/AlN/GaN MOS HEMTs.*

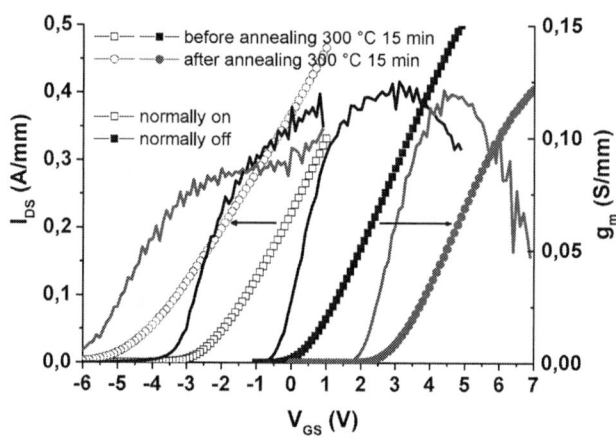

Figure 2 *Comparison of transfer and transconductance characteristics of virgin and and annealed E/D-mode MOS HEMTs integrated on the same chip at $V_{DS} = 10$ V. Arrows indicate evolution of transfer characteristics after PDA.*

Ni/Au metal together with the oxide was lifted-off to form the self-aligned MOS gates. By processing the D-mode HEMTs the gate recessing was skipped, see Figure 1. Sample was finished by post- deposition annealing at 300 °C in nitrogen atmosphere. Thermal stability was tested at selected temperatures up-to 150 °C.

3. Results

DC performance was measured by Keitley 4200 semiconductor characterization system. First we evaluated transistor properties before and after PDA, see Figs. 2 and 3. Before PDA we extracted the value of the threshold voltage V_T to be + 0.6 V and -2.4 V for E and D mode HEMT, respectively. The maximal drain current I_{dsmax} was about 0.45 A/mm for both of devices. This means that the gate recessing increased V_T by ~3 V without compromising I_D because of damage-free RIE. Similarly, transcondutances g_m of E/D-mode devices were almost the same (figure 2). After annealing at 300 °C in nitrogen atmosphere for 15 min the threshold voltage shift was observed. For normally-off transistor the threshold voltage increases as presented earlier [1] and reach value + 3 V. This shift after PDA was caused by reduced density of surface donors at the oxide/InAlN interface, which leaves polarization charge at the InAlN surface almost uncompensated [1]. However, for normally-on transistors the threshold voltage was decreased and reached a value of − 4.4 V. In this case we suppose

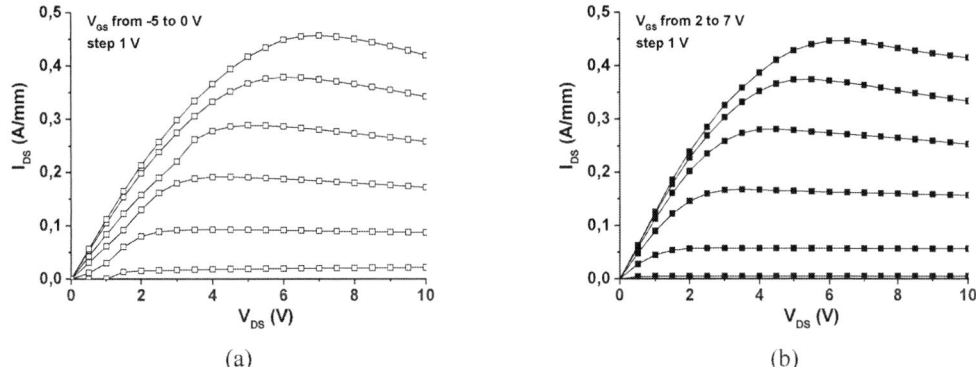

(a)   (b)

Figure 3 Output characteristics of E/D-mode MOS HEMTs at room temperature after PDA.

decreasing density of interface states at the oxide/GaN interface. We note that the high polarization charge at the InAlN surface which is not a case of the GaN surface is responsible for different behavior of E/D-mode devices. Next we evaluated temperature dependence of integrated E/D-mode HEMTs from

room temperature (20 °C) up-to 150 °C in ambient air. As temperature rises, for D-mode HEMT we observed a threshold voltage increase due to enhanced electron injection from the negatively biased gate with subsequent trapping in the oxide (Fig. 4a) [5]. For E-mode HEMT however the value of transconductance as well as of the threshold voltage decreases (see Fig. 4(b)). We suppose emission of electrons from the oxide/InAlN interface dominates in this case [6]. Finally we note that changes of the V_{th} due to temperature increase were reversible (not shown).

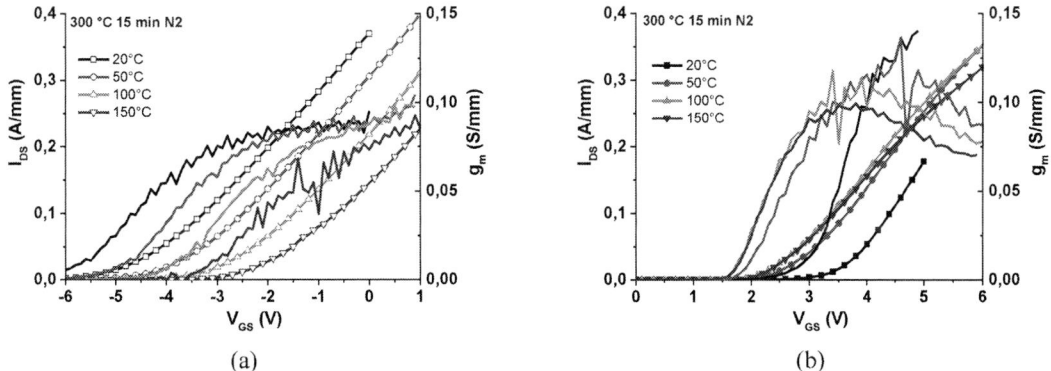

(a)   (b)

Figure 4 Transfer and transconductance characteristics of E/D-mode MOS HEMTs at different temperature at $V_{DS} = 10\ V$

4. Conclusions

We have proposed and developed technology of E/D-mode n^{++}GaN/InAlN/AlN/GaN MOS HEMTs integrated on the same chip, suitable for preparation of future fast mixed-signal electronic circuits. Threshold voltage of the E-mode HEMT is adjusted by the gate recessing while gate insulation of the planar HEMT provides D-mode operation. Only one resist mask was used for the plasma etching, oxide deposition and metallization of the gate, providing the self-aligned approach. We show that for the E-mode HEMT the PDA decreases density of surface donors at the InAlN surface and increases V_{th}. After the same treatment V_{th} of D-mode HEMT decreases due to decreased density of GaN/oxide interface states. On the other hand, device operation at elevated temperature increases V_{th} of D-mode HEMT due to injection of electrons from the gate contact, while V_{th} of E-mode device decreases due to emission of electrons from the interface states at the InAlN/oxide. Thermal stability measurement revealed considerable changes of threshold voltage, degree of which can be modified by changing the oxide thickness.

Acknowledgement

We acknowledge J.-F. Carlin, N. Grandjean from EPFL Lausanne and G. Konstantinidis from IESL FORTH Crete for a support. This work was supported by the Slovak Research and Development Agency under the contract No. 15-0673.

References

[1] Blaho M, Gregušová D, Haščík Š, Jurkovič M, Ťapajna M, Fröhlich K, Dérer J, Carlin J-F, Grandjean N, and Kuzmik J 2015 Self-aligned normally-off metal-oxide-semiconductor n++GaN/InAlN/GaN high electron mobility transistors Phys. Status Solidi A 212 1086-1090

[2] Palankovski V and Kuzmik J 2012 A Promising New n^{++}-GaN/InAlN/GaN HEMT Concept for High-Frequency Applications *ECS Transactions* **50** 291-296

[3] Blaho, M., Gregušová, D., Haščík, Š., Seifertová, A., Ťapajna, M., Šoltýs, J., Šatka, A., Nagy, L., Chvála, A., Marek, J., Carlin, J., Grandjean, N., Konstantinidis, G., Kuzmík, J.,: Technology of integrated self-aligned E/D-mode n++GaN/InAlN/AlN/GaN MOS HEMTs for mixed-signal electronics. Semicond. Sci Technol. 31 (2016) 065011

[4] M. Gonschorek, J. -F. Carlin, E. Feltin, M. A. Py, and N. Grandjean, Appl. Phys. Lett. 89, 062106 (2006).

[5] M. Ťapajna1, M. Jurkovič, L. Válik, Š. Haščík, D. Gregušová, F. Brunner, E.-M. Cho and J. Kuzmík: Bulk and interface trapping in the gate dielectric of GaN based metal-oxide-semiconductor high-electron-mobility transistors, Appl. Phys. Lett. 102, 243509 (2013)

[6] M. Miczek, C. Mizue, T. Hashizume and B. Adamowicz: Effects of interface states and temperature on the C-VC-V behavior of metal/insulator/AlGaN/GaN heterostructure capacitors, J. Appl. Phys. 103, 104510 (2008)

ASDAM 2016, The 11th International Conference on Advanced Semiconductor
Devices And Microsystems, November 13-16, 2016, Smolenice, Slovakia

Scaling and Traps Induced Degradation of Cutoff Frequency in GaN HEMT

B. C. Ubochi, S. Faramehr, K. Ahmeda, P. Igić, and K. Kalna

Electronic Systems Design Centre (ESDC), College of Engineering, Swansea University
Bay Campus, Fabian Way,Swansea SA1 8EN, Wales, United Kingdom

e-mail: brendan.ubochi.840649@swansea.ac.uk

The effects of electric field induced traps generation in the drain access region is studied using industry standard TCAD, Atlas by Silvaco [1]. We show that the reduction in the cut-off frequency of the device from 13.9 (GHz) to 11.25 (GHz) could be linked to the electric field induced traps. We have used acceptor traps at an energy level of $E_T = E_V + 0.9$ (eV), corresponding to substitutional carbon in GaN, and a concentration of $N_{IT} = 5 \times 10^{17} (cm^{-3})$ to model the induced traps. Although vertical scaling has been used to reduce short channel effects, we observe that this leads to a reduction in the current arising from the reduced ionised surface donors [2].

1. Introduction

Due to its excellent material properties, GaN HEMTs have a huge potential in radio frequency (RF) and power applications. However, the maximum RF power output is not always reproducible due to the existence of defects in device structure acting as traps which ultimately inhibits the wide commercial use of these devices. Various proposals have been made regarding the nature, location, and effects of traps in GaN device structures [3]. These electronic traps which occur in all the layers of the device are usually caused by the presence of impurities, lattice mismatch and thermal expansion coefficients between epitaxial layers during material growth and also from imperfect fabrication processes. Surface passivation and improvements in growth and fabrication processes could reduce the density of traps/defects in the device structure. Some studies have suggested that when subjected to high electric fields, the device degrades as a result of hot electron damage, trapping at the surface, barrier and buffer layers[4]–[6]. Recently, a correlation between the RF output power degradation and bulk traps have been observed [7]. The mechanism of traps generation in the drain access region under high electrical stress has been proposed [3] as opposed to the traditional explanation by a self-heating [8].

Calibration to experimental terminal characteristics is obtained from the knowledge of the device dimension, material constants and by applying suitable values for the transport parameters. Carrier transport is modelled using the drift-diffusion transport model. The cutoff frequency is obtained from an extraction routine that uses scattering parameters obtained over a range of frequencies [1]. The results presented in [9] suggests that GaN HEMTs are expected to have a $f_T \times L_G$ product of 13 GHz.μm. The 1 μm gate length GaN HEMT having a simulated cut-off frequency of 13.9 GHz shows very good agreement. Figure 1 illustrates the cross-section of an asymmetrical, 1 μm gate-length device modelled. It consists of a 2 nm GaN-cap, 20 nm AlGaN barrier, 1 nm AlN spacer, and 3 μm GaN substrate with source-to-gate and gate-to-drain separations of 2 μm and 3 μm, respectively. The cap layer is an improvement that leads to smaller surface charge, higher breakdown voltage, higher sheet carrier density and a reduction in the gate leakage current [10]. Additionally, the AlN spacer

978-1-5090-3084-2/16 $31.00 © 2016 IEEE   181

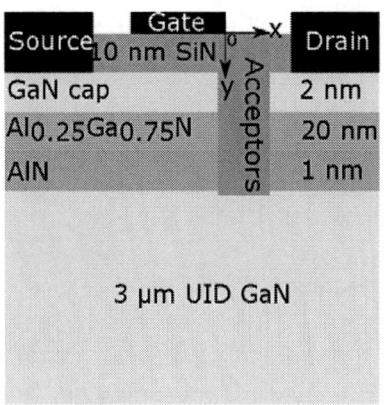

Figure 1: Schematic of the hetero-structure of the investigated 1(μm) gate length GaN HEMT.

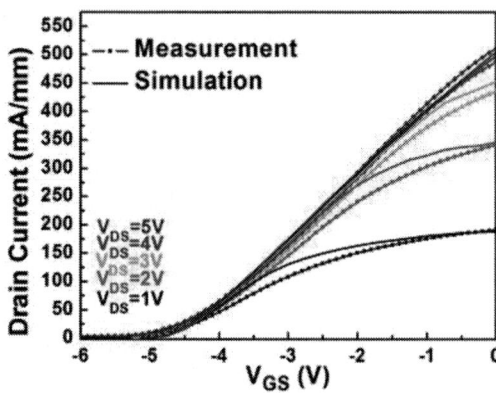

Figure 2: Transfer characteristics (I_D-V_{GS}) comparing calibrated simulations (full lines) to experimental measurements (dot lines) for V_{DS}=1V to V_{DS}=5V in step of 1V.

provides a better confinement of carriers in the channel and, consequently, an increase in density. In the calibration, we have used donor interface traps located at the surface of GaN cap with an energy, concentration and electron and hole capture cross-sections of $E_T=E_V+2.9$ eV [11], $N_D=4.2 \times 10^{19}$ cm^{-3} and $\sigma_{n,p}=1 \times 10^{-19}$ cm^2, respectively, in order to pin the Fermi level at the surface. The value of the electron capture cross-section has been used to model the onset of drain current. In order to control a leakage current through the buffer and thus to better confine electrons into the channel, acceptor traps are located in the GaN buffer with an energy, concentration and electron and hole capture cross-sections of $E_T=E_C - 2.2$ eV [12], $N_A=3.9 \times 10^{16}$ cm^{-3} and $\sigma_{n,p}=1 \times 10^{-15}$ cm^2 [12], respectively.

The free carrier distribution is modelled using the Boltzmann statistics. The device is uniformly n-type doped with a concentration of 1×10^{16} cm^{-3}. In order to model carrier velocity variation with field strength, we have used the parallel field dependent mobility model [13]. Fig. 2 shows a good agreement between the modelled and measured transfer characteristics of the device using parameters of Table 1.

Table 1: Material parameters used in simulations.

Material Properties	GaN	AlN
Electron Saturation Velocity	1×10^7 cm s^{-1}	1.4×10^7 cm s^{-1}
Electron Mobility	1070 cm^2 V^{-1}s^{-1}	150 cm^2 V^{-1}s^{-1}
Hole Mobility	30 cm^2 V^{-1}s^{-1}	14 cm^2 V^{-1}s^{-1}
β	2	2

2. Scaled Performance of the GaN HEMT

In order to predict improvements that could be obtained from lateral scaling, we scaled down the gate length and the access regions by one-half, one-quarter and one-eight corresponding to L_G= 0.5 μm, L_G=0.25 μm, L_G=0.125 μm, respectively. Fig. 3 shows the transfer characteristics (I_D-V_{GS}) at V_{DS}=5V for the laterally scaled down devices. Additional results indicate that the cut-off frequency of the 1 μm gate-length HEMT can be increased from 13.9 GHz to 90.7 GHz when the device is scaled down proportionally in lateral dimensions by a factor of 8. In Fig. 4, we observe a reduction in the maximum drain current

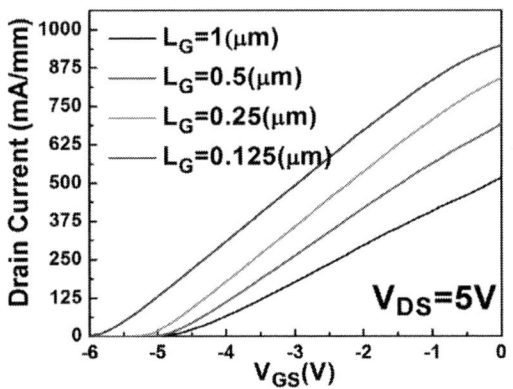

Figure 3: Transfer characteristics (I_D-V_{GS}) at V_{DS}=5V using a lateral scaling only for the scaled devices.

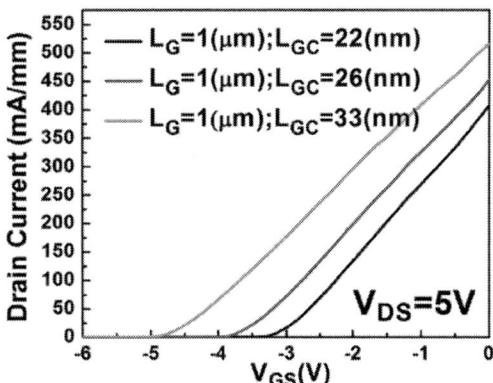

Figure 4: Transfer characteristics (I_D-V_{GS}) at V_{DS}=5V using a vertical scaling of the gate-to-channel distance (GC) only with a fixed gate length of 1 (µm).

when the gate to channel separation is reduced from 33 nm to 26 nm and 22 nm, respectively, whilst maintaining a fixed gate length of 1 µm. This observed reduction in the current is due to the reduction in the ionized surface states [2] which varies in proportion to the AlGaN barrier thickness. The reduction in the 2DEG density will result in a decrease in the cut-off frequency by 11.5% and 17.3% for a 25% and 50% increase in the aspect ratio, respectively.

3. Effects of Drain Access Electric Field Generated Traps

In order to model the electric field induced traps, we have located traps at an energy E_T=E_V+0.9 (eV) laterally from a position at L=-0.1 µm to L=2.4 µm where the gate edge at the drain side is set to be at zero (see Fig. 1). Fig. 5 shows the variation of the transfer characteristics with different trap densities that affect device behaviour by trapping phenomena in the drain access region [14]. The maximum drain current is seen to decrease as the trap density is increased. In addition, it can be seen that the threshold voltage is shifted positively with increasing trap densities. In comparison with the device without traps, we show that for an acceptor traps density of N_{IT}=5 ×10^{17} cm^{-3}, the cut-off frequency is reduced from 13.9 GHz to 11.25 GHz as shown in Fig. 6.

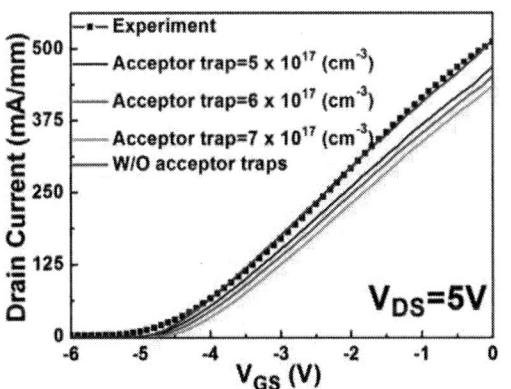

Figure 5: Simulated transfer characteristics (I_D-V_{GS}) at V_{DS}=5 V of the HEMT for indicated acceptor trap densities.

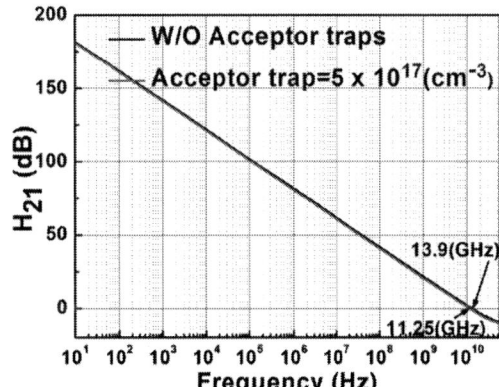

Figure 6: Simulated current gain versus frequency at V_{DS}=5V for devices without traps and with acceptor trap density of 5 (10^{17} cm^{-3}).

978-1-5090-3084-2/16 $31.00 © 2016 IEEE

4. Conclusion

Electric field induced degradation of the DC and RF characteristics of the GaN HEMTs have been studied based on the mechanism of traps generation in the drain access region. We have used donor interface traps in order to pin the Fermi level at the surface. In order to control a leakage current through the buffer and thus to better confine electrons into the channel, acceptor traps were located in the GaN buffer.

We have shown that the cut-off frequency of the 1 μm gate length GaN HEMT is reduced from 13.9 GHz to 11.25 GHz as a result of an acceptor trap density of $N_{IT}=5 \times 10^{17}$ cm^{-3}. The results from lateral scaling suggest that the cut-off frequency can be increased to 90 GHz when the device is scaled down in proportion laterally by a factor of 8. When a vertical scaling is applied in hope to maintain the gate control over carrier transport in the channel [15], which aims to reduce short channel effects, the maximum drain current and cut-off frequency are reduced as a result of the reduced ionised surface donors in GaN HEMTs[2].

Acknowledgements

This research is supported by the Sêr Cymru National Research Network in Advanced Engineering and Materials (NRN081). We thank Edward Wasige of the University of Glasgow for providing us with the experimental data.

References

[1]   Silvaco, *Atlas User 's Manual*, CA:Silvaco Inc., Santa Clara, 2015.

[2]   J. P. Ibbetson, P. T. Fini, K. D. Ness, S. P. DenBaars, J. S. Speck, and U. K. Mishra, *Appl. Phys. Lett.*, 77, 250, 2000.

[3]   M. Faqir, G. Verzellesi, G. Meneghesso, E. Zanoni, and F. Fantini, *IEEE Trans. Electron Devices*, 55, 1592–1602, Jul. 2008.

[4]   G. Koley, V. Tilak, L. F. Eastman, and M. G. Spencer, *IEEE Trans. Electron Devices*, 50, 886–893, Apr. 2003.

[5]   W. S. Tan, M. J. Uren, P. A. Houston, R. T. Green, R. S. Balmer, and T. Martin, *IEEE Electron Device Lett.*, 27, 1–3, Jan. 2006.

[6]   D. K. Sahoo, R. K. Lal, H. Hyungtak Kim, V. Tilak, and L. F. Eastman, *IEEE Trans. Electron Devices*, 50, 1163–1170, May 2003.

[7]   A. R. Arehart, A. Sasikumar, G. D. Via, B. Poling, E. R. Heller, and S. A. Ringel, *Microelectron. Reliab.*, 56, 45–48, 2016.

[8]   B. Benbakhti, A. Soltani, K. Kalna, M. Rousseau, and J.-C. De Jaeger, *IEEE Trans. Electron Devices*, 56, 2178–2185, Oct. 2009.

[9]   S. C. Binari, K. Ikossi, J. A. Roussos, W. Kruppa, D. Park, H. B. Dietrich, D. D. Koleske, A. E. Wickenden, and R. L. Henry, *IEEE Trans. Electron Devices*, 48, pp. 465–471, 2001.

[10]   S. Faramehr, P. Igic, and K. Kalna, *The Tenth International Conference on Advanced Semiconductor Devices and Microsystems*, 1–4, 2014.

[11]   H. K. Cho, C. S. Kim, and C.-H. Hong, *J. Appl. Phys.*, 94, 1485, 2003.

[12]   W. D. Hu, X. S. Chen, F. Yin, J. B. Zhang, and W. Lu, *J. Appl. Phys.*, 105, 84502, 2009.

[13]   D. M. Caughey and R. E. Thomas, *Proc. IEEE*, 55, 2192–2193, 1967.

[14]   S. Faramehr, K. Kalna, and P. Igić, *Semicond. Sci. Technol.*, 29, 25007, Feb. 2014.

[15]   K. Kalna and A. Asenov, *Semicond. Sci. Technol.*, 17, 579–584, 2002.

ASDAM 2016, The 11th International Conference on Advanced Semiconductor
Devices And Microsystems, November 13-16, 2016, Smolenice, Slovakia

Trap analysis of GaN-based heterostructures using current transients mesurements

M. Florovič[1], J. Škriniarová[1], D. Gregušová[2], J. Kováč[1], P. Kordoš[1]

[1] *Institute of Electronics and Photonics, Faculty of Electrical Engineering and Information Technology, Slovak University of Technology, Ilkovičova 3, 812 19 Bratislava, Slovakia*
[2] *Institute of Electrical Engineering, Slovak Academy of Science, Dúbravská cesta 9, 841 04 Bratislava, Slovakia*
e-mail: martin.florovic@stuba.sk

This work deals with the trap analysis of the GaN-based structures using current time response on the voltage pulse applied to Schottky and ohmic contacts containing AlGaN and InAlN layers with different composition and thickness. Monitoring of the current time evolution for the investigated samples allows identifying and comparing particular traps. The analysis of the current transients measured at various temperatures via fitting up to four exponentials is used to resolve traps located in the particular layers and their activation energies were determined.

1. Introduction

GaN-based high electron mobility transistors (HEMTs) have appeared as excellent devices for high power applications. GaN is a wide bandgap semiconductor, therefore there should be reached high operating temperatures [1]. In addition, the high carrier mobility combined with the two-dimensional electron gas (2DEG) density results in a very low on-state resistance [2]. Either thanks of high breakdown field of GaN, such HEMTs are appropriate to be used in high-power and high-speed applications. These designate GaN-based HEMTs for a massive adoption in the industrial, automotive and photovoltaic fields.

Despite the excellent performance of the devices, several technological aspects are still optimized to find wide GaN HEMT application in the market. Unfortunately, those devices still suffer from trapping/detrapping and early degradation mechanisms at the surface, inside the AlGaN, resp. InAlN layer and in the GaN buffer that affect the dynamic performance and the reliability of the transistors [3]. In addition hot electrons from the 2DEG channel can gain enough energy to overcome the barrier that exists in the extrinsic region of the structure and get trapped there. These trapped electrons change electrostatics and deplete the 2DEG concentration resulting in the channel current reduction [4].

This work is focused on the trap analysis of the GaN-based structures evaluating the current time response for reverse voltage pulse applied to Schottky contact (similar to the HEMT gate) and low forward voltage pulse applied to ohmic contacts (representing 2DEG channel in HEMT). The analysis offers two types of investigated structures containing AlGaN/AlN and InAlN layer respectively with different composition and thickness. There are currently various methods such as deep-level transient spectroscopy (DLTS), gate-lag and low-frequency noise measurements and conductance analysis to observe trapping/detrapping processes in semiconductor structures along which current transient spectroscopy can offer sufficient results by those already introduced methods [5].

978-1-5090-3084-2/16 $31.00 © 2016 IEEE

2. Experimental

GaN-based structures were prepared by low-pressure metal-organic vapour phase epitaxy. Sample 1 – AlGaN/AlN/GaN material structure was grown on a 4H–SiC substrate, Fe-doped GaN buffer layer was 1.5–1.7 μm thick followed by 1.25 nm thick AlN interlayer and $Al_{0.29}Ga_{0.71}N$ barrier layer of thickness 20 nm on the top. Sample 2 – InAlN/AlN/GaN material structure was prepared on a sapphire substrate, the GaN buffer layer was 2 μm thick, followed by 1 nm thick AlN interlayer and 8 nm thick $In_{0.21}Al_{0.79}N$ barrier layer.

The conventional processing steps were applied to prepare AlGaN/GaN HEMT devices [6]. Circular Schottky contacts of diameter 150 μm surrounded by ohmic contact area such as TLM ohmic contacts with the length of 75 μm and width of 150 μm with gap between two contacts ~ 20 μm were chosen for measurements.

The current-voltage (I-V) characteristics and current transients were measured using semiconductor parameter analyzer Agilent 4155C in the temperature range 25 – 150 °C. After the zero initial voltage, reverse voltage pulse ($V_D = -6$ V for Schottky contacts), resp. low voltage pulse ($V_D = +1$ V for TLM contacts) were applied in the time range of $\sim10^{-3}$ s to $\sim10^4$ s to measure the current transients. After each transient measurement the device state was recovered by white LED illumination for one minute at zero applied voltage.

3. Results and discussion

Typical I-V characteristics of investigated structures are shown in Fig. 1. In Fig. 1a reverse current $\sim10^{-4}$ A to $\sim10^{-3}$ A of Schottky contacts is depicted, under the threshold voltage of particular HEMT there is 2DEG channel depletion, current flows via surface leakage, non fully depleted space charge area, buffer and top layer traps and defects. Sample 1 exhibits more temperature sensitive I-V characteristics than sample 2. For low applied voltage $V_D = +1$ V TLM structure self-heating effect should be neglected as shown in Fig. 1b. However, hot electrons in the 2DEG channel can gain sufficient kinetic energy, therefore the charge trapping is possible and non-uniform active trap distribution in particular layers creates additional time dependent electrostatic field.

Normalized trapping transients of investigated structures are shown in Fig. 2. According to the thermodynamic theory trapping/detrapping process of particular trap decays in time in an exponential way, this assumption is proportional to a recovery process from a non-equilibrium state. Therefore the fitting function can be expressed as the sum of particular processes:

$$I_{fitted} = \sum_{i=1}^{n} a_i \exp(-t/\tau_i) + I_\infty \qquad (1)$$

Fitting of the measured data I_{data} was performed to minimize the sum of $|I_{data} - I_{fitted}|^2$ at the measured points. Time constants τ_i and magnitude a_i were evaluated from the fitting process, positive (negative) values of a_i correspond to the trapping (detrapping) processes, $[\tau_i, a_i]$ image allows to resolve traps (TP1,TP2,TP3,TP4) with particular time constants in the temperature range 25 – 150 °C, in this case up to four exponentials were used. Since reverse current transients of Schottky diode in Fig. 2b don't meet these criteria it wasn't further analysed. With these temperature-dependent measurements the activation energy of particular traps was extrapolated based on the Arrhenius plot (Fig. 3). In general there were observed traps with their energies in investigated structures – (TP1) ~ 0.55 eV, (TP2) ~ 0.74 eV, (TP3) ~ 0.60 eV, (TP4) ~ 0.36 eV.

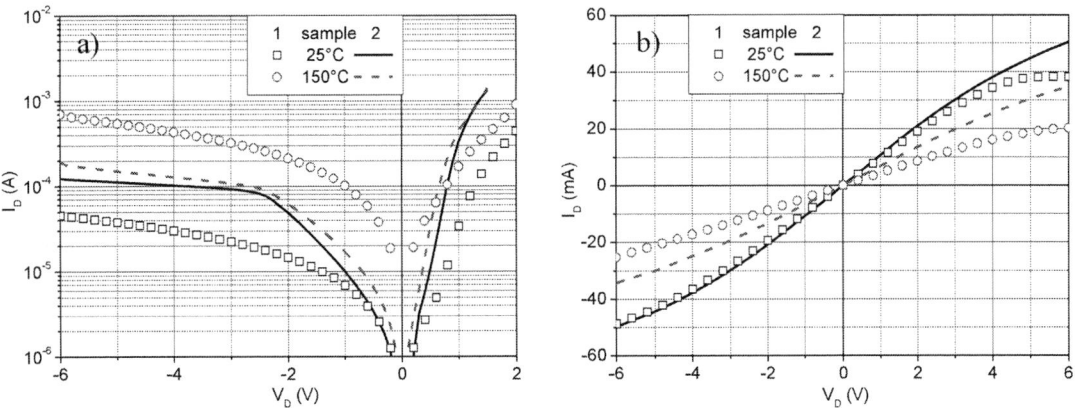

Fig. 1. I-V characteristics of a) Schottky and b) TLM – ohmic structure

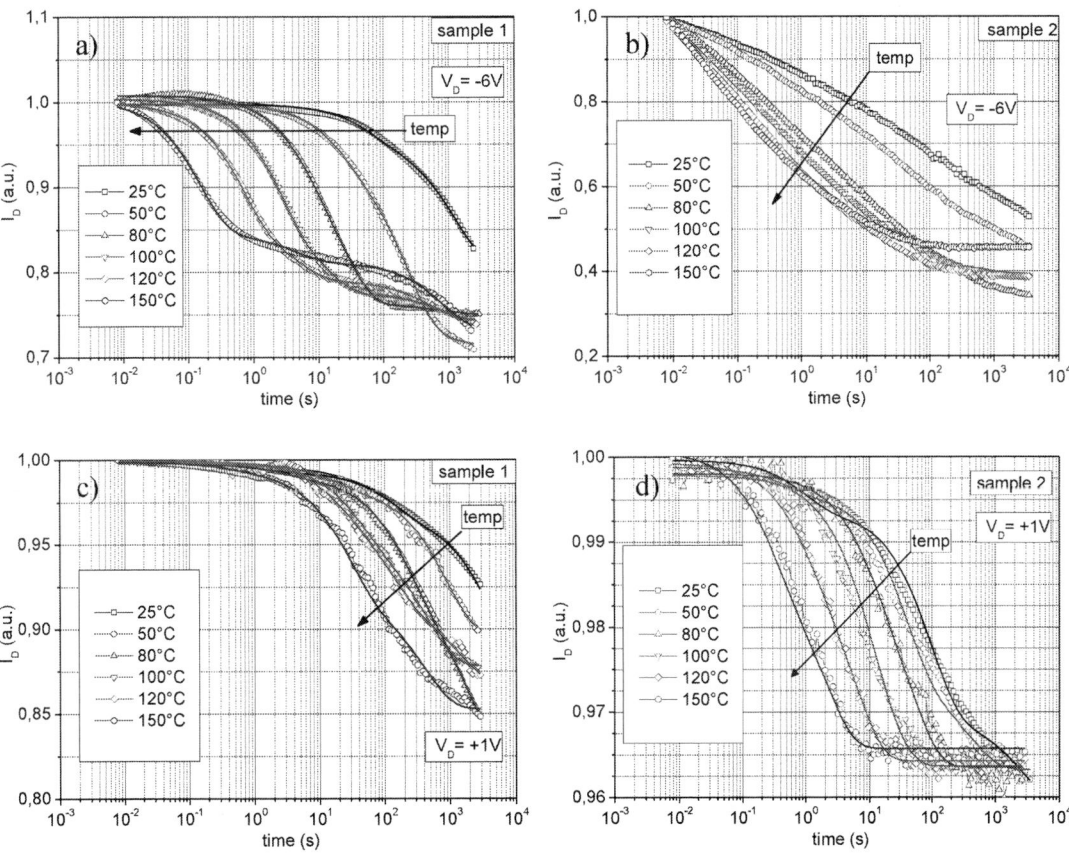

Fig. 2. Measured (dots) and fitted (lines) current transients of
a,b) Schottky and c,d) TLM – ohmic structure

The trap states energies similar to those evaluated in this article were reported also by the others. The traps with an energy $0.55 - 0.57$ eV (TP1) are considered as the main reason of high power GaN-based device degradation [3,7]. However, the physical origin of these traps is still controversial because the ~ 0.60 eV trap might be an intrinsic defect but the ~ 0.57 eV trap was assigned to Fe dopant and the $0.50 - 0.60$ eV trap to the nitrogen antisite point defects [3]. The similar trap energy (TP3) ~ 0.60 eV was obtained for the sample 1. The traps in the energy range 0.80 – 0.85 eV might be attributed to defects mainly associated with dislocations as commonly observed in MOCVD-grown structures [7]. Therefore the trap states ~ 0.74 eV (TP2) in the investigated structures can be connected predominantly with the GaN buffer near the interface defects associated with dislocations but traps with energy ~ 0.69 eV are connected with both AlGaN and InAlN layers [8]. Three deep traps in InAlN/GaN heterostructure with the energy 0.35 – 0.49 eV were inspected by DLTS [9], likewise for sample 2 there was observed the trap with the energy (TP4) ~ 0.36 eV.

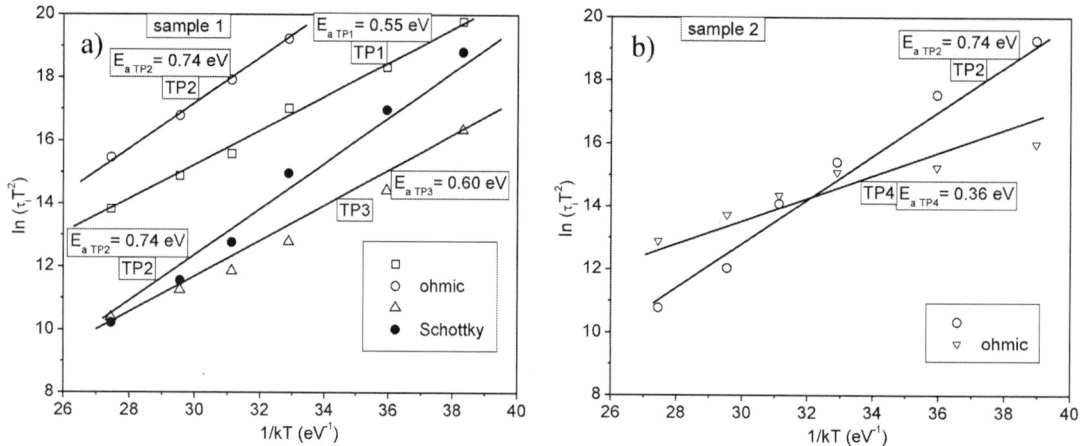

Fig. 3. Arrhenius plot with depicted trap energies for a) sample 1 and b) sample 2

4. Conclusions

Trapping effects of two different GaN-based structures were studied evaluating Schottky and ohmic contacts. There were compared I-V characteristics such as current transients of $Al_{0.29}Ga_{0.71}N/AlN/GaN$ and $In_{0.21}Al_{0.79}N/GaN$ devices at various temperatures. Based on the Arrhenius plot the trap activation energies (TP1) ~ 0.55 eV, (TP2) ~ 0.74 eV, (TP3) ~ 0.60 eV, (TP4) ~ 0.36 eV were extrapolated from the temperature dependent time constant positions. Only the trap energy of ~ 0.74 eV was observed in both types of structures, i.e. independently on the barrier composition, which confirms that these trap states are located in the GaN buffer. Additional experiments are needed to clear the origin of another trap states TP1, TP2 and TP4.

Acknowledgement

The results presented in this work were obtained with partial support of grants VEGA No. 1/0739/16, 1/0491/15 and 2/0105/13.

References:

[1] Maier, D., et al., IEEE Trans. Device Mater. Reliab. 10, 427–436 (2010).
[2] Moens, P., et al., IEEE Electron Device Meet. IEDM 2015 Technical Dig., 903–906 (2015).
[3] Hu, A., et. al., Appl. Phys. Lett. 108, 042107 (2016).
[4] Binari, S. C., et al., Proc. IEEE, Vol. 90, 1048-1058 (2002).
[5] Joh, J., et. al., IEEE Trans. Electr. Dev., Vol. 58 (No. 1), 132-140 (2011).
[6] Florovic, M., et. al., ADEPT 2013 Proc., 44-47 (2013).
[7] J.A.Y. Polyakov, et. al., Mater. Sci. Engn. 94, 1-56 (2015).
[8] A. Sasikumar, et al., Appl. Phys. Lett. 103, 033509 (2013).
[9] Z. Chen, et. al., Jpn. J. Appl. Phys. 50, 081001 (2011).

ASDAM 2016, The 11th International Conference on Advanced Semiconductor Devices And Microsystems, November 13-16, 2016, Smolenice, Slovakia

Temperature-dependent of sub-threshold slope of AlGaN/GaN MOSHFETs with HfO$_2$ gate oxide prepared by ALD

Roman Stoklas[1*], Dagmar Gregušová[1], Karol Frohlich[1], and Jan Kuzmík[1]

[1]Institute of Electrical Engineering, Slovak Academy of Sciences, SK-84104 Bratislava, Slovakia

*roman.stoklas@savba.sk

The sub-threshold slope (SS) of AlGaN/GaN MOSHFETs with HfO$_2$ gate oxide prepared by Atomic Layer Deposition (ALD) were investigated. The different SS values, like as 290 mV/dec and 100 mV/dec for MOSHFETs with and w/o oxygen-plasma pre-treatment (PHf- and Hf-MOS) respectively, were estimated. The analysis were realized in temperature range from RT to 300°C. From SS, the average D$_{it}$, like as 8.4x10^{12} eV^{-1}cm^{-2} and 1.5x10^{12} eV^{-1}cm^{-2} for Hf-and PHf-MOS, respectively, were evaluated. The gate leakage current has a strong effect on the SS. The gate leakage were reduced about two orders of magnitude for PHf-MOS (~10^{-8}A/mm at -10V) with oxygen-plasma pre-treatment. The improved SS obviously is also due to the large I$_{ON}$/I$_{OFF}$ ratio (10^8). An increase of SS values with temperature, were found. However, the PHf-MOS have exhibited an approximately half angle of slope of SS with temperature (0,37mV/°C) than Hf-MOS (0,63mV/°C). As a result from SS the D$_{it}$ for PHf-MOS were reduced approximately five times.

1. Introduction

AlGaN/GaN high electron mobility transistors (HEMTs) have shown outstanding performance on high power and high frequency applications. However, the GaN HFETs suffer from much higher sub-threshold slope (> 300mV/dec) than the theoretically limit of 60 mV/dec. Sub-threshold slope (SS) is very important in digital and analog applications like to assure excellent pinch-off, a low dissipated power and to achieve good power added efficiency and noise figure in power amplifiers [1]. Most crucially, subthreshold slope can be used to quantify trap density in the gate region of AlGaN/GaN HFETs, which degrades gate modulation efficiency, closely related to the SS and ultimately to high frequency performance [2]. The AlGaN/GaN metal-insulator-semiconductor HFETs (MISHFETs) with appropriate thin gate insulator such as Al$_2$O$_3$, SiO$_2$, or Si$_3$N$_4$ layer were therefore investigated to reduce the gate leakage current. The use of thin gate insulator is also required to achieve the high frequency operation for RF devices [3]. And also, HfO$_2$ has been intensively studied in III–V MOS devices [4], and applied in Si CMOS technology.

In this paper, we demonstrate effects of temperature on electrical properties of AlGaN/GaN MOS-HFETs with HfO$_2$ gate oxide prepared by ALD with and w/o oxygen-plasma pre-treatment. We found a strong correlation between sub-threshold slope, gate leakage current and I$_{ON}$/I$_{OFF}$ ratio.

2. Experiment

The AlGaN/GaN heterostructure used in our study was grown on the top of a SiC substrate. The active layer consisted of a GaN buffer, followed by a 17 nm Al$_x$Ga$_{1-x}$N with an AlN mole fraction of 29%. Finally, the structure was capped by 3-nm-thick GaN-cap layer. The conventional processing steps were used to fabricate HFET structures, like as: (1) Mesa isolation achieved with argon sputtering; (2) evaporation a Ti/Al/Ni/Au multiplayer for ohmic contact; (3) rapid thermal annealing of the multiplayer at 850 °C for 30 s in an N$_2$ ambient to

978-1-5090-3084-2/16 $31.00 © 2016 IEEE

prepare ohmic contacts. For MOSHFETs, the gate oxide was deposited before Schottky metallization on the source-drain access region, i.e. it served also as a passivation layer.

For analysis, two different types of MOS-HFETs were prepared by ALD at 300°C. The MOS-HFETs differ by using the oxygen-plasma (O2 plasma) pre-treatment before ALD deposition [5], and without pre-treatment (designated as PHf-MOS and Hf-MOS in the next, respectively). The gate oxide thickness was controlled by elipsometry measurement. After deposition of the gate oxide on the source-drain access region, i.e. it served also as passivation layer, the Schottky metallization consisted of a double Ni/Au layer was patterned by optical lithography. The transistor structures with different gate lengths (2 and 2.5 μm) and gate widths (60 and 120 μm) were prepared. Also processed along with the HFETs and MOSHFETs, there were large-gate "fat"-FET structures with a gate length of 100 μm and van der Pauw patterns with an active area of 0.3×0.3 mm^2. The devices were characterized by the temperature dependent Current-Voltage (TD-IV) measurements.

3. Results and Discussion

At first, the I-V measurement was performed at room temperature to investigate the threshold voltage V_{th}, current density I_{DS}, extrinsic transconductance $g_{m,ext}$ and leakage current I_{leak}. Fig. 1 shows the drain–current characteristics of MOSHFETs as a function of gate voltage. As shown in Fig. 1, the reduction of drain leakage current for PHf-MOS exceeds two orders of magnitude, with nearly the same drain saturation current I_{DS} (~ 0.6A/mm). However, the threshold voltage V_{DS} for Hf-MOS is a more negative, which could correspond with a higher density of oxide/semiconductor interface states (see Fig.1). By extrapolation from the $log(I_{DS})$ vs. V_{GS} curve, the SS of the MOS-HFETs were calculated as follows [6]:

$$SS = \frac{dV_{GS}}{d\ logI_{DS}} = \ln(10) * \frac{kT}{q} * \left(1 + \frac{C_q + C_{it}}{C_b}\right) \qquad (1)$$

in which C_q, C_{it} and C_b are the quantum, interface-trap and barrier capacitances, respectively. In the deep sub-threshold region, the C_q although finite at T>0 K owing to the Fermi–Dirac distribution, is three to four orders of magnitude smaller than C_b, i.e., few electrons are present in the channel. As a result, one can extract C_{it} and the interface-trap density D_{it} by neglecting C_q, as follows:

$$D_{it} = C_{it}/q \qquad (2)$$

We estimated different SS values, like as 290 mV/dec and 100 mV/dec for Hf- and PHf-MOS, respectively (see Fig. 1). The I-V curves at 300°C for both samples are also added in Fig. 1. The measurements were realized with a voltage ramp of 1V/s, approximately. The average D_{it} estimated at RT were 8.4x10^{12} eV^{-1}cm^{-2} and 1.5x10^{12} eV^{-1}cm^{-2} for Hf-and PHf-MOS, respectively. One reason for high interface states is the low quality of oxide/semiconductor interface, which affects adversely an electric field distribution due to hot electron mechanism at the gate edges and in G-D access region. Due to a long emission time of oxide/semiconductor interface states, the AlGaN/GaN interface states are also affected and SS could be estimated.

In the next, a temperature dependent I-V (TD-IV) measurement was used to evaluate the SS. From the TD-IV, the SS slope for Hf-MOS increased markedly at elevated temperature (in the range of 290-460 mV/dec) with the ramp of 0,63mV/°C. On the other side, the SS slope for PHf-MOS only slightly increased at elevated temperature (in the range of 100-200 mV/dec) with a ramp of 0,37mV/°C, see in Fig. 2. As a result, the PHf-MOS sample is a much more stable with the temperature than the Hf-MOS.

Moreover, the gate leakage current I_{leak} has a strong effect on the subthreshold slope (shown in Fig. 3). The lower I_{leak} of PHf-MOS, with a half angle of slope than Hf-MOS approximately, can be explained by a larger band offset at oxide/GaN interface than that of Hf-MOS due to O2 plasma, where a thin oxidized GaN cap-layer was formed.

The improved SS obviously is also due to the large I_{ON}/I_{OFF} ratio (for PHf-MOS in our case), see in Fig. 4. The values of I_{ON} and I_{OFF} were calculated at $V_{GS} = 1V$ and from the pinch-off region, respectively. There is a strong linear dependence between I_{ON}/I_{OFF} and I_{leak} and the use of O2 plasma is very important to both lower I_{leak} ($< 10^{-8}$ A/mm) and increase I_{ON}/I_{OFF} ratio ($\sim 10^8$).

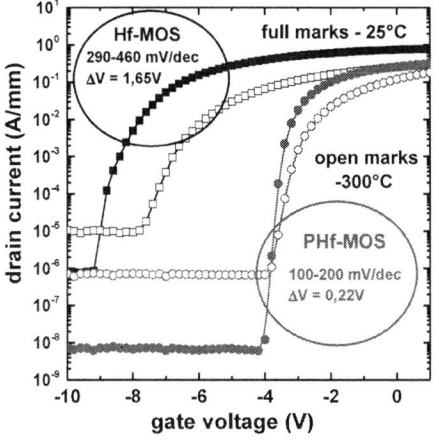

Fig. 1 SS slope for MOSHFETs

Fig. 2 Temperature dependencies of SS slope

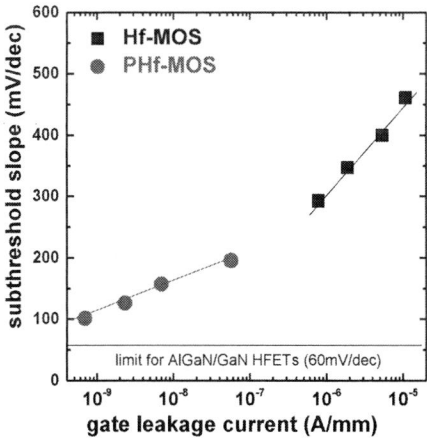

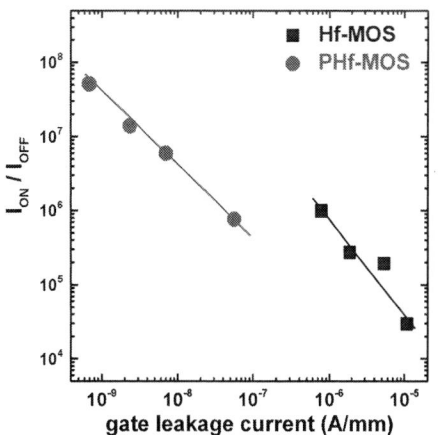

Fig. 3 Linear dependence of SS and gate leakage current

Fig. 4 Linear dependence between I_{ON}/I_{OFF} ratio and gate leakage current (I_{leak})

The D_{it} evaluated from the SS (estimated at different temperatures) for PHf-MOS is shown in Fig. 5. The density of states for Hf-MOS (not shown here) have been approximately five times higher than that for PHf-MOS. A similar D_{it} for a different SS at elevated temperature corresponds with a similar ratio of SS slope and kT/q (SS vs. kT/q ratio $\sim$ 3.8 and 11 for PHf- and Hf-MOS respectively). From the SS analysis, we proposed a relatively simple way to estimate interface trap density. The understanding of these interface traps is critical to optimize the gate modulation efficiency of MOSHFETs and to maximize their high frequency performance.

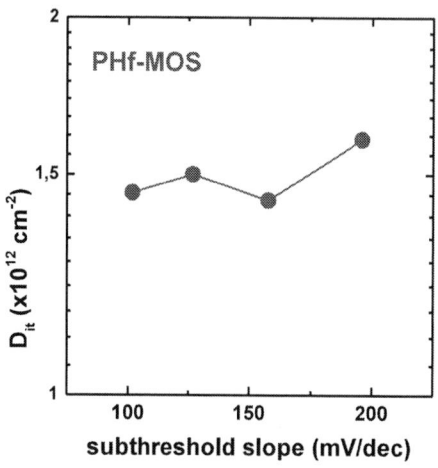

Fig. 5 D_{it} vs. SS of PHf-MOS

4. Conclusions

In summary, an influence of oxygen-plasma pre-treatment of MOS-HFETs with HfO_2 dielectric layer on their electrical properties, were analysed. To estimate D_{it}, the SS slope obtained from the IV measurements, were used. Nearly the same maximum drain current densities for both types of MOSHFETs were evaluated, but about two order of magnitude lower leakage current for PHf-MOS were observed. The measurements were realized from RT to 300°C. The SS value increase at evaluated temperature for both samples. However, the PHf-MOS have exhibited an approximately half angle of slope of SS with temperature (0,37mV/°C) than Hf-MOS (0,63mV/°C). The PHf-MOS also shown a lower shift of ΔV_{th} at 300°C (0.2V and 0.8V for PHf- and Hf-MOS, respectively) and higher I_{ON}/I_{OFF} ratio ($\sim 10^8$). All improvements mentioned above are responsible for the reduction of D_{it} for PHf-MOS approximately five times. We suppose, an formation of interfacial layer at oxide/GaN interface due to chemical reactions between the precursors and GaN during ALD. And even the oxygen vacancies (V_O) and their energy levels in HfO_2 may contribute to the formation of leakage paths [4].

Acknowledgement

This work was supported by the Slovak Grant Agencies VEGA (Grant Nos. 2/0105/13 and 2/0138/14) and APVV (Grant. No. 15-0637), as well as by the Centre of Excellence for New Technologies in Electrical Engineering project, ITMS code 26240120011, supported by the Research and Development Operational Programme funded by the ERDF.

References
[1] C. Sanabria, et al., IEEE Electron Device Lett. 27, 19 (2006)
[2] D. K. Kim, et al., J. Semicond. Technol. Sci. 14, 601 (2014)
[3] D. Gregušová, et al., Semicond. Sci Technol. 24, 075014 (2009)
[4] Y.C. Chang, et al., Appl. Phys. Lett. **92**, 072901 (2008)
[5] D. Gregušová, et al., Appl. Phys. Lett. 104, 013506 (2014)
[6] J. Chung, et al., IEEE Electron Device Letters 29 (11), 1196, (2008)

ASDAM 2016, The 11th International Conference on Advanced Semiconductor
Devices And Microsystems, November 13-16, 2016, Smolenice, Slovakia

Nanocrystalline diamond films for electronic monitoring of gas and organic molecules

A. Kromka[1,2], T. Izak[1], M. Davydova[1], B. Rezek[1,3]

[1]Institute of Physics, Czech Academy of Sciences,
Cukrovarnická 10, 162 00 Prague, Czech Republic
[2]Faculty of Civil Engineering and [3]Faculty of Electrical Engineering,
Czech Technical University in Prague, Prague, Czech Republic
e-mail: kromka@fzu

Nowadays, implementation of specific materials for (bio) sensors is one of the most rapidly developing micro-electronic field. Synthetic diamond thin films exhibit an extraordinary combination of intrinsic properties which make it an attractive material for investigation of solid state interaction with (bio) molecules or complex biological systems. Hydrogen terminated intrinsic diamond films reveal a phenomenological property - induced p-type surface conductivity which has been found as suitable for fabrication of various electronic devices, mainly impedance elements and field effect transistors. This study draws on research conducted by employing hydrogen terminated nanocrystalline diamond as the functional layer for (bio) sensoric uses and point out recent models for the detection principles.

1. Introduction

Diamond exhibits an extraordinary combination of intrinsic properties which make it as a promising material for the interdisciplinary oriented community - material engineering, thin film physics, plasma chemistry, civil engineering, life science, etc. [1,2]. Especially, its p-type surface conductivity (i.e. two dimensional hole gas) makes the diamond thin film a promising material for fabrication of electronically active devices and sensor uses [3,4]. Such sensoric devices use diamond as a functional layer which is employed as either electrically passive or active layer, or as both.

The diamond surface reveals a distinct feature - its surface can be covalently terminated with different atoms (molecules) resulting in tailored wetting properties, i.e. hydrophobic or hydrophilic, respectively. The water contact angle can vary from extremely low values ($<5°$) up to as high as values of $110°$ (super-hydrophobic surface) and is controlled on the atomic level by the diamond surface termination (Fig. 1). Among of variety of atomic or molecular terminations, oxygen and hydrogen terminated diamond are mostly investigated and employed surfaces.

2. Hydrogen and oxygen termination of diamond

2.1 Hydrogen termination

The hydrogen terminated diamond surface is hydrophobic with the contact angle varied from 60 to $100°$ depending on the size of crystals, their orientation and surface morphology. It is naturally built during the diamond CVD growth. The H-terminated diamond surface induces a hole accumulation in close proximity to the surface. The hole accumulation mechanism is still under discussion. Due to electronegativity differences of hydrogen and carbon (2.1 vs. 2.5, respectively), formed C–H dipoles and adsorbates are essential for hole

978-1-5090-3084-2/16 $31.00 © 2016 IEEE   193

accumulation [5]. The C–H dipoles lead in a spontaneous polarisation and after exposing such surface to ambient air, the surface conductivity occurs due to the surface transfer doping effect [6].

The first diamond electronic device based on the p-type induced surface conductivity should be dedicated to Kawarada [7] who introduced the diamond based metal-semiconductor field effect transistor (FET).

2.2 Oxygen termination

Diamond surface termination with oxygen results in a hydrophilic surface (CA <10°) keeping electrically insulating properties. Oxygen termination can be simply achieved by UV/ozone treatment [8], wet chemistry (like $H_2SO_4+KNO_3$, 200°C, 30 min) or by dry plasma [9]. Oxygen atoms can be bonded in the form C-O-C (ether) or C=O (carbonyl). In some cases, more complex bonds can occur, i.e. carboxylic anhydride bonds (O=C-O-C=O).

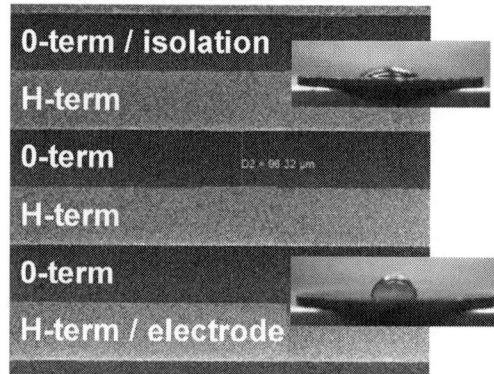

Fig. 1: Scanning electron microscopy of oxygen (O-term) and hydrogen (H-term) terminated diamond surfaces and appropriate contact angle images (insets).

Generally, O- and H-terminated diamond surfaces are not resolvable under an optical microscope. After their exposing to electron irradiation (e.g. to scanning electron microscopy, SEM), both these individual atomic terminations become well resolvable. The H-terminated diamond surface exhibits negative electron affinity (NEA) and thus, it can emit electrons easier than oxygen terminated surface (darker regions, Fig. 1 – on the left). The H-terminated diamond patterns are represented by brighter regions in the SEM image due to the NEA.

3. Intrinsic diamond as sensor

Diamond as a biocompatible semiconductor with the functional surface (i.e. dangling bonds just ready for a covalent binding of atoms/molecules) is an attractive material for merging research on the interaction of solid state with biological systems. The preliminary studies on the diamond functionalization with DNA and proteins motivated continuing research on developing diamond based biosensors employing the p-type surface conductivity induced even on the intrinsic diamond. The p-type surface conductivity can be used either as an optically transparent planar electrode during impedance measurements or as a conductive channel in field effect transistors. In these planar devices, resistive O-terminated surface electrically isolates and separate the H-terminated electrodes/channels. It means that O-terminated patterns substitute the *MESA*-like isolation known from the standard silicon devices. There is no doubt that hydrogen and oxygen terminated surface micro-patterns are highly relevant for electronic applications of intrinsic (undoped) diamond thin films.

3.1 Gas sensors

The operating principle of diamond gas sensors is based on a modulation of the induced surface conductivity due to reactions with adsorbed gas molecules. One of the first studies used the Pd/intrinsic diamond/boron doped-diamond configuration for the detection of H_2 [10]. Later on, the same configuration was used for the detection of benzene and toluene [11]. The undoped monocrystalline diamond gas sensors for detection of reducing (NH_3, CO, H_2) and oxidising gases (O_2, NO_2, HCl) were presented in work [12]. To improve the sensor response and detection limit, diamond was plasmatically structured in nanocones [13]. Nevertheless, most of such devices have been realised on monocrystalline diamonds which are technologically limited in the deposition area.

As an alternative solution to monocrystalline diamond, we have developed the sandwich structure of intrinsic H-terminated diamond layer/metal interdigitated electrodes/insulating ceramic, see Fig. 3a [14–16]. The new design used built-in interdigitated electrodes in the diamond layer. We found that the O-terminated diamond surface did not exhibit any response to the tested gases, as expected. On the other hand, the H-terminated sensor showed a positive response to the tested gases: humidity, carbon dioxide (CO_2), ammonia (NH_3), and phosgene ($COCl_2$). Moreover, the H-terminated diamond layers exhibited a peculiar selectivity towards phosgene as the oxidising gas. It was observed that the diamond surface conductivity increased up to 31 times for $COCl_2$ increase from 5 to 20 ppm.

Representative impedance gas sensitivity to various $COCl_2$ concentrations measured at 25°C and 105°C is shown in Fig. 3b. First, the gas sensor was exposed to synthetic air (SA) and then to phosgene. As observed, room temperature measurements exhibited only a slight increase of the surface conductivity from 3.9×10^{-5} S to $4.9 \ 10^{-5}$ S for $COCl_2$ from 5 to 20 ppm. A quite dynamic response expressed by an increase in the conductivity is observed for the elevated temperature of 105°C. A pronounced increase of surface conductivity (from 7.4×10^{-6} S to 1.7×10^{-5} S) is found when the sensor was exposed to 20 ppm of $COCl_2$ at 105°C. A good enough sensitivity of diamond sensor to phosgene was attributed to the phosgene dissociation in the monolayer of water adsorbed on the diamond and consequent more effective charge transfer due to the multiplication effect of the increased concentration of H_3O^+.

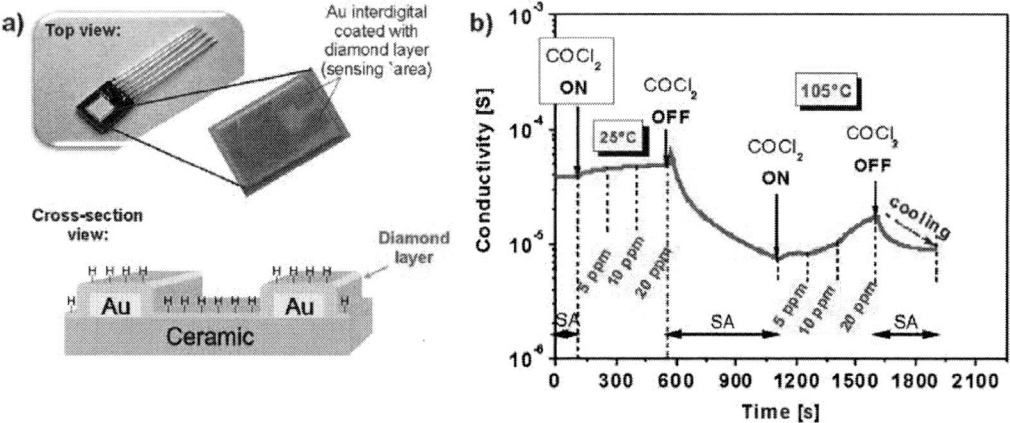

Fig. 3: a) Digital photo of diamond based gas sensor and its schematic cross-section view. b) Time dependence of the surface conductivity of the H-terminated diamond coated sensor to phosgene, i.e. 5, 10 and 20 sccm, respectively. During measurements, samples were kept at 25°C and 105°C. For a review see also works [14–16].

3.2 Electrical monitoring of proteins and living cells

Non-invasive label-free techniques for *in vitro* monitoring of cell growth are of high interest due to simplicity, fast response in real time and excluding experiments with animals. Presently, there are two analytic approaches which employ either *optical* or *electronic* signal processing [17]. Both these approaches reached state of the art in a transducer type which converts a stimulus-induced cellular response into a quantifiable signal (i.e. biosensor signal) [18]. From the broad family of electronic systems, diamond based impedance and field effect transistor devices seem to be the most used devices.

Impedance sensors for cell monitoring: Impedance measurements seem to be one of the simplest and still a powerful method for monitoring of the cellular signals. The monitored impedance signal is sensitive not only to ionic currents but also to cell growth stages (i.e. cell attachment, spreading, shape, proliferation, differentiation and communication) [19]. The typical impedance sensors use gold interdigital electrodes which are deposited on an electrically isolating material (glass, ceramics or plastic). In some cases, gold electrodes may limit optical monitoring of cultivated cells from the sensor backside. Here, electrodes made of intrinsic diamond are optically transparent and provide a suitable biocompatible substrate for cell cultures. Impedance measurements employing conductive H-terminated interdigitated electrodes isolated by O-terminated regions revealed an increase of absolute impedance as well as of cell index [20]. The impedance increase in the first hours was be attributed to the coverage of the interdigitated electrodes with osteoblast-like MG 63 cells.

FET sensors for cells monitoring: The solution-gated field effect transistor (SG FET) design is more suitable for miniaturization than impedance sensors, or others. The first diamond based FETs cell biosensor should be related to the work Rezek et al. [21]. The SG FET was fabricated from H-terminated nanocrystalline diamond films. The active area of the H-terminated channel was $20 \times 60\ \mu m^2$. As the gate, Ag/AgCl reference electrode was used.

It was observed that exposing the transistor channel to fetal bovine serum proteins (i.e. the part of the cultivation medium) decreases the surface conductivity. Fig. 4 shows transistor output characteristics of diamond SG FET in McCoy's 5A medium, device after FBS adsorption and adsorbed device after rinsing by water. In comparison to the reference conditions (i.e. device loaded with McCoy's 5A medium), the transistor transfer characteristic reflected the negative gate potential shift approx. −50 mV for FBS medium. The transistor with adsorbed FBS proteins was washed by its rinsing in phosphate-buffered saline (PBS)

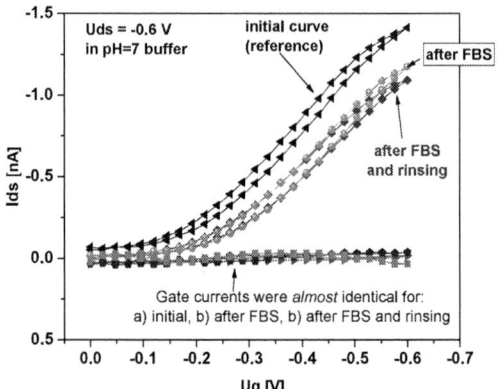

Fig. 4: Transistor output characteristics and gate currents of diamond SG FET in McCoy's 5A medium (reference), after FBS adsorption and after FBS adsorption rinsed in water, for review see also [21].

and deionized water. The "cleaning" procedure did not influence the transfer characteristic, which was nearly identical before and after the transistor cleaning. Gate currents were almost the same for all 3 combinations (i.e. the reference, adsorbed FBS and FBS rinsed in water. It should be noted that the diamond based SG FET transistor revealed stable and repeatable characteristics for measurements under physiological conditions and its exposing to mediums required approx. 25 s to stabilise the signals due to new equilibrium conditions.

The direct electrostatic contribution from adsorbed proteins, which have negative net charge, on the channel current modulation was refused due to observed an opposite effect – i.e. decrease of the channel current. Therefore, the decrease of channel current was attributed to sticking of proteins to close vicinity of the diamond surface and repelling the counter-ions. A similar effect has also been observed after adsorbing lipid bilayers to the diamond FETs [22].

Izak et al. used the diamond SG FET for real time monitoring of the SAOS-2 cell growth [23]. They found that gate currents increased nearly by two orders during the cell cultivation. Once cell de-adhered (delaminated), the SG FET transistors recovered to initial values, i.e. to the low gate currents. The observed effect, also labelled as transistor "triggering", was repeatable and reproducible measure of cell adhesion.

Last but not least feature of the diamond SG FET is its electronic stability after the gamma irradiation at least up to 5 Gy [24]. The diamond SG FETs were stable themselves and their transfer characteristics and gate currents reflected modification of proteins as well as cell cultures (adhesion, delamination) after the γ-irradiation at different doses.

4. Conclusions

Diamond, especially as a thin film, is a promising functional material for fabrication of biosensors which should be simply downsized. Electronic surface properties of intrinsic undoped diamond films were simply controlled by the terminating atoms (hydrogen vs. oxygen). Our findings demonstrated the electronic functionality of the diamond surface even for the film consisting of nano-sized crystals (<150 nm). The H-terminated diamond gas sensing element should find industrial uses, especially for detection $COCl_2$. The optically transparent diamond electrodes were suitable for impedance based sensors and real-time monitoring of cultivated cells. Electrical measurements on the SG FETs confirmed surface channel modulation by the presence of proteins and living cells. The sensitivity of induced p-type surface conductivity to surrounding environment has opened attention on a variety of chemical and biological sensors, microfluidic systems for in vitro cultivation and testing new drugs, or even for providing cytotoxic experiments.

Acknowledgement

We would like to thanks Martin Stuchlik (TU Liberec) for gas sensor testing, Marie Kratka and Lenka Michalikova for FETs experiments, Marie Kalbacova and Lucie Bacakova for cell cultivations, Ondrej Rezek and Pavla Bauerova for technical support. This work was supported by the projects P108/12/G108 (AK, BR) and postdoc project 14-06054P (MD).

References

[1] J.A. Carlisle, Nat. Mater. **3** (10), 668–669 (2004).

[2] C.E. Nebel, B. Rezek, D. Shin, H. Uetsuka, N. Yang, J. Phys. Appl. Phys. **40** (20), 6443–6466 (2007).

[3] W. Yang, R.J. Hamers, Appl. Phys. Lett. **85** (16), 3626 (2004).

[4] T. Izak, T. Sakata, Y. Miyazawa, T. Kajisa, A. Kromka, B. Rezek, Diam. Relat. Mater. **60** 87–93 (2015).

[5] K. Hirama, H. Takayanagi, S. Yamauchi, J.H. Yang, H. Kawarada, H. Umezawa, Appl. Phys. Lett. **92** (11), 112107 (2008).

[6] F. Maier, M. Riedel, B. Mantel, J. Ristein, L. Ley, Phys. Rev. Lett. **85** (16), 3472 (2000).

[7] H. Kawarada, Surf. Sci. Rep. **26** (7), 205–206 (1996).

[8] X.F. Wang, M. Hasegawa, K. Tsugawa, A.R. Ruslinda, H. Kawarada, Diam. Relat. Mater. **24** 146–152 (2012).

[9] O. Babchenko, B. Rezek, J. Stuchlík, A. Kromka, J.-C. Arnault, P. Bergonzo, Adv. Sci. Eng. Med. **6** (7), 802–808 (2014).

[10] Y. Gurbuz, W.P. Kang, J.L. Davidson, D.L. Kinser, D.V. Kerns, Sens. Actuators B Chem. **33** (1–3), 100–104 (1996).

[11] Y. Gurbuz, W.P. Kang, J.L. Davidson, D.V. Kerns, Sens. Actuators B Chem. **99** (2–3), 207–215 (2004).

[12] S. Ri, K. Tashiro, S. Tanaka, T. Fujisawa, H. Kimura, T. Kurosu, M. Iida, Jpn. J. Appl. Phys. **38** (Part 1, No. 6A), 3492–3496 (1999).

[13] Q. Wang, S.L. Qu, S.Y. Fu, W.J. Liu, J.J. Li, C.Z. Gu, J. Appl. Phys. **102** (10), 103714 (2007).

[14] M. Davydova, A. Kromka, P. Exnar, M. Stuchlik, K. Hruska, M. Vanecek, M. Kalbac, Phys. Status Solidi A **206** (9), 2070–2073 (2009).

[15] A. Kromka, M. Davydova, B. Rezek, M. Vanecek, M. Stuchlik, P. Exnar, M. Kalbac, Diam. Relat. Mater. **19** (2–3), 196–200 (2010).

[16] M. Davydova, M. Stuchlik, B. Rezek, K. Larsson, A. Kromka, Sens. Actuators B Chem. **188** 675–680 (2013).

[17] Y. Fang, Int. J. Electrochem. **2011** 1–16 (2011).

[18] Y. Fang, Drug Discov. Today Technol. **7** (1), e5–e11 (2010).

[19] I. Giaever, C.R. Keese, Proc. Natl. Acad. Sci. **81** (12), 3761–3764 (1984).

[20] T. Ižák, K. Novotná, I. Kopová, L. Bačáková, B. Rezek, A. Kromka, Phys. Status Solidi B **250** (12), 2741–2746 (2013).

[21] B. Rezek, M. Krátká, A. Kromka, M. Kalbacova, Biosens. Bioelectron. **26** (4), 1307–1312 (2010).

[22] P.K. Ang, K.P. Loh, T. Wohland, M. Nesladek, E. Van Hove, Adv. Funct. Mater. **19** (1), 109–116 (2009).

[23] T. Izak, M. Krátká, A. Kromka, B. Rezek, Colloids Surf. B Biointerfaces **129** 95–99 (2015).

[24] M. Krátká, O. Babchenko, E. Ukraintsev, J. Vachelová, M. Davídková, M. Vandrovcová, A. Kromka, B. Rezek, Diam. Relat. Mater. **63** 186–191 (2016).

ASDAM 2016, The 11th International Conference on Advanced Semiconductor
Devices And Microsystems, November 13-16, 2016, Smolenice, Slovakia

Modelling the I-V-T characteristics of 4H-SiC DMOSFET in presence of SiO₂/SiC interface traps and fixed oxide

Gian-Domenico Licciardo, Luigi Di Benedetto and Alfredo Rubino

Department of Industrial Engineering, University of Salerno,
Via Giovanni Paolo II, 132, 84084, Fisciano (SA), Italy
e-mail: gdlicciardo@unisa.it, ldibenedetto@unisa.it and arubino@unisa.it

A new analytical model of the 4H-SiC DMOSFET is proposed that is capable to predict the forward operation of the device in a wide range of temperature, by including in its DC current-voltage characteristics the effects of the parasitic resistances, of the insulator-semiconductor interface traps on the threshold voltage and channel mobility, as well as their temperature dependences. The accuracy of the model has been verified by comparisons with numerical simulations using interface trap density varying in the range $[0;10^{14}]cm^{-2}eV^{-1}$ and a temperature operation up to 500K. Comparisons with experimental data taken on 1.2kV commercial devices validate the model.

1. Introduction

Nowadays, Double-implanted Metal-Oxide-Semiconductor Field-Effect-Transistor (Fig.1) is one of the most attractive devices in 4H-SiC due to its good figure-of-merits for high and medium power applications [1]-[2]. To date, the quality of the gate oxide is one of the major limitations of such device, because the presence of interface traps at the semiconductor/oxide interface affects the operation of the device [3]-[4]. Indeed, the density of defects significantly modifies the charge distribution in the channel and in the accumulation region and, as consequence, the threshold voltage can shift [5], the channel and accumulation layer mobility can vary as well as the sub-threshold slope. Due to the difficulties related to the analytical formulation of the problem, the existent literature does not provide physical models, specifically derived for 4H-SiC, which quantitatively predict the variations of the characteristics induced by the defects.

In this paper, differently than the available literature [6], we show a physically-based analytical model of the 4H-SiC DMOSFETs that is able to describe the effect of the interface traps on the V_{TH} and the channel mobility, μ_{CH}, in a wide temperature range, as well as their impact on the I_D-V_{DS} and I_D-V_{GS} curves, by using closed-form expressions and avoiding adjusting parameters. Numerical simulations [7] and experimental data are used to validate

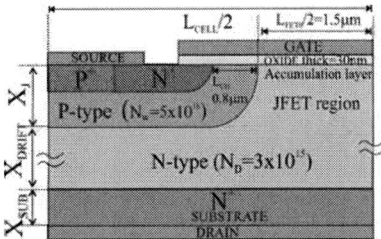

Fig. 1. Cross-section view of a DMOSFET structure with the main physical and geometrical parameters.

978-1-5090-3084-2/16 $31.00 © 2016 IEEE   199

the model.

2. Analytical model

In order to derive the traps contribute, the temperature-dependent trapped charge density, $Q_{a,it}(T,V_{GS})$, has been calculated as:

$$Q_{a,it}\left(T,V_{GS}\right)=-q\int_{E_V}^{E_C}D_{it,TA}\left(E\right)f\left(T,E\right)dE= \tag{1}$$

$$=-qD_{it,T0}WTA\left[2F1\left(1,\frac{kT}{WTA};\frac{kT+WTA}{WTA};-\frac{N_C}{n_S}\right)-e^{\frac{E_V-E_C}{WTA}}2F1\left(1,\frac{kT}{WTA};\frac{kT+WTA}{WTA};-\frac{N_C}{n_S}e^{\frac{E_V-E_C}{kT}}\right)\right]$$

where $2F1\left(a,b;c;z\right)=\sum_{i=0}^{\infty}\frac{(a)_i(b)_i}{(c)_i}\frac{z^i}{i!}$ is the Gauss hypergeometric function. One can evaluate

the threshold voltage from (1) as $V_{TH}\left(T,V_{GS}\right)=V_{TH,0}\left(T\right)-C_{OX}^{-1}Q_{a,it}\left(T,V_{GS}\right)$ [2]. The channel mobility expression $\mu_{CH}\left(T\right)=\mu_{FE}\left(T\right)//\mu_{CS}\left(T\right)$ has been obtained, by using the Matthiessen's rule, as the superposition of the field-effect mobility

$\mu_{FE}\left(T\right)=\mu_{n0,FE}\left(\dfrac{T}{300}\right)^{-1.5}\left[1+\left(\dfrac{\mu_{n0,FE}E_m}{v_{SAT}}\right)^{\beta}\right]^{-\frac{1}{\beta}}$ and the Coulomb scattering contribution due

to the interface traps, $\mu_{CS}\left(T\right)=32\varepsilon^2\hbar kT\left(m^*q^2\left|Q_{a,it}\right|\right)^{-1}$ [8]. The above equations have been used in the current-voltage expressions of the SPICE model 1 to obtain the forward characteristics of the DMOSFET:

$$I_{DS}=\begin{cases}\mu_{CH}C_{OX}\dfrac{W}{L_{CH}}\left[2\left(V_{GS,Int}-V_{TH}\right)V_{DS,Int}-V_{DS,Int}^2\right] & :V_{DS,Int}\leq V_{GS,Int}-V_{TH}\\[2mm]\mu_{CH}C_{OX}\dfrac{W}{2L_{CH}}\left(V_{GS,Int}-V_{TH}\right)^2\left(1+\lambda V_{DS,Int}\right) & :V_{DS,Int}>V_{GS,Int}-V_{TH}\end{cases} \tag{2}$$

where the voltages applied to the channel, $V_{GS,Int}\approx V_{GS}$ and $V_{DS,Int}=V_{DS}-R_{TOT}I_D$, have been obtained by depurating the voltage terminals from the parasitic resistance $R_{TOT}=R_{ACC}+R_{JFET}+R_{DRIFT}+R_{SUB}$, due to the accumulation layer, R_{ACC}, the JFET region, R_{JFET}, the epilayer, R_{DRIFT}, and the substrate, R_{SUB}, whose expressions are as follows [9]:

$$R_{ACC}=\frac{L_{FET0}}{WC_{OX}\mu_{na}\left(V_{GS}-V_{TH}\right)};R_{SUB}=\frac{W_{SUB}}{q\mu_{n,sub}N_{SUB}WL_{CELL}}$$

$$R_{JFET}=\frac{X_J+W_{SCV}}{q\mu_{nn}N_{EPI}WL_{FET}};R_{DRIFT}=\frac{1}{q\mu_{nn}N_{EPI}WL_{CELL}}\left[X_{DRIFT}+\frac{L_{FET}-L_{CELL}}{2}\right]$$

3. Results

In Fig. 2-4 comparisons with numerical simulation results [7] are reported to validate the model. DMOSFET structure of Fig.1 has X_J=0.6µm, N_W=5·10^{16}cm^{-3}, t_{OX}=30nm,

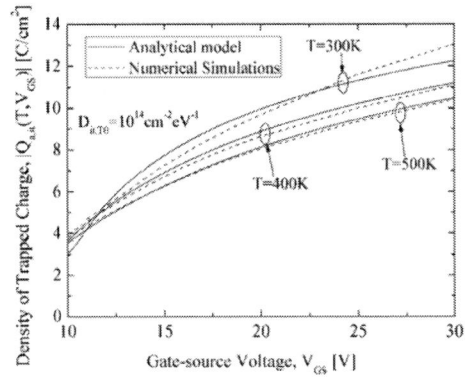
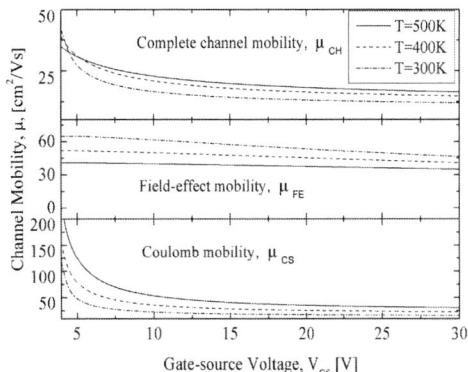

Fig. 2: Trapped charge density as a function of the V_{GS} at different temperature.

Fig. 3: Temperature dependency of μ_{CH}, μ_{FE} and μ_{CS} as function of V_{GS}.

$L_{CH}=0.8\mu m$, $N_D=3\cdot10^{15}cm^{-3}$, and $qWF=4eV$, which are often used in 4H−SiC DMOSFETs [10]–[13]. Furthermore, physical models include field and doping dependent carriers mobility, SRH recombination [14]-[16], incomplete ionization and bandgap narrowing with parameters taken from [17]-[24].

Fig. 2 shows the accuracy of (1) to describe the V_{GS} dependence of the trapped charge density at the oxide/semiconductor interface, when the temperature varies from $300K$ to $500K$ and $D_{it,T0}=10^{14}cm^{-2}eV^{-1}$.

Fig. 3 shows the V_{GS} dependence of the overall channel mobility and its two components in the temperature range 300-$500K$.

In Fig.4 the proposed model is proved by comparing the I_D-V_{DS} and I_D-V_{GS} characteristics of numerical simulations and experimental measurements of the commercial device in [25], at $T=300K$ and $T=423K$. It is clear as the model accurately describes both electrical and thermal behaviors of the device.

4. Conclusions

The proposed model analytically describes the static curves of 4H-SiC DMOSFET under different temperature values and under interface traps effects. Comparisons with numerical simulations and experimental measurements validate the model.

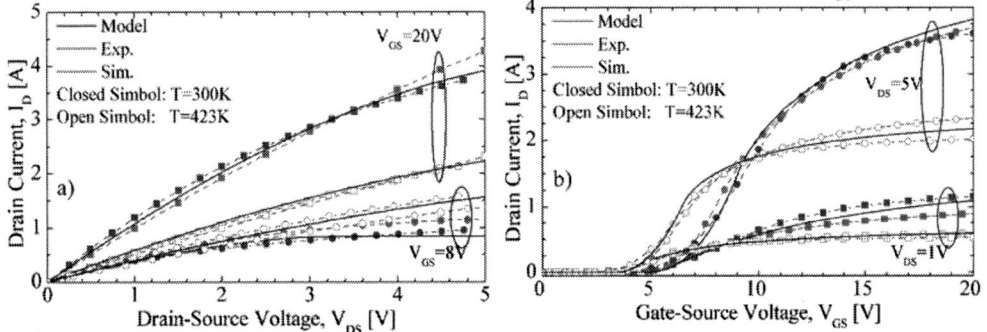

Fig. 4: Comparisons between analytical, experimental and simulated a)I_D-V_{DS} and b) I_D-V_{GS} curves of CREE C2M1000170D.

978-1-5090-3084-2/16 $31.00 © 2016 IEEE

References

[1] K. Hamada *et al.*, *IEEE Trans. on Electron. Devices*, **62**, 2, 278-285, 2015.

[2] S. Allen *et al.*, *Materials Science Forum*, **821–823**, 701–704, 2015.

[3] G.D. Licciardo *et al.*, *IEEE Trans on Power Elect.*, **30**, 10, 5800-5809, 2015.

[4] G.D. Licciardo *et al.*, *Materials Science Forum.*, **858**, 825-828, 2016.

[5] A. Castellazzi *et al.*, *Microelectronics Reliability,* **52**, 9, 2414–2419, 2012.

[6] R. Fu *et al.*, *IEEE Trans. on Industry Applications*, **48**, 181-190, 2012.

[7] SILVACO Int., Santa Clara, CA, ATLAS User's Manual, 2005, Ver. 5.10R.

[8] S. Potbhare *et al.*, *IEEE Trans. Electron Devices*, **55**, 8, 2029-2040, 2008.

[9]L. Di Benedetto *et al.*, *IEEE Electr. Dev. Lett.*, **37**, 2016. *Doi: 10.1109/LED.2016.2613821*

[10] A. Saha *et al.*, *IEEE Trans. on Electron Dev.*, **54**, 10, 2786–2791, 2007.

[11] G. D. Licciardo *et al.*, *IEEE Trans. on Electron. Devices*, **63**, 4, 1783-1787, 2016.

[12] K. Matocha *et al.*, *Materials Science Forum,* ***778–780***, *903–906, 2014.*

[13] L. Di Benedetto *et al.*, *IEEE Trans. on Electron. Devices*, **63**, 9, 3795-3799, 2016.

[14] S. Bellone *et al.*, *IEEE Trans. on Electron Dev.*, **56**, 12, 2902-2911, 2009.

[15] S. Bellone *et al.*, *IEEE Trans. on Instrum. and Meas.*, **57**, 6, 1112-1117, 2008.

[16] S. Bellone *et al.*, *IEEE Trans. on Electron Dev.*, **54**, 11, 2998-3006, 2007.

[17] S. Bellone *et al.*, *Solid State Electron.*, **120**, 6–12, 2016.

[18] S. Bellone *et al.*, *Solid State Electron.*, **109**, 17–24, 2015.

[19] M. L. Magherbi *et al.*, *Solid State Electron.*, **109**, 12-16, 2015.

[20] S. Bellone *et al.*, *Procedings of the International Semiconductor Conference (CAS),* **2**, 405-408, 2010.

[21] L. Di Benedetto *et al.*, *IEEE Electr. Dev. Lett.*, **35**, 2, 244-246, 2014.

[22] S. Bellone *et al.*, *IEEE Trans. on Electron Dev.*, **59**, 9, 2546-2549, 2012.

[23] S. Bellone *et al.*, *IEEE Trans on Power Electron.*, **29**, 1, 514–521, 2014.

[24] S. Bellone *et al.*, *IEEE Trans on Power Electron.*, **29**, 5, 2174–2179, 2014.

[25] CREE model C2M1000170D

ASDAM 2016, The 11th International Conference on Advanced Semiconductor
Devices And Microsystems, November 13-16, 2016, Smolenice, Slovakia

The Effect of Self-Heating and Electrical Stress induced Polarization in AlGaN/GaN Heterojunction Based Devices

K. Ahmeda[1,*], S. Faramehr[1], P. Igić[1], K. Kalna[1], S. J. Duffy[2], A. Soltani[3], B. Benbakhti[2]

1. Electronic Systems Design Centre (ESDC), College of Engineering, Swansea University Bay Campus, Fabian Way, Swansea, SA1 8EN, Wales, United Kingdom. *717828@swansea.ac.uk
2. Department of Electronics and Electrical Engineering, Liverpool John Moores University, Byrom Street, L3 3AF, Liverpool, United Kingdom.
3. LN2, Sherbrooke, Canada and IEMN, Lille, France.

The effect of self-heating and polarisation is studied in AlGaN/GaN Transmission Line Measurement (TLM) structures with varying the contact spacing between the source and drain. The measurement results of the I-V characteristics are calibrated and investigated by TCAD Atlas-Silvaco. The self-heating simulations show a hotspot at the vicinity of the drain side. The electrical stress that is applied on the Ohmic contacts decreases the polarisation as the source-drain distance is reduced, causing the inverse piezoelectric effect.

1. Introduction

Gallium Nitride (GaN) semiconductor material possesses unique material properties that make it a promising candidate for high-power and high-frequency applications [1]. Low resistance Ohmic contacts of AlGaN/GaN based devices are essential to achieve forecasted device performance in RF switching functionality whilst enforcing device reliability [2, 3]. In this work, we study $GaN/Al_{0.32}Ga_{0.68}N/AlN/GaN/Al_{0.1}Ga_{0.9}N$ TLM (Transmission Line Measurement) structures grown on a p-type doped (5×10^{18} cm^{-3}) Si (111) substrate. The device structure consists of a 1.7 μm thick $Al_{0.1}Ga_{0.9}N$ back barrier for a better confinement of electrons in the 2DEG channel, followed by a 15 nm GaN buffer, a 1 nm AlN spacer, a 25 nm $Al_{0.32}Ga_{0.68}N$ barrier and a 1 nm GaN cap layer. The TLM structures have various contact spacing of L_1= 4 μm, L_2= 8 μm, L_3=12 μm and L_4= 18 μm (Fig. 1(a)). Fig. 1(b) illustrates energy band diagram overlapped with electron concentration in the structure cross-section.

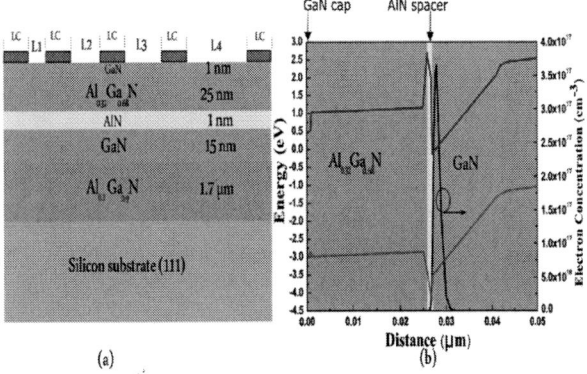

Figure 1: (a) Schematic cross section of TLM structures grown on the Si (111) substrate. (b) Energy band diagram and electron concentration of the structure.

978-1-5090-3084-2/16 $31.00 © 2016 IEEE

2. Simulation Approach

I-V characteristics were simulated via a 2D drift-diffusion transport model using Fermi-Dirac statistics and Shockley-Read-Hall recombination model provided by Silvaco TCAD. Thermal models were used to study the self-heating effects. The measurement I-V characteristics of the TLM structures are simulated by combining two mobility models, the analytic low field mobility model and the high field mobility using an electron low field mobility of 800 cm/V.s and a saturation velocity of 1.9×10^7 cm/s. The effect of self-heating during the simulation was taken in consideration by using Giga module in Atlas which solves the lattice heat flow equation in addition to the overall carrier transport simulations to make the simulations electro-thermal. The thermal conductivity model for the device layers is assumed to have a power function dependence and have been calibrated against experimental data taken from [4,5]. The Poisson and continuity equations are solved self-consistency in all simulations [6].

3. Results

Figs. 2 and 3 show the simulated I-V characteristics of the AlGaN/GaN TLMs without and with self-heating effect, respectively. Carbon traps with an energy level of $E_a = E_V + 0.9$ eV and a concentration of 1×10^{17} cm^{-3} are considered in the GaN buffer [7] while Iron traps at $E_a = E_V + 0.6$ eV with a concentration of 4×10^{18} cm^{-3} in the $Al_{0.1}Ga_{0.9}N$ back barrier. A larger difference between the measurement results and the simulations data that is caused by high electric field is seen for the shortest contact spacing of L1= 4 μm at V_{DS}=15V. When the distance between the source-drain is increased, the agreement between calibrated simulations at low electric fields and the measurement results is improving because self-hearting effects on carrier transport play a less role in the larger TLM structures as expected. Fig. 3 compares the simulation results obtained from electro-thermal simulations. The drain current reduction due to enhanced electron scattering with increased lattice temperature resulting in a mobility degradation caused by the self-heating shown in the Fig. 3. A very good agreement between electro-thermal simulations and measurement data can be observed. The agreement of the simulations improves as the distance between the source and drain becomes larger which indicates that calibrations of thermal material parameters for a bulk GaN was accurate but non-linear thermal effects start to occur in smaller TLM structures when scaled from 18 μm down to 4 μm.

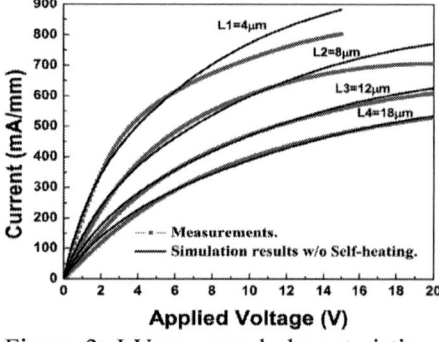

Figure 2: I-V measured characteristics of TLM (red lines). The simulations were calibrated against measured values (black lines) without including the self-heating.

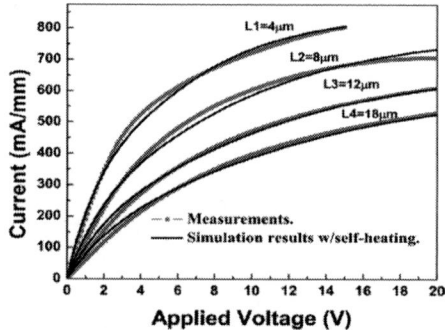

Figure 3: I-V measured characteristics of TLM (red lines). The simulations were calibrated against measured values (black lines) while including the self-heating.

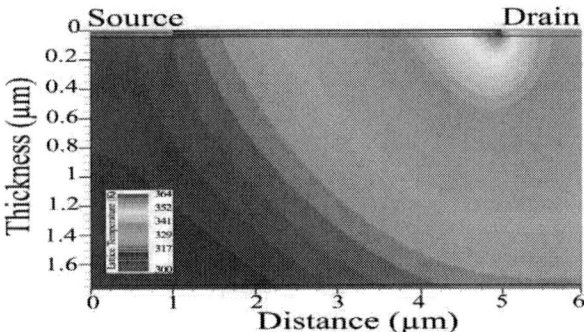

Figure 4: 2D distribution of the lattice temperature for a contact spacing distance of L1= 4 μm at applied bias of 15 V.

Fig. 4 shows that the increase in temperature is due to high electric field occurring between the contacts, particularly near the drain side. Fig. 5 illustrates the temperature profile of the 2DEG channel at V=15 V. The hot spot appears at the drain side. The shortest structure has the highest channel temperature at around 362 K even though the maximum applied voltage to this structure is 15 V, a smaller than that of the other structures. When the distance between the contacts is increased, the temperature is gradually reduced until negligible self-heating is observed, as shown in Figs. 3 and 5.

When an external force is applied, for example, an electrical stress on the AlGaN/GaN TLM structure by applying voltage via contacts, the wurtzite crystal structure of III-Nitrides suffers a when an electrical stress is applied. This affects the polarization along with various spacing between the contacts. This phenomenon is known as the inverse piezoelectric effect [8]. We have found that the inverse piezoelectric effect in the TLM structures has increased with the spacing (Fig. 6) resulting in decrease of the total polarisation (piezoelectric and spontaneous) by 7 %, 10 %, 17% in the 12 μm, 8 μm, and 4 μm TLM structures, respectively, when compared to the largest 18 μm TLM structure. This decrease in polarisation may arise from the electrical stress on Ohmic pads during measurements. Fig. 6 illustrates hypothetical I-V

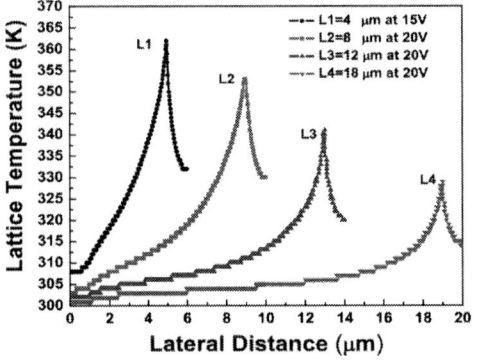

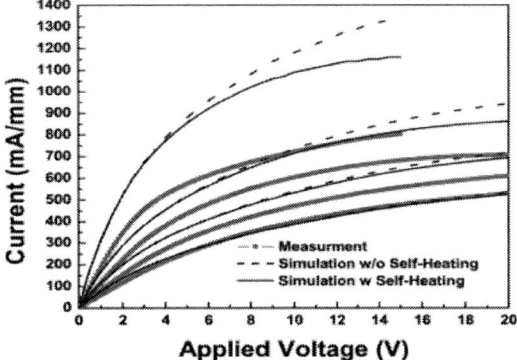

Figure 5: Channel temperature profile of the TLM structures at V=15 V for L1 and V=20 V for L2, L3 and L4.

Figure 6: Measured I-V characteristic of TLM structures (red lines) vs. the hypothetical low-field simulations (black dashed lines without the self-heating, black lines with the self-heating) assuming fixed polarization used for the L4=18 μm.

characteristics for when the polarisation factor is fixed at a value calibrated for the TLM structure with L4=18 µm, which would hugely overestimate the measured current even at a low electric field.

4. Conclusion

We have investigated $GaN/Al_{0.32}Ga_{0.68}N/AlN/GaN/Al_{0.1}Ga_{0.9}N$ TLM structures grown on a *p*-type doped Si (111) substrate. The I-V measurement results were calibrated by 2D drift-diffusion transport model using low and high mobility models, Fermi-Dirac statistic and Shockley-Read-Hall recombination model by commercial tool Atlas-Silvaco. The electro-thermal model with calibrated thermal conductivity against experimental measurements of bulk GaN was employed to study the self-heating effects on the TLM structure. We have found that the current becomes significantly limited by increase in a lattice temperature with the increase in applied bias and that this limitation occurs sooner in shorter structures. The maximum temperature (362 K) was predicted at a vicinity of the drain. In addition, we have observed that, by applying electrical stress (voltage) on the Ohmic contacts, the total polarization value decreases by 7 %, 10 %, 17% in the 12 µm, 8 µm, and 4 µm TLM structures, respectively, when compared to the largest 18 µm source-to-drain TLM structure. The inverse piezoelectric effect changing the polarization in the heterostructure thus affecting a 2DEG in the channel.

References

[1]   M. Ishida, Y. Uemoto, T. Ueda, T. Tanaka, and D. Ueda, *The 2010 International Power Electronics Conference (IPEC)*, 1014–1017, Jun. 2010.

[2]   B. P. Luther, S. E. Mohney, T. N. Jackson, M. A. Khan, Q. Chen, and J. W. Yang, *Appl. Phys. Lett.*, 70, 57-59, Jan. 1997.

[3]   A. Motayed, R. Bathe, M. C. Wood, O. S. Diouf, R. D. Vispute, and S. N. Mohammad, *J. Appl. Phys.*, 93, 1087–1094, Dec. 2002.

[4]   R. Quay, and V. Palankovski, Springer-Verlag, Wien 2004.

[5]   H. Shibata, Y. Waseda, H. Ohta, K. Kiyomi, K. Shimoyama, K. Fujito, H. Nagaoka, Y. Kagamitani, R. Simura, and T. Fukuda, *Materials Transactions, The Japan Institute of Metals*, 48, 2782-2786, 2007.

[6]   Silvaco, *Atlas User' s Manual*, CAD:Silvaco Inc., Santa Clara, 2016.

[7]   M. J. Uren, J. Moreke, and M. Kuball, *IEEE Trans. Electron Devices*, 59, 3327–3333, Dec. 2012.

[8]   U. Chowdhury, J. L. Jimenez, C. Lee, E. Beam, P. Saunier, T. Balistreri, S. Y. Park, T. Lee, J. Wang, M. J. Kim, J. Joh, and J. A. de Alamo, *IEEE Electron Device Lett.*, 29, 1098-1100, 2008.

ASDAM 2016, The 11th International Conference on Advanced Semiconductor
Devices And Microsystems, November 13-16, 2016, Smolenice, Slovakia

Effect of HCl pretreatment on the oxide/semiconductor interface state density in AlGaN/GaN MOS-HEMT structures with MOCVD grown Al_2O_3 gate dielectric

M. Ťapajna, K. Hušeková, O. Pohorelec, L. Válik, Š. Haščík, F. Gucmann, K. Fröhlich, D. Gregušová, and J. Kuzmík

[1]Institute of Electrical Engineering, SAS, Dúbravská cesta 9, 841 04 Bratislava, Slovakia,
e-mail: milan.tapajna@savba.sk

Suppression of surface donors (SDs) in AlGaN/GaN MOS-HEMTs represents a promising approach towards realization of normally-off switching devices with high threshold voltage. In this work, density of oxide/barrier interface traps (D_{it}) was determined in AlGaN/GaN MOS-HEMT structures with different SDs density (N_{DS}), resulted from HCl pre-treatment variation. The results suggest deteriorated interface quality for sample without HCl cleaning. D_{it} was found to increase from $\sim 10^{12}$ to $\sim 10^{13}$ $eV^{-1}cm^{-2}$ in the energy range of 0.8 to 1.1 eV below the conduction band edge for structures with and without HCl cleaning step, respectively. On the other hand, our analysis indicates negligible contribution of interface traps to observed threshold voltage in thermal equilibrium. This indicates the nature of SDs to be different from that of interface traps.

1. Introduction

GaN based high electron mobility transistors (HEMTs) are very attractive for high-power switching applications due to low ON-state resistance and high breakdown in the OFF state. Due to safety reasons, normally-off switching device operation with threshold voltage (V_{th}) close to 5 V is necessary in the applications [1]. Although number of concepts for normally-off GaN HEMTs has been proposed in the literature [2], V_{th} of these devices typically does not exceed 3 V. It has been demonstrated that metal-oxide-semiconductor (MOS) gated HEMT (MOS-HEMT) can overcome this limitation, providing that sufficiently high negative net charge exist at the oxide/III-N interface [3]. This can be achieved by suppression of the so-called surface donors SD, which compensates III-N barrier surface polarization charge [3]. Recently, we have reported that SD density (N_{DS}) can be manipulated in GaN MOS-HEMTs using HCl pre-treatment applied prior to deposition of Al_2O_3 by MOCVD [4]. On the other hand, different surface treatment can result in deteriorated oxide/barrier interface quality and thus reliability issues related to V_{th} instabilities.

In this work, we analyze the impact of HCl pre-treatment on the distribution of oxide/semiconductor interface traps density (D_{it}) in GaN/AlGaN/GaN MOS-HEMTs with MOCVD-grown Al_2O_3 at 600 °C. It was found that for samples without HCl pretreatment, giving reduced N_{DS}, D_{it} increases from 10^{12} to 10^{13} $eV^{-1}cm^{-2}$ in the energy range of 0.8 to 1.1 eV below GaN conduction band edge (CBE).

2. Experimental details

GaN/$Al_{0.29}Ga_{0.71}$N/GaN (3 nm/17 nm/1.5 μm) heterostructures were grown on sapphire substrates by metal-organic chemical vapor deposition (MOCVD). After Ti/Al/Ni/Au Ohmic contacts evaporation and annealing (850 °C, 1 min), Al_2O_3 gate dielectrics were grown by

978-1-5090-3084-2/16 $31.00 © 2016 IEEE   207

MOCVD at 600°C using Al acetylacetonate precursor diluted in toluene using O_2 as reactant and Ar as carrier gas. Prior to the oxide deposition, HCl treatment was applied to part of the wafer (w/ HCl in the following) to etch away the native oxide, while part of the sample was cleaned without HCl treatment (w/o HCl). After post-deposition annealing at 700 °C for 60 min in N_2 ambient, Ni/Au gate metallization was evaporated and patterned. Electrical characterization of MOS HEMT devices was performed on circular diode structures with diameter of 80 μm.

Three complementary techniques were used for D_{it} determination: (i) D_{it} in the energy range of E-E_C=0.5 to 1.1 eV was measured using V_{th}-transients, deduced from the capacitance transients monitored at $V_g \approx V_{th}$-0.5 V at temperatures ranging from 25 to 275 °C. Only relative change (ΔV_{th}) between V_{th} transients measured after 'filling' pulse V_F=0 (no filling) and V_F=4 – 8 V (interface traps filling) was used for D_{it} determination as exemplified in Fig. 1(a). This way, other parasitic capture/emission processes such as oxide bulk trapping reported in [5] can be cancelled-out. $D_{it}(E)$ is determined from the time derivative of ΔV_{th} transient. (ii) Filtered light-assisted V_{th} transient technique was used to measure D_{it} distribution of very deep interface states in the range of 1.75 to 3.25 eV. Here, the difference between V_{th} transient measured in the dark and after light exposure was taken into account, as exemplified in Fig. 1(b) for light energy cantered around 2.1 eV. The light from broad-band Xe-lamp (SP, ASB-XE-175) was filtered using band-pass metal interference filters with FWHM spectral transmittance ranging from 17.5 to 5 nm for light energy of 1.75 to 3.25 eV, respectively. In contrast to [5], the light was exposed after certain time (30 s) and kept until the transient termination. Transient needs to be long enough to reach saturation between V_{th} shifts with and without light exposure. It was found that 200 s is sufficient to reach the saturation. (iii) For shallower traps (E-E_C=0.2 – 0.3 eV), C-V curve frequency dispersion technique was used, as proposed in [6]. This technique relies on the frequency response of interface traps capable to capture and emit electrons when aligned with the Fermi level, i.e. at positive gate voltages (V_g-s). Depending on the measurement frequency, difference in capacitance onset voltages, ΔV, is related to interface traps density, while the trap energy range is obtained from the

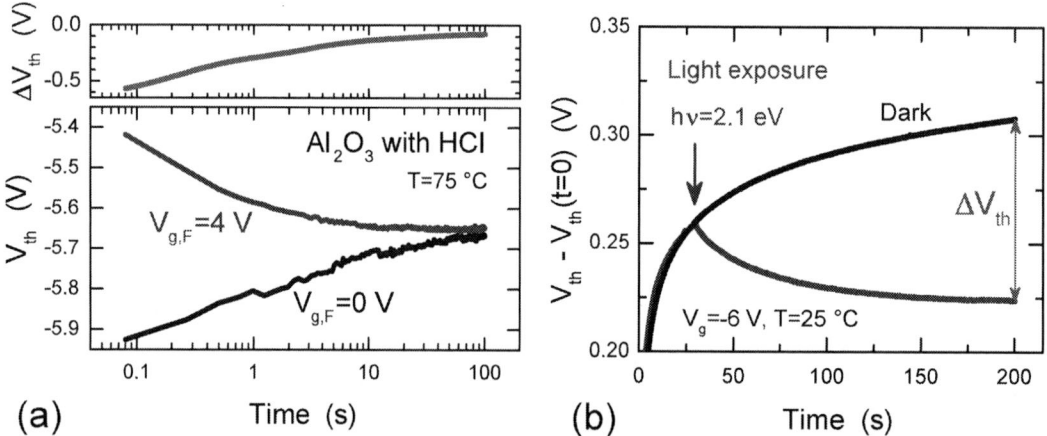

Fig. 1 (a) Typical V_{th} transients (extracted from C transients) of MOS-HEMT with HCl after V_g step from filling voltage ($V_{g,F}$) to depletion part (V_g=V_{th}-0.5). ΔV_{th}, calculated from the difference between V_{th}-t using $V_{g,F}$=0, (i.e. no filling of traps located above E_F in equilibrium) and V_g=4 V is then used for D_{it} determination, while emission time at given temperature relates to trap energy. (b) Typical V_{th} transients monitored after V_g step from 0 to V_{th}-0.5 at room temperature in the dark and after filtered light exposure. After saturation of both transients, ΔV_{th} is extracted graphically and used for D_{it} calculation while light energy is used for E_{it} measured from E_C.

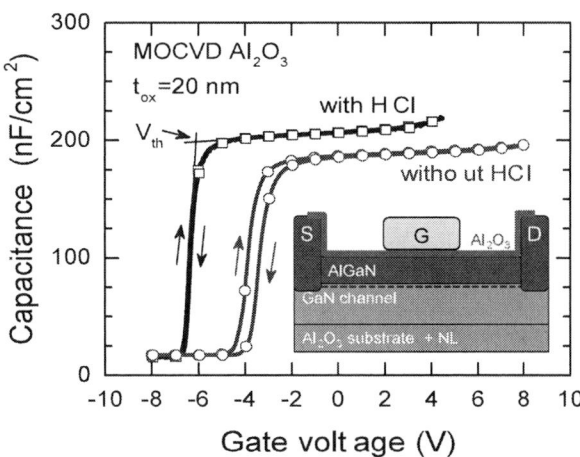

Fig. 2 CV hysteresis of MOS-HEMTs structures with and without HCl pretreatment prior to Al_2O_3 deposition. Measurement frequency of 1 MHz and sweep rate of 0.2 V/s was used.

frequency difference, i.e. time constant constraints.

3. Results and discussion

Figure 2 compares typical capacitance-voltage (CV) double-sweep characteristics of investigated MOS-HEMT structures. V_{th} (determined graphically) increases from -6 V to -3.5 V for structures with 20-nm thick Al_2O_3 with and without HCl treatment, respectively. This was attributed to change in N_{DS}, as deduced from V_{th}-t_{ox} dependences reported elsewhere [4]. On the other hand, structures with HCl pre-treatment show negligible hysteresis (<100 mV), while it increases to 0.55 V for device without HCl pre-treatment. As the positive hysteresis can be attributed to capture of electrons by interface states at positive V_g-s, increased hysteresis indicates higher D_{it} in the relevant energy range.

Experimentally determined D_{it} distributions are summarized in Fig 3. Both structures show typical U-shaped distribution with similar D_{it} ranging from 3×10^{11} to 8×10^{12} $eV^{-1}cm^{-2}$ for $E>E_C$-1.5 eV. As expected, sample w/o HCl treatment show increased D_{it} featuring broad peak in the energy range E_C-0.8 to 1.1 eV below E_C, reaching D_{it} of 1.5×10^{13} $eV^{-1}cm^{-2}$ that is an order of magnitude higher than that for sample w/ HCl. The increased D_{it} for structure w/o HCl is fully consistent with the observed CV hysteresis. During positive CV sweep, interface states above Fermi level at V_g=0 ($E_{F,0}$) can capture electrons resulting in positive shift in the CV curve. Electrons trapped in deep traps (>0.5 eV below CBE) cannot be re-emitted back to the barrier CB due to much higher time constant compared to measurement time, giving positive CV hysteresis.

Finally, let us discuss possible relation between N_{DS} and interface trapped charge N_{it} contributing to V_{th}. Also shown in Fig. 3 are energy band diagrams for both structures calculated by 1-D Poisson equation using theoretical polarization charges and the net oxide/III-N interface charge determined from V_{th}-t_{ox} dependences [4]. Using standard model for interface trap nature, N_{it} can be calculated from the integration of $D_{it}(E)$ from $E_{F,0}$ to E_{CNL}, giving positive N_{it} of 2.3×10^{11} and 1.35×10^{12} cm^{-2} for sample with and without HCl treatment, respectively. This gives positive ΔN_{it} of 1.12×10^{12} cm^{-2} that is an order of magnitude lower than the difference in N_{DS} between the two samples being 1.33×10^{13} cm^{-2}. Moreover, reduction of positive net charge is necessary to explain observed V_{th} shift between the

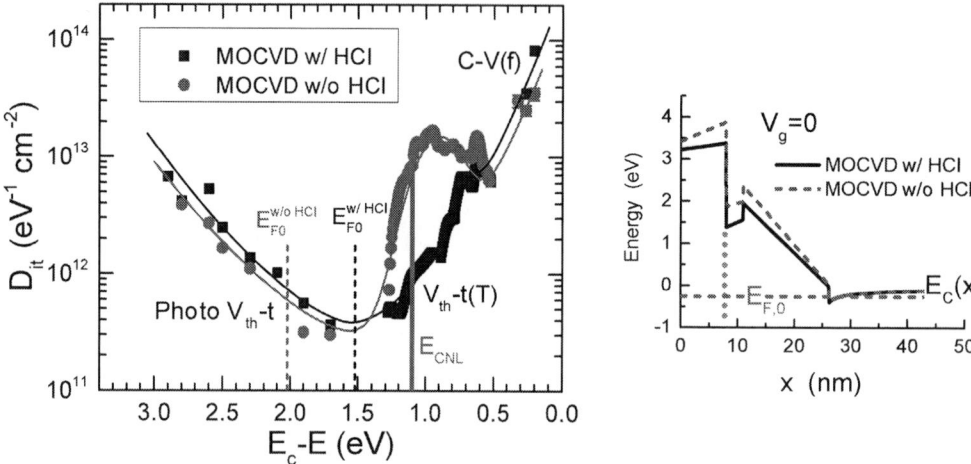

Fig. 3 (a) $D_{it}(E)$ distributions of MOS-HEMT structures with and without HCl pretreatment prior to Al_2O_3 deposition determined using three complementary techniques. Also shown is a position of theoretical charge neutrality level E_{CNL} and position of E_F in equilibrium ($E_{F,0}$) at oxide/III-N interface taken from the calculated band diagrams for both structures (b).

analyzed structures. Our analysis therefore indicates the nature of SD to be different from that of interface traps.

4.   Conclusions

We determined D_{it} distribution in Al_2O_3/GaN/AlGaN/GaN MOS-HEMT structures with different N_{DS}, resulted from HCl cleaning step prior to the oxide deposition. As expected, the data suggest deteriorated interface quality when HCl cleaning is not applied on the III-N surface. D_{it} was found to increase from $\sim10^{12}$ to $\sim10^{13}$ eV^{-1}cm^{-2} in the energy range of 0.8 to 1.1 eV below GaN CBE. Our analysis indicates N_{it} related shift in V_{th} in opposite direction compared to the observed V_{th} shift, indicating independence between N_{DS} and D_{it}.

Acknowledgement

This work was supported by V4-Japan joint call on advanced materials (project SAFEMOST) and national projects APVV 15-0031 and VEGA 2/0138/2014.

References

[1]   M. A. Briere, *Power Semicond.* **4**, 30, 2013.
[2]   T. Kachi, *IEICE Electron. Exp.* **10**, 1, 2013.
[3]   M. Blaho, D. Gregušová, *et al., Phys. Status Solidi* A **212**, 1086, 2015.
[4]   M. Ťapajna, *et al., in Proceedings of the Int. Workshop Nitride Semicond. IWN 2016*, Orlando, USA, 2016, C0.2.04.
[5]   M. Ťapajna, M. Jurkovič, *et al., J. Appl. Phys.* **116**, 104501, 2014.
[6]   Y. Hori, C. Mizue, and T. Hashizume, *Japan. J. Appl. Phys.* **49**, 080201, 2010.

ASDAM 2016, The 11th International Conference on Advanced Semiconductor
Devices And Microsystems, November 13-16, 2016, Smolenice, Slovakia

RF Performance Optimization of SiGe:C Transistors via Ge-B Doping Strategies and RTP

A. Emre Yarımbıyık, Duygu İşler Öksüz and Enes Cesur

TÜBİTAK (The Scientific and Technological Research Council of Turkey), UEKAE (Electronics Cyrptology Inst.) YİTAL (Semiconductor Technologies Research Lab.), Annibal Street, 41400, Gebze-Kocaeli / TURKEY
e-mail: emre.yarimbiyik@tubitak.gov.tr, duygu.isler@ tubitak.gov.tr

This paper focuses on the optimization of Germanium and Boron profiles at the epitaxial SiGe:C layer of a self-aligned SiGe:C transistor structure with 0.13 µm emitter width as well as the last significant thermal step (RTP) which follows the in-situ phosphorus doped poly-Si deposition of the emitter electrode. At first, critical process steps affecting the doping profile at the base such as emitter epitaxy, selective implant of collector (SIC) and the last RTP (950°C, 5 seconds) have been conserved and 6 different Germanium and Boron doping strategies were applied at the 250Å SiGe:C epitaxial growth which is followed by a 150 nm Si-CAP deposition. The epi-growth strategy with the best RF performance outcome was chosen for further optimization with the RTP step. As a result of these two successive optimizations, simulated transistors exhibit f_t , f_{max} , BVceo and β of 192GHz, 211GHz, 2.09V and 1195 respectively.

1. Introduction

SiGe:C process flow corresponds to one of the most complicated and cumbersome fabrications in planar semiconductor device production. For this reason, trial and error is not a good approach for optimizing the key steps of the process flow. SENTAURUS semiconductor simulation tools have been used to assist the process planning for 0.13 micron technology SiGe:C transistors fabricated in TÜBİTAK YİTAL (Semiconductor Technologies Research Laboratory).

The doping profile of the base region is the dominant factor affecting the performance of the SiGe:C transistors. N-type doping levels of the emitter and the collector are set with emitter epitaxy and SIC respectively. The doping profile of the base region is affected by these levels, SiGe:C epitaxial growth recipe which also includes the Boron deposition information, Si-CAP thickness and the thermal steps, especially the RTP (rapid thermal processing) step that follows the emitter epitaxy. In order to optimize the SiGe:C epitaxy and RTP all other processes were kept the same. SiGe:C epitaxy and RTP were then optimized in turn as follows: 1) optimization of SiGe:C epitaxy with an RTP of 950°C for 5 seconds, 2) optimization of RTP with the SiGe:C recipe choice of the first optimization.

Epitaxial deposition of SiGe:C instead of Si reduces the energy level of the conduction band at the base region. This increases the collector current allowing more electrons to be transferred when subjected to the same emitter-base voltage. In addition, a graded concentration of Germanium (increasing in the direction of collector) creates a built-in electric field which accelerates the electrons across the base, enabling the transistor to work at higher frequencies [1]. Trapezoidal or hybrid (box + trapezoid) Germanium profiles have also been a choice for researchers and fabs [2, 3, 4]. Boron concentration also plays an important role for the performance of the transistor. This study includes an examination of different (box-

978-1-5090-3084-2/16 $31.00 © 2016 IEEE

triangular-trapezoidal-hybrid) Germanium profiles along with box Boron profiles changing in width (limited by our processing capabilities) and position.

2.   Ge-B doping and RTP optimization

A self-aligned SiGe:C transistor structure with 0.13 µm emitter width is taken as the starting point [5]. To examine different Ge-B doping strategies, all process steps except the epitaxial growth of the SiGe:C layer are kept the same. Phosphorus concentration at the emitter is 5×10^{20} and SIC (selective implant of collector) is performed with a dose of 10^{14} ions/cm^2 at 150keV. At this stage of optimization, the RTP step which follows the emitter epitaxial deposition is at 950°C for 5 seconds. Six different Germanium-Boron doping strategies were applied at the 250Å SiGe:C epitaxial growth. In all simulations 0.2% Carbon was deposited along with Boron to suppress its diffusion [6], SiGe:C deposition is followed by a 150 nm Si-CAP deposition.

Figure 1 shows 6 Ge-B doping strategies which include equal amount of total Boron doping and profiles of Germanium varying from a usual box and triangle to trapezoid and hybrid (box + trapezoid). Table 1 includes the performance data (current gain, unity gain cut-off frequency, maximum oscillation frequency, emitter-collector open base breakdown voltage) of 6 different transistors corresponding to these doping recipes. Here $f_t \times BVceo$ is a useful metric, because it shows a combination of how fast and flexible (in the sense of wider operation range) the transistor is together. Ge-B profile #1 and #6 are the best transistors according to this metric. Using Table 1, one can choose a doping profile that fits the desired performance. Here our primary goal is to fabricate a SiGe:C transistor with the highest possible f_t value with at least a moderate breakdown voltage; therefore we choose #6 as our doping strategy (doping profile before the final thermal process, RTP).

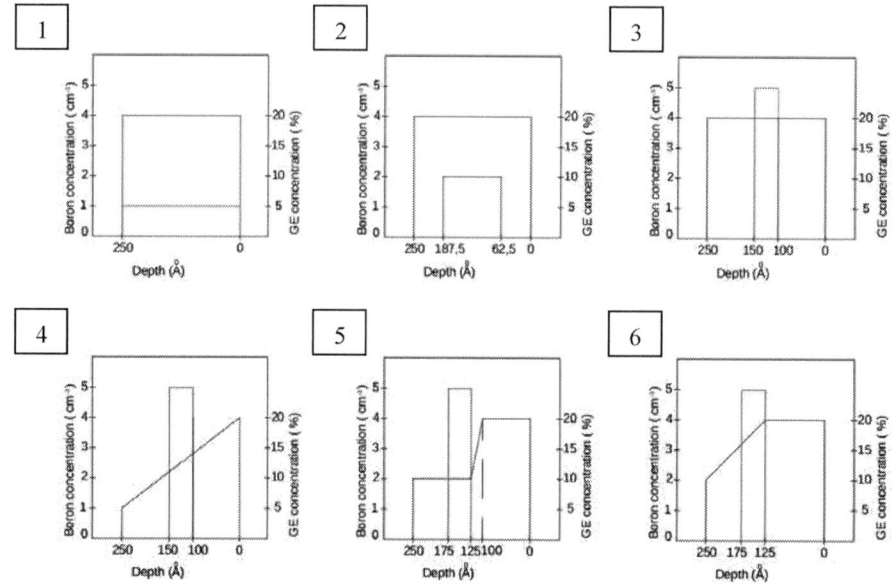

Figure 1- Six Ge-B doping profiles (before RTP) which include equal amount of B.

Having chosen doping profile #6 as the goal to reach in the epitaxial growth of the SiGe:C layer, we went on to look for the optimum temperature and time for the RTP that follows the emitter epitaxial deposition. The RTP times between 5 and 20 seconds were examined for 900°C, whereas for 950°C and 1000°C 5 to 15 seconds were simulated. Figure 2

Ge-B Profile #	β	f_t (GHz)	f_{max} (GHz)	BVceo (V)	$f_t \times$ BVceo (GHz V)
1	1737	137	162	2.96	405.5
2	2611	173	181	1.91	330.4
3	1962	159	173	2.11	335.5
4	480	159	177	2.16	343.4
5	327	155	179	2.37	367.4
6	1195	192	211	2.09	401.3

Table 1– Perfomance data of 6 doping profiles with an RTP of 5seconds at 950°C
(interface traps are not taken into account)

and Figure 3 show the cutoff frequency (f_t) and breakdown voltage (BVceo) vs. RTP time respectively. The best f_t result comes out of an RTP at 950°C for 5 seconds. BVceo values for all trials are higher than 2V, which is good enough for our purposes. Looking at 900°C and 950°C curves, one sees that as the thin, 50 Å thick Boron roughly in the middle of the 250 Å SiGe:C layer diffuses, base region starts to widen, breakdown voltage increases but the transistor starts to get slower. The reason for the dramatic difference in the 1000°C curves is the diffusion of Phosphorus at the emmitter turning the trend backwards around 8 seconds. Current gain (β) for all the data points except (950 °C, 5 seconds, β=1195) and (1000°C, 15 seconds, β=1043) are between 290 and 358 (interface traps which reduce the current gain by increasing the recombination rate [7] are not taken into account in this computation).

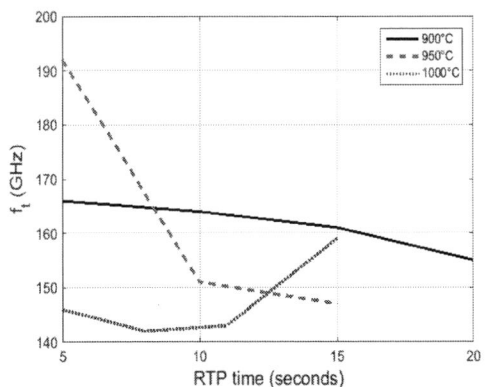

Figure 2- Cut-off frequencies (f_t) vs. t_{RTP}

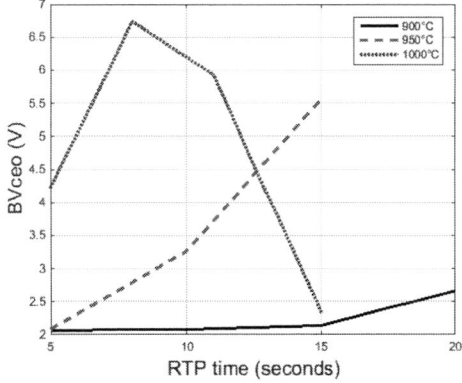

Figure 3- Breakdown voltages (BVceo) vs. t_{RTP}

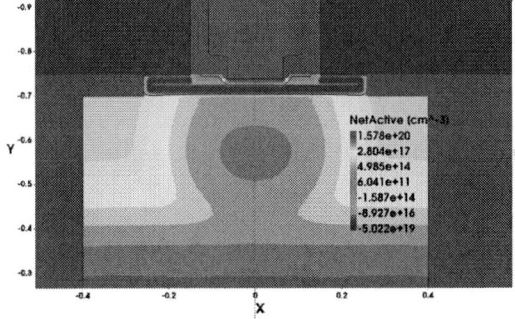

Figure 4- Net doping concentration.

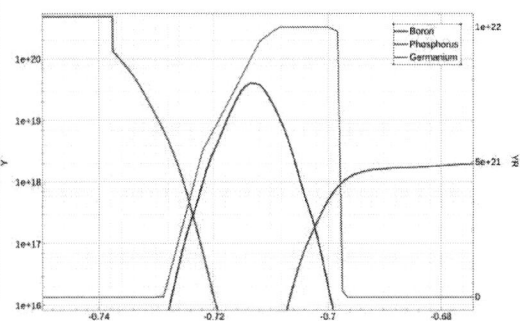

Figure 5- Doping profile along the line in Fig.4

Figure 5 shows the doping profile along the line in Figure 4, passing through the base region of the transistor which was simulated with the 6th doping profile of Figure 1 and a final RTP of 5 seconds at 950 °C. B and P concentrations are plotted in logarithmic left y axis, whereas the Ge concentration is plotted in the linear right y axis (10^{22} stands for 20% as the density of atoms in the crystalline Si structure is 5×10^{22} atoms/cm^3 [7]).

3   Conclusion

We have optimized our SiGe:C epitaxial deposition and rapid thermal processing (RTP) steps with SENTAURUS simulation tools to reach our fabrication goal of cut-off and maximum oscillation frequencies approaching 200GHz: f_t=192 GHz, f_{max} = 211GHz, BVceo = 2.09V. Reduced emitter width, better control over Boron doping width and RTP profile are the keys for further performance improvement.

Acknowledgement

We would like to thank to all members of the YİTAL group, for making this work possible.

References

[1]   G.A.M. Hurkx, *"Bipolar transistor physics"* - *Bipolar and Bipolar-MOS integration*, edited by Hart P.A.H., Elsevier Science B.V., 1994.
[2]   S. Decoutere et.al., Proceedings of the Bipolar/BiCMOS Circuits and Technology Meeting, IEEE, 2008, p9.
[3]   T. Hashimoto et.al., *Journal of the Elect. Dev. Soc.,vol:2, no:4, 50-58, 2014.*
[4]   J. Böck et.al., International Electron Devices Meeting Tech. Dig., IEEE, 2004, p255.
[5]   A.E. Yarımbıyık, Proceedings of the National Conference on Modeling and Simulation in Defence Applications, METU, 2015, p238.
[6]   H. Rücker et.al. , Appl. Phys. Letters 73, 1682-1684, 1998.
[7]   R. F. Pierret, *Advanced Semiconductor Fundamentals Vol.4*, Prentice Hall, NJ, 2003.

ASDAM 2016, The 11th International Conference on Advanced Semiconductor
Devices And Microsystems, November 13-16, 2016, Smolenice, Slovakia

Silicon carbide thin films deposited by PECVD technology for applications in photoelectrochemical water splitting devices

J. Huran[1], P. Boháček[1], V. Sasinková[2], A. Kleinová[3], M. Mikolášek[4], A.P. Kobzev[5],
L. Hrubčín[1,5], J. Arbet[1] and M. Sekáčová[1]

[1] Institute of Electrical Engineering, Slovak Academy of Sciences, Dúbravská cesta 9, 84104 Bratislava, Slovakia

[2] Institute of Chemistry, Slovak Academy of Sciences, Slovak Academy of Sciences, Dúbravská cesta 9, 84541 Bratislava, Slovakia

[3] Polymer institute, Slovak Academy of Sciences, Dúbravská cesta 9, 84538 Bratislava, Slovakia

[4] Faculty of Electrical Engineering and Information Technology, Slovak University of Technology, Ilkovičova 3, 81219 Bratislava, Slovakia

[5] Joint Institute for Nuclear Research, Joliot-Curie 6, 141980 Dubna, Russian Federation

e-mail: jozef.huran@savba.sk

Amorphous silicon carbide films were prepared by PECVD technology on Si substrate with the aim to use SiC films in the technology of photoelectrochemical water splitting devices. Concentration of elements in the films was determined by RBS and ERD method. FTIR, Raman and I-V measurement were used before and after immersion of samples to aqueous sulfuric acid electrolyte. Differences between Raman or FTIR spectra and differences between I-V characteristics before and after immersion to electrolyte are discussed.

1. Introduction

The use of wide band gap semiconductors enables the operation of devices and electronics in harsh environments. Photoelectrochemical (PEC) water splitting devices require semiconductor photoelectrode material fulfilling a number of primary requirements such as band gap, band edge alignment and corrosion resistance to electrolyte, as well as durability. Due to their excellent chemical stability and their back-end compatible deposition temperature (Td<400 $^{\circ}$C), amorphous silicon carbide thin films deposited by PECVD technology are promising candidates for applications in PEC water splitting devices [1, 2].

In this paper, PECVD technology was used for deposition of SiC films using SiH_4, CH_4, H_2 on silicon substrate. Structural and electrical properties were analysed by RBS, ERD, FTIR, Raman methods and IV measurements of structure Al/SiC/Si/Al, before and after immersion to aqueous pH 2.0 and pH 1.0 sulfuric acid electrolyte.

2. Experiment

Amorphous silicon carbide films (a-SiC:H) were deposited on Si substrates by plasma enhanced chemical vapor deposition (PECVD) technique using SiH_4, CH_4 and H_2 gas as precursors at various gas flow rates. P-type silicon wafer with resistivity 6-10 Ωcm and (100) orientation was used as the substrate. Technological parameters for samples were: substrate

978-1-5090-3084-2/16 $31.00 © 2016 IEEE

temperature was 300 °C and gas mixture was for wafer AS1 (SiH$_4$-5 sccm, CH$_4$-30 sccm, H$_2$-100 sccm), and for wafer AS2 (SiH$_4$-5 sccm, CH$_4$-4sccm, H$_2$-100 sccm), respectively. For all samples were RF power 100 W at 13.56 MHz and pressure 100 Pa. The concentration of elements in the films was determined by RBS and ERD analytical method simultaneously. Raman spectroscopy study of the SiC films were performed by using a Raman microscope with 532 nm laser and chemical compositions were analyzed by Fourier transform infrared spectroscopy (FTIR). Raman and FTIR spectroscopy were performed before and after immersion of samples to aqueous pH 2.0 and pH 1.0 sulfuric acid electrolyte (200 h durability test). Electrical properties of SiC films before and after immersion of samples to aqueous pH 2.0 and pH 1.0 sulfuric acid electrolyte were studied by measurement of the *I-V* characteristics on structure Al/SiC/Si/Al. Circular electrodes of Al (200 nm thick) as a contacts with diameter 0.5 mm were formed using metal masks on the side with SiC film on each sample. The other side of samples was fully covered by Al ohmic contact (~300 nm thick).

3.   Results and discussion

RBS and ERD spectra of samples AS1 and AS2 are shown in figure 1. Concentrations of elements were for film AS1(silicon-35 at.%, carbon-33, hydrogen-30 at. %, oxygen-2 at.%) and for film AS2(54,15,29,2), respectively.

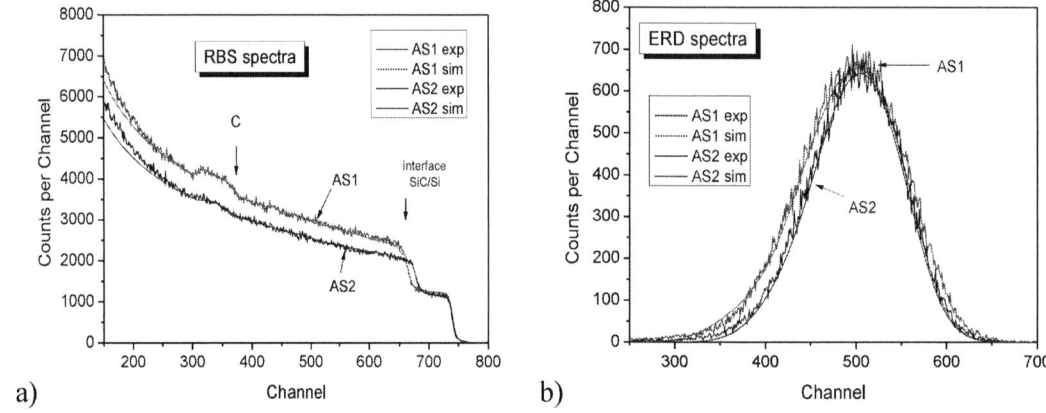

a)   b)

Fig. 1 Experimental and simulated RBS a) and ERD b) spectra for AS1, AS2 samples.

FTIR results, figure 2, showed the presence of Si-C, Si-H, C-H bonds. From the IR spectra of films were determined the main vibration frequencies. The band at 2800 to 3000 cm^{-1} is attributed to stretching vibration of the CH$_n$ group, sp^2 (2880 cm^{-1}) and sp^3 (2920 cm^{-1}) configurations. The band at 2100 cm^{-1} is due to SiH$_m$ stretching vibrations and C-H$_2$ wagging mode from the Si-C-H$_2$ complex at 1000 cm^{-1}. The band at 790 cm^{-1} can be assigned to Si-C stretching vibration. In the case of sample AS2 is clear intensity at 660 cm^{-1} that is assigned to symmetric Si-C stretching. Significant differences between FTIR spectra before and after immersion to electrolyte are not observable. Figure 3 shows Raman spectra of samples AS1 and AS2 before and after immersion to aqueous pH 2.0 and pH 1.0 sulfuric acid electrolyte. Raman intensity at 520 cm^{-1} originates from the lattice vibration of silicon substrate. The Raman band between 930 cm^{-1} and 990 cm^{-1} is created by acoustical and optical phonon modes of cubic or one of the hexagonal polytypes of SiC. The Raman intensity, band between 1300 cm^{-1} and 1700 cm^{-1}, is assigned to diamond like carbon (DLC). It is also observed relatively weaker band around 830 cm^{-1} which is typical for amorphous SiC structure.

978-1-5090-3084-2/16 $31.00 © 2016 IEEE   216

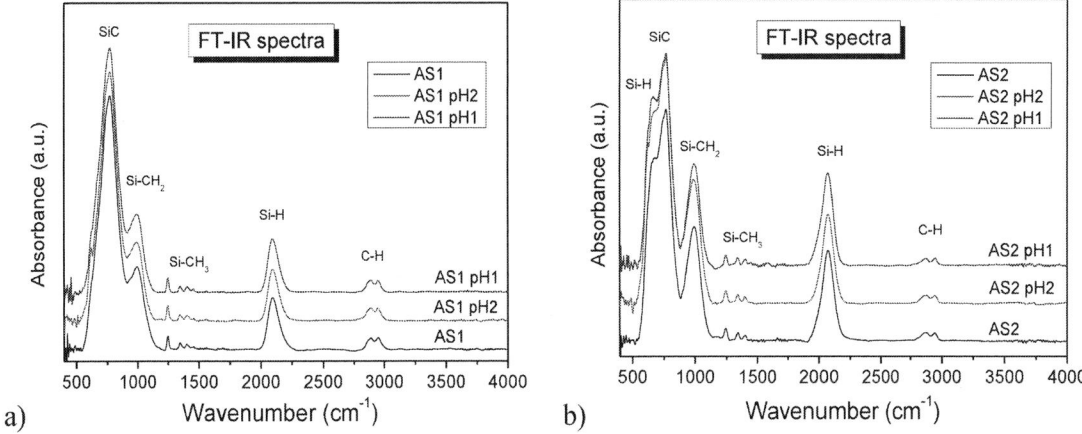

a)   b)

Fig. 2 FTIR spectra of samples AS1 a) and AS2 b) before and after immersion to aqueous pH 2.0 and pH 1.0 sulfuric acid electrolyte.

In the case of sample AS2 is clear intensity, band between 60 cm^{-1} and 260 cm^{-1}, that is typical for amorphous Si and intensity at 470 cm^{-1} that is assigned to Si-Si vibrational mode. Raman intensities of SiC and Si-Si vibrations decreased after immersion of samples to electrolytes with pH2 and pH1. It can be explain by slightly oxidation of SiC film after immersion to electrolyte, result in, SiC films starts more amorphous than before immersion to electrolyte and amorphisation continued with immersion to stronger electrolyte.

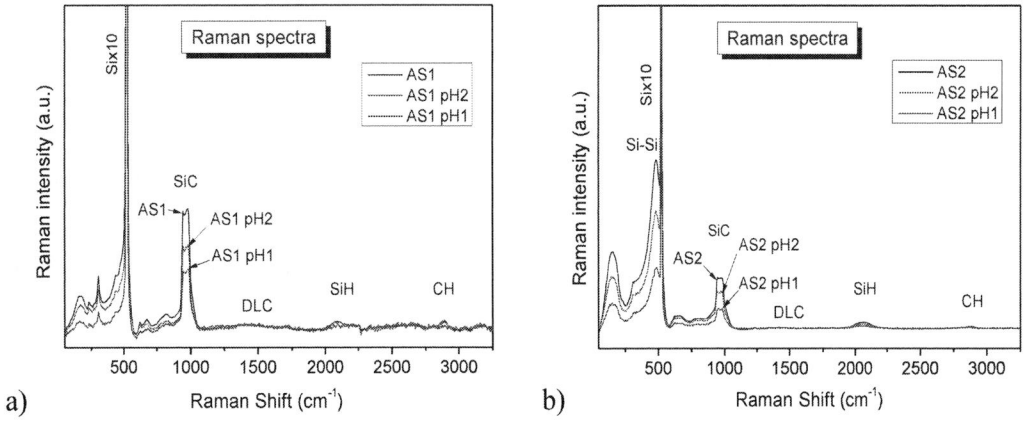

a)   b)

Fig. 3 Raman spectra of samples AS1 a) and AS2 b) before and after immersion to electrolyte.

Figure 4 shows *I-V* characteristics of structure Al/SiC/Si/Al with AS2 film before and after immersion of samples to aqueous pH 2.0 and pH 1.0 sulfuric acid electrolyte. *I-V* dependencies shown linear behavior for both polarity of the bias voltage ranging from 0.01 to 0.08 V for all measured samples. The measured *I-V* characteristics of sample AS2 are similar but great quantitative differences (more than one order) in the conductance are observable in the linear part of characteristics. The lowest calculated conductance of SiC film is about 3×10^{-11} S, rises up to 2×10^{-10} S for the samples immersed to aqueous pH 2.0 and pH 1.0 sulfuric acid electrolyte, respectively. The highest calculated conductance is about 4×10^{-10} S for the sample before the immersion to the electrolyte. The measured structures exhibit rising rectification ratio of 2, 8, and 33 at ±4 V of the bias voltage for the samples before and after immersion to aqueous pH 2.0 and pH 1.0 sulfuric acid electrolyte. At low bias voltage, the transport is probably governed by a multistep tunneling hole emission process through the a-SiC/c-Si interface. At still higher reverse bias voltages (>0.2 V), space-charge-limited hole conduction in the presence of traps in the a-SiC film limits transport [3].

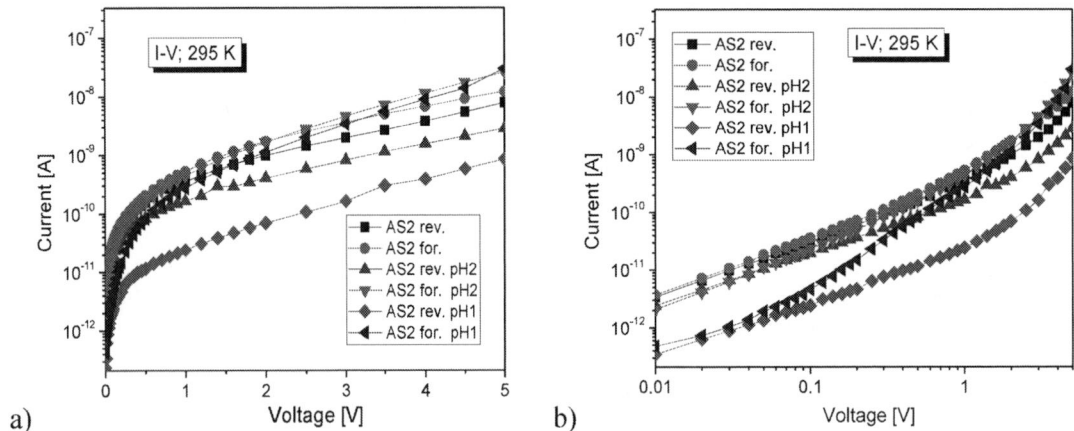

Fig. 4 I-V characteristics of structure Al/SiC(P)/Si/Al –sample AS2 in the semi-logarithmic (a) and in the logarithmic (b) scale, respectively.

4. Conclusions

We have investigated the structural properties of amorphous SiC films and electrical properties of structure Al/SiC/Si/Al before and after immersion of samples to pH2 and pH1 electrolyte. The RBS and ERD results showed main concentrations of Si, C and H. The FT-IR results of the SiC films showed the presence of the Si-C, Si-H, and C-H bonds. Significant differences between FTIR spectra before and after immersion to electrolyte are not observable. Raman spectra of SiC films showed interesting band intensities between 930 cm^{-1} to 990 cm^{-1}. SiC film after immersion to electrolyte starts more amorphous than before immersion to electrolyte. The measured *I-V* characteristics exhibited rising rectification ratio of 2, 8, and 33 at ±4 V of the bias voltage for the samples before and after immersion to aqueous pH 2.0 and pH 1.0 sulfuric acid electrolyte. The highest calculated conductance was about 4×10^{-10} S for the sample before the immersion to the electrolyte.

Acknowledgement

This research has been supported by the Slovak Research and Development Agency under the contracts APVV-0443-12 and by the Scientific Grant Agency of the Ministry of Education and Sport of the Slovak Republic and Slovak Academy of Sciences, No. 1/0651/16.

References

[1] M. G. Walter, E. L. Warren, J. R. McKone, S. W. Boettcher, Q. Mi, E. A. Santori, and N. S. Lewis, *Chem. Rev.* **110**, 6446–6473, 2010
[2] L.M. Peter, *Electroanalysis* **27**, 1 – 9, 2015
[3] A. N. Nazarov, Ya. N. Vovk, V. S. Lysenko, V. I. Turchanikov, V. A. Scryshevskii, and S. Ashok, *J. Appl. Phys.* **89**, 4422, 2001.

ASDAM 2016, The 11th International Conference on Advanced Semiconductor
Devices And Microsystems, November 13-16, 2016, Smolenice, Slovakia

Investigation of Metal Contacts on Semi-Insulating GaAs: Physics, Technology and Applications

F. Dubecký[1], G. Vanko[1], D. Kindl[2], P. Hubík[2], E. Gombia[3], P. Boháček[1], M. Sekáčová[1] and B. Zaťko[1]

[1] Institute of Electrical Engineering, SAS, Dúbravská cesta 9, 841 04 Bratislava, Slovakia
[2] Institute of Physics v.v.i., AS CR, Cukrovarnická 10, CZ-162 00 Praha 6, Czech Republic
[3] IMEM-CNR Institute, Parco Area delle Scienze 37/A, 43010 Fontanini, Parma, Italy
e-mail: elekfdub@savba.sk

The work reports on a study of the symmetric metal/SI GaAs/metal (M-S-M) diodes in order to demonstrate the effect of contact metal work function. We compare the high work function Pt contact versus the low work function Mg contact. The Pt-S-Pt, Mg-S-Mg and Mg-S-Pt structures are characterized by the current-voltage measurements. The Mg-S-Pt structure show a significant current decrease at low bias while the Mg-S-Mg structure shows saturation current at high voltages more than an order of magnitude lower with respect to the Pt-S-Pt reference. The phenomena observed in Mg-containing samples are explained by the presence of insulating MgO layer at the M-S interface. The reported findings have potential applications in sensors based on SI GaAs.

1. Introduction

Semi-insulating (SI) GaAs has become an important candidate for the fabrication of X- and γ-ray detectors [1] due to its high resistance to radiation damage, good physical characteristics, quality of the base material and well developed device technology. Applications in the field of ultraviolet [2] detection and terahertz devices [3] have been also investigated. Apart from the base material [1], the performance of SI GaAs-based devices significantly depends on structure and technology of the electrodes, which are of our interest here.

Focusing on radiation detectors, the standard electrode setup uses a small-area barrier Ti/Pt/Au contact on one side of the SI GaAs wafer and full-area ohmic n^+ contact on the opposite side. An improvement of the SI GaAs radiation detector performance was achieved by the "non-alloyed" ohmic contact introduced by Alietti et al. [4]. Evaporated metal-SI GaAs contact represents another, more economical and simple solution, investigated here. In case of undoped n-type SI GaAs, a metal with a sufficiently low work function, Φ, in comparison with the semiconductor (SI GaAs: Φ = ~4.5 eV [5]), should, in ideal case, cause anti-blocking band bending and thus behave as "non–injecting" quasi-ohmic contact. This is, however, not an easy task to achieve in general, for the usual presence of the Fermi level (FL) pinning at the metal-semiconductor interface.

In this study, we investigate the metal-SI GaAs-metal (M-S-M) structures with a conventional Pt as high Φ metal and less explored Mg in a role of low Φ metal with identical geometry on both sides to show the effect of the contacts in a straightforward way. Three distinct types of structures with full-area contacts, deposited on the identical "detector-grade" bulk SI GaAs material, are compared: (i) Pt-S-Pt, (ii) Mg-S-Pt, and (iii) Mg-S-Mg. The prepared structures are characterized by the current-voltage (*I-V*) measurements performed at room temperature. We find that the transport characteristics of the considered structures strongly depend on the contact configuration.

2. Technology and experimental details

The M-S-M structures were prepared from Freiberger bulk undoped SI GaAs 3" wafer grown by vertical gradient freeze with (100) crystallographic orientation and dislocation density of about 3000 cm^{-2}. The wafer was polished from both sides down to (180±10) μm. Resistivity and the Hall mobility measured by the van der Pauw method at 300 K (RT) give 5.3×10^7 Ωcm and 5950 cm^2/Vs, respectively. The measured values fulfil the key requirements for a "detector-grade" bulk SI GaAs [1]. The wafer was segmented into fragments used for the fabrication of three sets of different samples. Samples with Mg/Au contacts on both sides, Mg/Au and Pt/Au contacts, and Pt/Au

978-1-5090-3084-2/16 $31.00 © 2016 IEEE   219

contacts on both sides of the wafer fragments were fabricated. Just before metal evaporation the surface oxides were removed by dipping the sample in a solution of HCl:H$_2$O = 1:1 at RT for 30 sec. The low Φ Mg (Φ = 3.68 eV) and high Φ Pt (Φ = 5.65 eV) metal contacts with topside/backside thickness of 10/40 nm covered in situ by 5/60 nm Au were evaporated in a dry high-vacuum system using electron gun. After the evaporation, samples with dimensions of about 3×3 mm^2 were cleaved from the wafer fragments. I-V characteristics of the prepared structures were measured using a Keithley 237 source controlled by personal computer directly on the cleaved samples using tip on the top contact. Measurements were performed in slow, steady-state regime at RT in the dark using an electrically shielded probe station. The measurement in opposite polarity started till the residual discharge current reached value of a detectable limit.

3.   Results and discussion

The I–V characteristics of the fabricated structures measured at RT in the dark are shown in Fig. 1. The dashed straight line corresponds to the calculated linear-ohmic dependence controlled by the bulk material resistivity and the structure geometry giving a resistance of $R_{OBL} = 9.6 \times 10^7$ Ω. The measured lower current (labelled hereafter as "reverse" – this label is related to the lower current branch in all cases; in the case of the Mg-S-Pt structure it corresponds to minus polarity applied to the thin Mg contact) characteristics typically consist of four regions: (i) initial linear part, (ii) soft saturation region, (iii) current increase and (iv) high field saturation region. The initial linear part of the measured characteristics of particular samples give resistances $R_{Mg-Mg} = 1.9 \times 10^9$ Ω, $R_{Mg-Pt} = 6.6 \times 10^9$ Ω and $R_{Pt-Pt} = 9.8 \times 10^7$ Ω. The sublinear, "soft saturation" observed in a bias voltage region of about 0.1 V - 5 V (Mg-S-Mg, Mg-S-Pt) and 0.05 V - 0.5 V for the sample Pt-S-Pt is obviously related to the saturation current due to the thermionic emission [6]. This region is followed by a current increase being superlinear for samples with Mg contact, but linear for the sample with Pt contacts, reaching a second saturation region at a bias over about 50 V. The sublinear part observed in Mg-S-Pt sample (reverse branch) is followed by an injection region ($I \sim V^2$) in a bias range of 2 V - 20 V. The bias voltage >80 V represents the usual operation range of a radiation detector, which is characterized by a dominant transport mechanisms controlled by the saturation drift velocity of charge carriers in high electric fields (>10^4 Vcm^{-1} [7]). Even if the "reverse" breakdown voltage of all samples exceeds 100 V we stopped the measurement at 100 V because the cleaved samples could be influenced by possible leakage through the edge surface. As for the "higher" current ("forward") branches, the initial linear part is followed by slight superlinear region above ~0.02 V. The parts of the "forward" characteristics of samples Mg-S-Pt and Pt-S-Pt above about 1 V correspond to those of the "reverse" direction, but are shifted to higher currents. However, the sample Mg-S-Mg shows serious asymmetry: instead of two qualitatively similar characteristics due to the same metal contacts on both sides, we observed strong unsymmetrical characteristics with qualitatively different transport behaviour in 0.02 V – 10 V range.

Regarding the transport characteristics, the M-S-M structures with identical metal on both sides, one would expect symmetrical I-V characteristics. Nevertheless, we observed partial deviation from the symmetry in the case of Pt-S-Pt sample and quite unexpectedly large difference in the case of Mg-S-Mg sample. We can speculate that the reason is related to the different thickness of the evaporated metallization, a role of the surface state prior evaporation or different interface formation after finishing the evaporation process. In case of Au/MgO/n-GaAs, the effect of slight Mg-oxide thickness variation influences the electrical transport considerably [8]. In addition, in the case of Mg-S-Mg structure, we could also speculate about the technology-induced asymmetry due to the sequential evaporation of contacts. In such case, the strong influence of Mg on one side could induce altering of the base material properties in the vicinity of the interface, or the second metallization process could lead to formation of a contact with different characteristics although the material and technology on both sides remain the same.

The most important feature of the sample with two Mg contacts is the observed high field saturation (over ~20 V) at current more than 20 times lower comparing with other two samples. We must remind the reader that Baldini et al. [6] in their study of transport characteristics of varying SI

978-1-5090-3084-2/16 $31.00 © 2016 IEEE

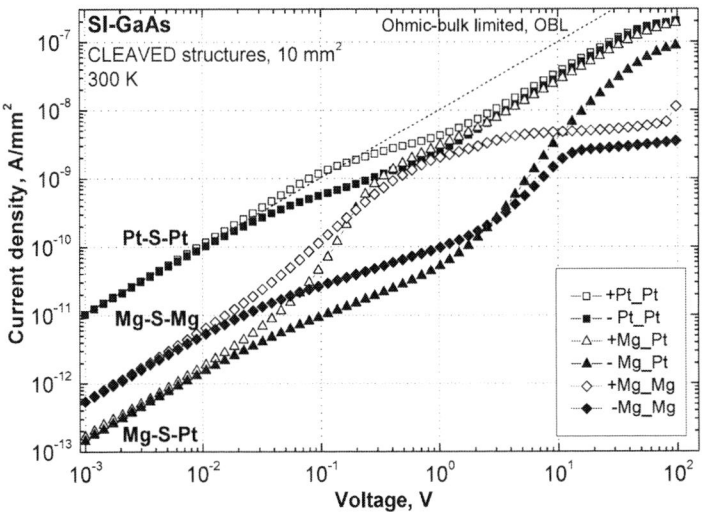

Fig. 1. *I-V* characteristics of the fabricated M-S-M structures in the „forward" (open symbols) and „reverse" (full symbols) directions.

GaAs materials with fixed contact shape and technology showed that the high-electric field transport regime is almost solely dictated by the properties of SI GaAs base material. Our observation, that Mg contact (in both, Mg-S-Mg and Mg-S-Pt samples) is able to markedly affect transport characteristics confirms the presence of "contact induced bulk effect" that was tentatively assumed in our previous work [9]. Specifically, it indirectly suggests the depletion of the GaAs bulk by the presence of MgO interface layer(s). The significant saturation current lowering at high voltage (>50 V) has a serious application potential in SI GaAs-based radiation M-S-M detectors that operate at high electric fields, and possibly also in other electronic devices, dur to the improvement in signal/noise ratio, that limits the energy resolution [10, 11], and lowered power consumption.

The Mg-S-Pt structure shows the lowest current density at rather low bias voltages (less than about 10 V), of about 100× lower in comparison with the Pt-S-Pt structure, which may be important in development of photodetectors or another physical sensors operating at low bias. The XPS results observed with the Au/Pt-S and Au/Mg-S [12] show that the AuMg-S interface is best characterized as AuMgO-S heterojunction with about 10 nm thick MgO layer. MgO is wide band gap material (7.8 eV [13]) and in contact with GaAs, it creates high conduction and valence bands offsets (in n-type GaAs, 2.2 eV and 4.2 eV, respectively [13]). We assume that the presence of MgO layer causes an effective elimination of the metal induced gap states [14] and therefore facilitates depinning of the FL and variable band bending at SI GaAs/MgO interface.

The increased effective resistance of the structures with MgO at low bias can be explained either by a reduction of the bulk free charge concentration in SI GaAs, or by different properties of Au/(Mg)/MgO contacts with respect to the Au/Pt ones. In the former case the bulk can be depleted from free electrons by the interface states and/or positively charged localized centres in the partially disordered MgO film. Let us recall that the RT free electron concentration in our SI GaAs is $\sim 2 \times 10^7$ cm^{-3}, which implies the FL position at about 0.62 eV below conduction band. In order to obtain the observed $\sim 80 \times$ reduction of low-bias current with respect to the OBL value, a FL shift of about 0.11 eV is necessary. Such a shift, and consequently bulk depletion, is possible without a substantial change of the occupation of the EL2 level (which lies ~ 0.75 eV below conduction band). Limited change of the EL2 (or other level) occupation, e.g. by 1×10^{15} cm^{-3} corresponding to a sheet density change of $\sim 2 \times 10^{13}$ cm^{-2} (at 180 µm thickness), can be easily accommodated by interface states of usual density, cf. the case of Si or Al$_2$O$_3$ on n-GaAs [15, 16]. The explanation of the increase of effective resistance in the structures with Au/Mg metallization by localization of bulk electrons at/in the MgO film seems therefore plausible. A detailed investigation of band bending at MgO/SI GaAs interface and its role in modification of the effective resistance, including the influence of possible dipole moments (cf. [9]), or a fixed positive charge, is still required to confirm this model.

A simple model describing the M-SI GaAs-M structure as a series connection of the bulk resistance and two Schottky-like diodes with opposite polarisations have proved as successful in explanation of the experimental photocurrent spectra [17]. The same approach applied to the dark currents shows that a reduction of the diode saturation current, which is a critical quantity, leads to the rise of the low-bias effective resistivity, while a difference in the saturation currents of top and bottom diodes causes the splitting of the I-V curves of opposite polarities. A competition between positively and negatively polarized diodes results then in sub- and superlinear parts of I-V curves, whereas the upper limit of observable current is the OBL value as measured for Pt-S-Pt structures. Even if we cannot directly involve the influence of the MgO layer in such an approach, it seems reasonable to assume that the effect of this layer can be effectively included via the lower saturation current of the corresponding reversely biased diode (contact).

4. Conclusion

The presented comparative study of Mg and Pt contacts on SI GaAs with the same geometry lead us to the following conclusions:

(i) the structure Mg-S-Pt demonstrates the highest decrease of the low-bias current,

(ii) the structure Mg-S-Mg demonstrates the lowest current in high bias voltage region before breakdown with relatively large difference following the bias polarity, while

(iii) the Pt-S-Pt structure shows the highest current in overall bias region with the initial resistance

corresponding to the measured bulk resistivity and relatively with the lowest influence of the bias polarity, so only this structure corresponds to the ohmic-bulk limited, OBL, transport mechanisms including the original base material resistivity.

The Mg-containing samples show that low work function/oxide forming contacts may affect properties of SI GaAs bulk and the especially high-field saturation current lowering observed in symmetric Mg-S-Mg could result in improvement of radiation detectors. In addition, the observed huge current decrease at low bias could be utilized in other SI GaAs based devices like photodetectors, photovoltaic devices and different sensors.

Acknowledgement

This work was partially supported by the Slovak Grant Agency for Science through grants Nos 2/0167/13 and 2/0152/16, by the Slovak Research and Development Agency under contract No APVV-0321-11, and by the Project Research and Development Centre for Advanced X-ray Technologies (ITMS code 26220220170) of the Research & Development Operational Program funded by the European Regional Development Fund (0.7).

References:

[1]   F. Dubecký, et al., *Nucl. Instr. and Meth. in Phys. Res.* **A 576,** 27, 2007.
[2]   M. Caria, et al., *Appl. Phys. Lett.* **81,** 1506, 2002.
[3]   J. Lloyd-Hughes, et al., *Appl. Phys. Lett.* **89,** 232102, 2006.
[4]   M. Alietti, et al., *Nucl. Instr. and Meth. in Phys. Res.* **A 362,** 344, 1995.
[5]   J. Massies, P. Devoldere, N. T. Linh, *J. Vac. Sci. and Technol.* **16,** 1244, 1979.
[6]   R. Baldini, et al., *Nucl. Instr. Meth. in Phys. Res.* **A 449,** 268, 2000.
[7]   J.J. Mareš, J. Krištofik, V. Šmíd, K. Jurek, S. Pospíšil, J. Kubašta, *Solid-State Electr.* **31,** 1309, 1988.
[8]   J.C. Breton, et al., *Appl. Phys. Lett.* **91,** 172112, 2007.
[9]   F. Dubecký, et al., *Solid-State Electr.* **82** (2013) 72-76.
[10] B. Zaťko, et al., *Nucl. Instr. Meth. in Phys. Res.* **A 531,** 111, 2004.
[11] T.E. Schlessinger, R.B. James, *Semiconductor for room temperature nuclear detector application,* Semiconductors and Semimetals, **Vol. 43**, Academic Press, San Diego, 1995.
[12] F. Dubecký, et al., *Appl. Surf. Sci.* 2016, in press.
[13] Y. Lu, J. C. Le Breton, P. Turban, B. Lépine, P. Schieffer, G. Jézéquel, *Appl. Phys. Lett.* **88,** 042108, 2006.
[14] B.E. Coss, W.Y. Loh, R.M. Wallace, J. Kim, P. Majhi, R. Jammy, *Appl. Phys. Lett.* **95,** 222105, 2009.
[15] S.M. Sze, *Physics of Semiconductor Devices*, Second edition, J. Wiley&Sons, New York 1981.
[16] J. Hu, A. Nainani, Y. Sun, K.C. Saraswat, H.S. Philip Wong, *Appl. Phys. Lett.* **99,** 252104. 2011.
[17] F. Dubecký, et al., *Solid-State Electr.* **118,** 30, 2016.

ASDAM 2016, The 11th International Conference on Advanced Semiconductor Devices And Microsystems, November 13-16, 2016, Smolenice, Slovakia

Nanostructuring by femtosecond laser ablation and RIE for MEMS and microfluidic systems fabrication

J. Zehetner[1], G. Vanko[2], J. Dzuba[2], T. Lalinsky[2]

[1]Research Centre for Microtechnology, University of Applied Sciences, Hochschulstrasse 1, 6850 Dornbirn, Austria

[2]Institute of Electrical Engineering, Slovak Academy of Sciences, Dubravska cesta 9, 841 04 Bratislava, Slovakia

e-mail: gabriel.vanko@savba.sk and johann.zehetner@fhv.at

AlGaN/GaN based sensors should be integrated into micro-electro-mechanical-systems (MEMS) and microfluidic devices used in biotechnology. The creation of appropriate diaphragms, surface structures or a combination of them is important for the fabrication of devices required in biotechnology and interdisciplinary research. Laser ablation as a direct mask writing procedure and deep reactive ion etching were used for nanostructuring of silicon nanopillars from the opposite site of the sensor device.

1. Introduction

An emerging interest in fabrication of high aspect ratio Si micropillars for use in biotechnology can be observed in recent years. Various applications such as capillary electrophoresis for DNA analysis [1], 3D cell culture [2] or electrochemical sensing [3] have been reported. However, several issues during the patterning of such structures can occur [4, 5, 6]. Recently, we demonstrated that femtosecond laser ablation in combination with RIE is a powerful tool to generate a variety of structures for possible utilization of AlGaN/GaN based high electron mobility transistors (HEMTs) [7]. This approach can be applied also in the technology for microfluidic devices and biotechnology. Controlling the laser polarization is an effective method to influence the interaction of the laser radiation with the substrate material used for device fabrication [8]. Using of Si facilitates the integration of additional materials and topological features within the same RIE process used to shape a device contour and a local surface pattern. In addition, this provides options to add our previously developed MEMS pressure sensor components to a fluidic device [9]. GaN-based MEMS sensors provide high temperature stability (useful in catalytic activated chemical and gas sensors [10]) and can be monolithically integrated with GaN electronics in a common chip for wireless signal transfer.

2. Experimental Methods

In this paper, we present possible approaches to fabricate various forms of AlGaN/GaN diaphragms covert on the backside with surface micro/nano structures to be integrated in microfluidic systems. The proposed MEMS structure of the sensor is depicted in Figure 1. It consists of an AlGaN/GaN heterostructure grown on Si supporting substrate. The thickness of the MOCVD grown low stressed GaN and AlGaN barrier layers are 4.2 µm and 20 nm, respectively. The Al mole fraction is nominally set to be 20%. The ohmic (Ti/Al based) and Schottky (Ir based) electrodes of the CHEMT are fabricated by means of electron beam evaporation followed by patterning lift off technique and appropriate annealing

978-1-5090-3084-2/16 $31.00 © 2016 IEEE

procedure. The micro-pillars are prepared using the combination of laser ablation in a role of micromasking and deep reactive ion etching (RIE).

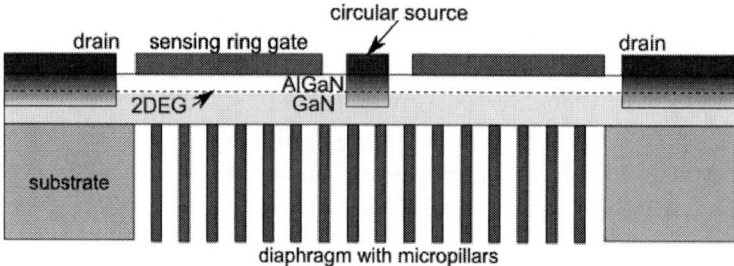

Figure 1. Schematic cross-section of the proposed MEMS device with integrated Si pillars

We use the SPIRIT from HighQ Laser, delivering 4W average power in the fundamental 1040nm wavelength at 200 kHz repetition rate and 350fs pulse length. For the second harmonic at 520nm, the maximum power is 1,6W after the 100mm focal length scanner optic. Typical parameters are 1000 mm/sec scanning speed and a laser power in the range of 30 to 120mWat at the 520nm line. Such a low laser power was sufficient as we did not ablate the Si-bulk material.

3. Results and Discussion

We demonstrate that 300µm long silicon needles can be produced on a 10µm thick membrane to provide a relevant design element for a microfluidic filter system (biomimetic surface derived from the baleen of whales) with integrated pressure sensor (Figure.1; 2a).

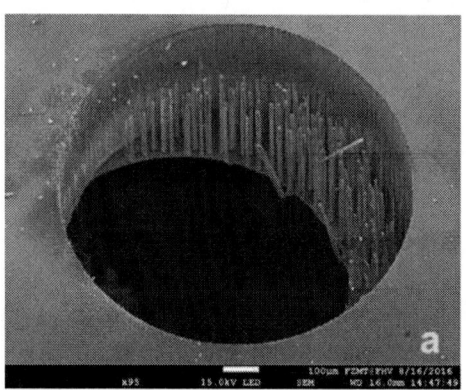
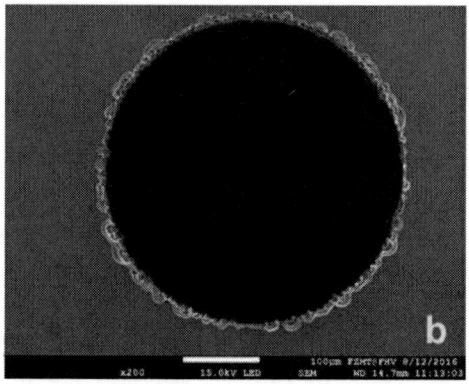

Figure 2.: 300 µm long Si micropillars on a 10 µm thick membrane (a), increased RIE under etching due to laser damage of the Si crystal structure at the contour corners [11] (b)

To fabricate RIE masks for sub micrometre features we investigated different types of coating materials on Si. Aluminium is an ideal mask material for RIE to produce micropillars. Laser ablation (60 to 120 mW) generated only pinholes in the 200 nm thick masking layer varying in sizes between 100 nm and 1 µm. With growing laser power or increasing number of scans (typically 1 to 5 scans) the density and size of the pinholes is growing (Figure 3) finally leading to the formation of islands. The final structure after RIE changes from bores to needles (Fig. 2a, 4a). Within one device contour a variety of needle array can be produced compassing a photolithography step. With the laser we scribed the contour of a specific device layout into a 100 to 200nm thick metal layer. The metal mask layer was not always totally

978-1-5090-3084-2/16 $31.00 © 2016 IEEE   224

removed. Intentionally the laser parameters were set to create pinholes, islands or ripples (Figure 3; 4; 6) in the metal layer inside the device counter. In such a manner we generated in one process step the etch mask contour of a device and the required surface structure for the subsequent reactive ion etch process (RIE). Islands generated in the aluminium (Al) mask layer produced micropillars after RIE (Figure 4a; 5).

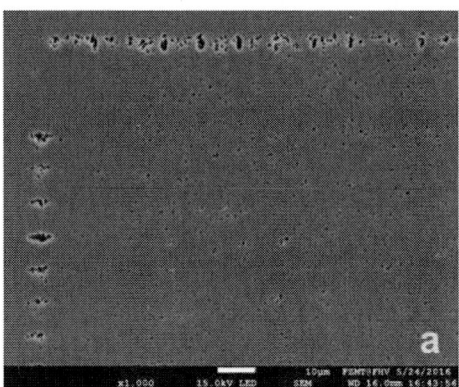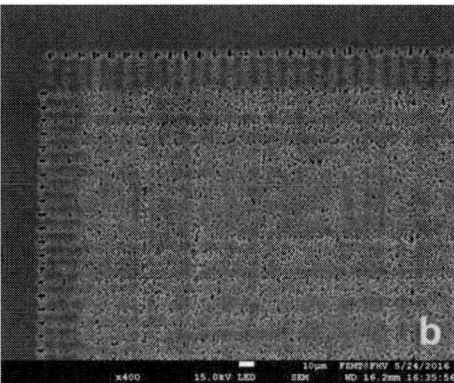

Figure 3.: Sub micron pinholes in 200nm Al mask, one xy laser scan (a), two xy laser scans at the same laser power and hatch distance of 5 μm (b)

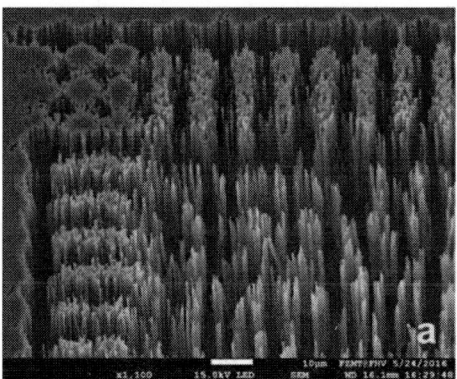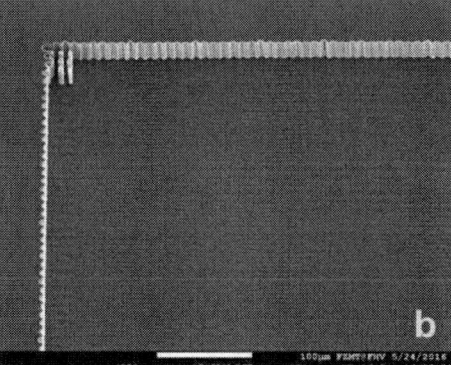

Figure 4.: Four laser scans at the same laser power as in figure 3 but different areal xy hatch distance, 5μm low needle density array and 10μm for the block structure (a), eight scans (b)

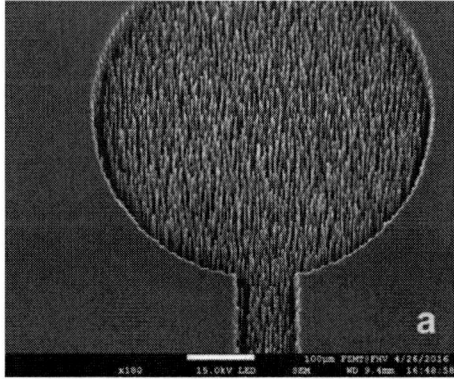

Figure 5.: Mimetic baleen structure in a microfluidic filter device (a), enlarged Si structure (b)

978-1-5090-3084-2/16 $31.00 © 2016 IEEE

Titan (Figure 6a), iron (Figure 6b) and nickel are developing ripples or pinholes depending on the number of scans and the laser power. The laser parameter and choice of material influences the 2D pattern and independently hereof, the RIE process parameters affect the 3D structure. This opens a way to generate a great diversity of structures e.g. for gas sensors or with hydrophobic/hydrophilic properties for science and applications.

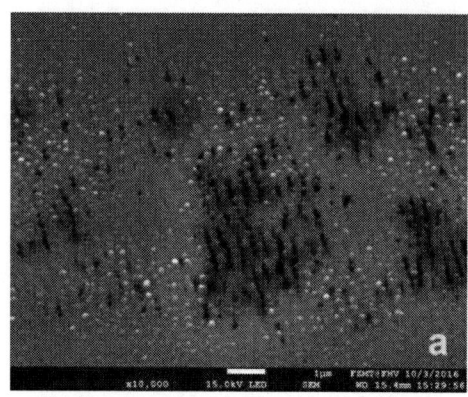 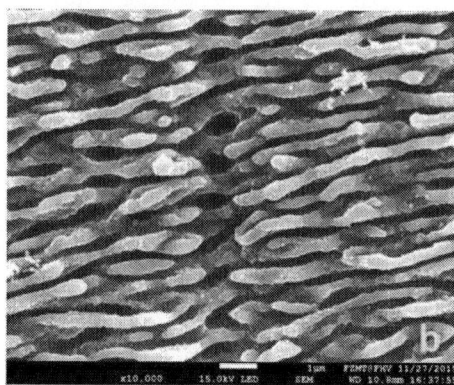

Figure 6.: Laser generated sub μm slots in a titan RIE etch mask (a), ripples and in steel (b)

4.   Conclusion

Femtosecond laser ablation in combination with RIE is a feasible process to fabricate membranes and surface structures in Si and SiC for MEMS pressure sensor, microfluidic devices, catalytic activated chemical and gas sensors. With ablation supported RIE we produced a biomimetic baleen structure on top of 10μm thick membranes for microfluidic filter systems with optional integrated pressure sensor.

References

[1]   Y. Ch. Chan, Y.-K. Lee and Y. Zohar, J. Micromech. Microeng. 16 (2006) 699–707

[2]   J. L. Tan, J. Tien, D. M. Pirone, D. S. Gray, K. Bhadriraju, C. S. Chen, Proc. Natl. Acad. Sci. U. S. A. 2003, 100, 1484

[3]   Y. Yoon, J.-B. Lee, Sensors 13 (2013) 16672-16681.

[4]   K. Miller, M. Li, K. M. Walsh and X.-A. Fu, J. Micromech. Microeng. 23 (2013) 035039.

[5]   H. M. Hegab, M. Soliman, S. Ebrahim and M. O. de Beeck, J Biosens Bioelectron 4 (2013) 1000140

[6]   F. Bai, M. Li, R. Huang, D. Song, B. Jiang and Y. Li, Nanoscale Research Letters 7 (2012) 557

[7]   G. Vanko, P. Hudek, J. Zehetner, J. Dzuba, P. Choleva, V. Kutiš, M. Vallo, I. Rýger, T. Lalinský, Microelectronic Engineering 110 (2013) 260-264

[8]   J. Zehetner, S. Kraus, M. Lucki, G. Vanko, J. Dzuba, T. Lalinsky, Microsystem Technologies 22 (2016) 1883.

[9]   J. Dzuba, G. Vanko, M. Držík, I. Ryger, V. Kutiš, P. Lobotka, J. Zehetner and T. Lalinský, Applied Physics Letters 107 (2015) 122102.

[10]   M. Haruta, Journal of New Materials for Electrochemical Systems 7 (2004) 163-172.

[11]   J. Zehetner, G. Vanko, P. Choleva, J. Dzuba, I. Ryger, T. Lalinsky, in Proc. of ASDAM, 20 to 22 Oct. 2014, Smolenice Castle, Slovakia.

ASDAM 2016, The 11th International Conference on Advanced Semiconductor
Devices And Microsystems, November 13-16, 2016, Smolenice, Slovakia

Strain induced response of AlGaN/GaN high electron mobility transistor located on cantilever and membrane

J. Dzuba[a], G. Vanko[a], O. Babchenko[a], T. Lalinský[a], F. Horvát[b], M. Szarvas[b], T. Kováč[b], B. Hučko[b]

[a]Institute of Electrical Engineering, Slovak Academy of Sciences, Dúbravská cesta 9,
841 04 Bratislava, Slovakia
[b]Institute of Applied Mechanics and Mechatronics, Faculty of Mechanical Engineering,
Slovak University of Technology in Bratislava, Nám. Slobody 17, 812 31 Bratislava,
Slovakia

e-mail: jaroslav.dzuba@savba.sk and gabriel.vanko@savba.sk

The III-Nitrides, especially AlGaN/GaN devices can be widely used in micro-electro-mechanical sensors thanks to their excellent mechanical and electric properties. In this work, we investigate influence of the applied mechanical load on the source-drain current of AlGaN/GaN C-HEMTs located on the clamped edge of the cantilever beam and the membrane, respectively. A linear dependence of the output current on the calculated strain is observed.

1. Introduction

Group III-Nitrides (III-Ns) are very attractive for various micro-electro-mechanical systems (MEMS) or sensors thanks to their excellent mechanical and electric properties which are preserved at high temperatures or in harsh environments [1, 2]. Several approaches are reported to investigate the piezoelectric response of III-N devices based on a bulk design. Proposed sensors work on the action of hydrostatic pressure which alters Ni/Au/GaN and Ni/AlGaN Schottky barrier height [3, 4], respectively, the internal fields in GaN/AlGaN/GaN heterostructures [5], and the polarization in AlGaN/GaN heterostructure [6]. In the other approach, the AlGaN/GaN high electron mobility transistor (HEMT) process technology is performed on a bulk sapphire or silicon carbide (SiC) substrate. The substrate with integrated HEMT as a strain sensor was then cut into a bulk cantilever which was exposed to bending [7-9]. Previously we reported the fundamental piezoelectric readout of such circular HEMT (C-HEMT) mechanical sensor integrated on the surface of a cantilever beam [10, 11]. Proposed MEMS structure exhibits piezoelectric induced charge as a response to the external mechanical forces applied to free end of the cantilever beam. This design is commonly considered as inappropriate for direct sensing of pressure. Therefore we developed more suitable circular shaped diaphragm design with the integrated C-HEMT [12]. The piezoelectric charge was measured when the diaphragm was loaded by the dynamic air pressure.

In this work, we investigate the output transistor characteristics of both previously manufactured C-HEMT cantilever-like [10, 11] and membrane-like devices [12] at various bias conditions. At first, no force neither pressure is applied to the cantilever and the membrane, respectively, and the idle output current is measured. Then, a dynamic load is applied on both structures and the resulted current changes in the channel are compared.

2. Processing technology

A schematic cross-section view of the HEMT strain sensing devices is depicted in Fig. 1. An undoped 25 nm/2 μm $Al_{0.25}Ga_{0.75}N$/GaN heterostructure is grown by metal-organic chemical vapor deposition (MOCVD) on a 300 μm thick 4H-SiC substrate. The C-HEMT processing

978-1-5090-3084-2/16 $31.00 © 2016 IEEE   227

technology was employed to form source and drain ohmic contacts using Nb/Ti/Al/Ni/Au metallic system alloyed at 850 °C. To form the ring gate contact, Ni/Au electron beam evaporation and lift-off were carried out. Subsequently, Ti/Au top-contact metallic layer was patterned on all C-HEMT contacts to improve device bonding. Finally, the completed device on SiC substrate was sawn to form cantilever beam of the MEMS strain sensor with area 12 x 4 mm² (Fig. 1, a). The C-HEMT with distance between ohmic contacts 180 µm, ring gate inner and outer radii 70 µm and 230 µm, respectively, is situated near the clamped edge of the cantilever to measure mechanical strain.

For the second investigated MEMS sensor, a similar heterostructure with 20 nm thin $Al_{0.25}Ga_{0.75}N$ and 4.2 µm thick GaN was grown on the 600 µm thick silicon (Si) substrate thinned down to 300 µm. Above mentioned front-side processing was employed to form C-HEMT ohmic contacts with distance 690 µm [11]. The inner and outer radii (710-730 µm) of the gate electrode were set to cover membrane surface near its outer rim (Fig. 1, b). The membrane is formed using the back-side micro-machining of the Si substrate by deep reactive ion etching (DRIE) which resulted in the vertically etched walls. Such membrane-like design is primarily used for the AlGaN/GaN MEMS pressure sensor previously presented [12].

3. Experiment

Proposed MEMS structures sense the external mechanical forces by converting the mechanical strain into change of the C-HEMT output (source-drain) current due to present direct piezoelectric effect. To investigate the piezoelectric response of the cantilever beam device (Fig. 1, a), the cantilever was fixed into a Lucite block. The free end of the cantilever was dynamically loaded by the known force to obtain desired tip displacement. While the altering mechanical load is applied, the piezoelectric charge generated in the heterostructure changes the inner electrical field and thus influences the conductivity of the transistor's channel. As a result, the generated piezoelectric charge, or (alternatively) in this case a fluctuation of the output current can be measured. The induced piezoelectric charge of the AlGaN/GaN/SiC cantilever strain sensor was investigated previously as a function of the C-HEMT geometry and the frequency of the loading force [10, 11].

In this work, we focus on characterization of the strain induced output current dependency of the C-HEMT positioned at the clamped edge of the cantilever beam. The source-drain current was measured by Hewlett Packard HP 4145B Semiconductor Parameter Analyzer while deformation of the cantilever tip was measured using of laser Doppler vibrometer (LDV) system Polytec OFV 303. The mechanical strain at the clamped edge of the cantilever beam with rectangular cross-section was calculated from the tip deformation.

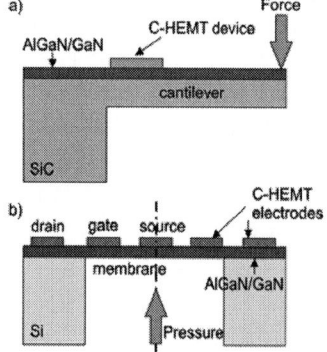

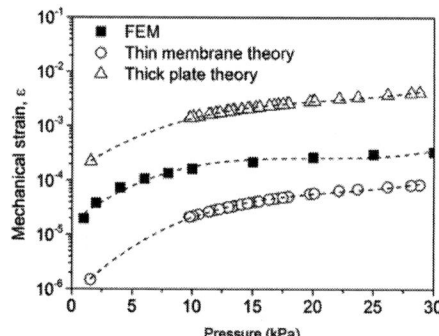

Fig. 1. The C-HEMT strain sensor located on the cantilever beam (a), and the C-HEMT pressure sensor integrated in a membrane (b).

Fig. 2. Different strains calculated in the investigated C-HEMT membrane sensor.

In contrast, the membrane-like MEMS design (Fig. 1, b) is loaded by uniformly distributed dynamic air pressure of the known and properly high amplitude to avoid the membrane cracking which could be observed when the concentrated force is applied. The dynamic air pressure was produced by the inductive piston into the piping system connected to the membrane of the C-HEMT strain sensor. The resulting output current changes were measured by aforementioned semiconductor analyzer. The mechanical strains in such membrane sensors were discussed earlier [13]. It has been shown that the membrane of our pressure sensor exhibits transition between the ideal plate-like and membrane-like behavior. Here, this behavior was also expected and confirmed by the finite element method (FEM) analysis (Fig. 2). Mechanical strains calculated at the position of the sensing electrode are compared with the known thin membrane and thick plate theory. It is clear that analytical expressions of the strains using the membrane-plate theory would lead to underestimated or overestimated values. Hence, we calculated the strains in the membrane-like design of our sensor (Fig. 1, b) using FEM analysis.

4. Results

The output current characteristics of both C-HEMT devices were measured while various bias conditions were set. Firstly, no force neither pressure was applied to free end of the cantilever and the membrane, respectively. After the idle output current (I_0) was measured, the dynamic load was applied on both structures and the current changes were compared. In the cantilever beam design, relatively small force causing less than 50 μm displacement was applied to the free end of the cantilever to ensure the small deflection and strains. Change of the output current (I_{amp}) relative to the idle current (I_0) depends linearly on the strain (ε) induced near the clamped end of the cantilever beam (Fig. 3). The optimal working condition of the C-HEMT and so the highest sensitivity was observed at drain-source voltage set to V_{ds}=6V and gate voltage V_g=-3V, respectively. The area and position of the gate electrode were previously optimized [11]. In case of the membrane-like C-HEMT design, the pressure range 8-30 kPa was applied to the membrane to obtain strains comparable with those calculated within the cantilever. Similarly high sensitivity was observed even with less power supply voltage V_{ds}=2V (Fig. 4.). Additionally when geometry optimization was performed, the electrode would be located in the region where the highest strains are observed while the membrane is loaded. It has been shown previously that mechanical strains and stresses strongly rely on the geometry and thus the sensor's sensitivity strongly depends on the optimally located electrodes [13].

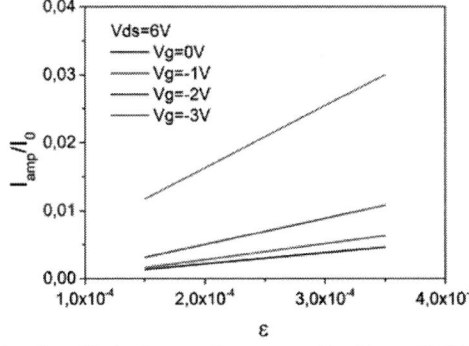

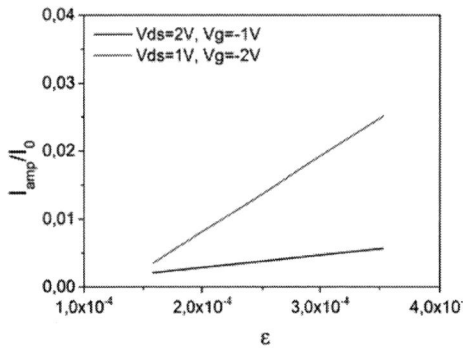

Fig. 3. Relative change of the C-HEMT output current located at the clamped end of the cantilever.

Fig. 4. Relative change in output current of C-HEMT located on membrane at various drain-source voltage V_{ds} and gate voltage V_g.

Acknowledgement

This work was supported by the Slovak Research and Development Agency under the Contract Nos. APVV-14-0613 and APVV-0455-12 and VEGA project 1/0712/14.

References

[1] V. Cimalla, J. Pezoldt, O. Ambacher, *Journal of Physics D: Applied Physics* **40**, 20, p. 6386, 2007.

[2] S. J. Pearton *et al.*, *Journal of Physics: Condensed Matter* **16**, 29, p. R961, 2004.

[3] Y. Liu *et al.*, *Applied Physics Letters* **84**, 12, p. 2112, 2004.

[4] Y. Liu *et al.*, *Applied Physics Letters* **88**, 2, p. 22109, 2006.

[5] Y. Liu *et al.*, *Journal of Applied Physics* **99**, 11, p. 113706, 2006.

[6] Y. Liu, P. P. Ruden, J. Xie, H. Morkoç, K.-A. Son, *Applied Physics Letters* **88**, 1, p. 13505, 2006.

[7] M. Eickhoff, O. Ambacher, G. Krötz, M. Stutzmann, *Journal of Applied Physics* **90**, 7, p. 3383, 2001.

[8] B. S. Kang *et al.*, *Applied Physics Letters* **83**, 23, p. 4845, 2003.

[9] Chia-Ta Chang, Shih-Kuang Hsiao, E. Y. Chang, Chung-Yu Lu, Jui-Chien Huang, Ching-Ting Lee, *IEEE Electron Device Letters* **30**, 3, p. 213, 2009.

[10] T. Lalinský *et al.*, *Microelectronic Engineering* **88**, 8, p. 2424, 2011.

[11] G. Vanko *et al.*, *Sensors and Actuators A: Physical* **172**, 1, p. 98, 2011.

[12] J. Dzuba *et al.*, *Applied Physics Letters* **107**, 12, p. 122102, 2015.

[13] J. Dzuba *et al.*, *Journal of Micromechanics and Microengineering* **25**, 1, p. 15001, 2015.

ASDAM 2016, The 11th International Conference on Advanced Semiconductor
Devices And Microsystems, November 13-16, 2016, Smolenice, Slovakia

Electrical Properties of Back-to-Back Metal/Insulator/Semiconductor Diodes Based on AlGaN/GaN Heterostructure

J. Osvald

Institute of Electrical Engineering, Slovak Academy of Sciences,
Dúbravská cesta 9, 841 04 Bratislava, Slovakia
e-mail: elekosva@savba.sk

AlGaN/GaN based heterostructure on which transistor active elements are widely studied may be also used for preparation of devices which protect the circuits against external voltage surge. Such devices are called varactors and consist of two back-to-back connected Schottky diodes. We have studied current - voltage and capacitance - voltage characteristics of two back-to-back connected MIS (Metal Insulator Semiconductor) diodes. It is shown that the C-V curves of such a varactor structure are influenced by the I-V curves since the direct current polarization of both diodes are dependent on the relation between forward and reverse currents flowing through the diodes.

1. Introduction

Aluminum gallium nitride/gallium nitride (AlGaN/GaN) heterostructure transistors reached relatively high level of maturity. Certainly many problems are still to be solved. One of them is a protection of these devices against a voltage surge, either man or nature made [1]. We have studied in this contribution electrical properties of two back-to-back connected Schottky diodes with two-dimensional electron gas (2DEG) based on AlGaN/GaN heterostructures. Capacitance properties of such a connection of two Schottky diodes are used, *e. g.,* in so called varactor. This device is intended to use for circuits protection against voltage surge. The varactor has high capacitance (low impedance) for low polarization voltages and its the capacitance decreases considerably (impedance increases) by accidental voltage surge. The whole voltage then drops on the varactor and the circuits in series are protected against the voltage increase. At which voltage the capacitance of the varactor changes its value depends on the *I-V* curves of the diodes forming the varactor. For the case of the ideal Schottky diodes the diode reverse current is very low. Consequently also the voltage drop on the forwardly polarized diode is low and the transition voltage of the varactor practically equals to the threshold voltage of the diodes forming the varactor. In real contacts situation is not so ideal [2,3], since there is practically always present some leakage current in reverse direction. Electrical properties of back-to-back connected diodes are interesting themselves, especially for non-equal Schottky contacts of the diodes and were studied recently. It was shown that a circuit with two such Schottky diodes may have current-voltage *(I-V)* characteristic as a circuit with only one Schottky diode and the second contact Ohmic [4].

Different situation occurs if a thin dielectric layer is used between the metal and the semiconductor forming thus MIS (Metal Insulator Semiconductor) diode instead of Schottky diode [5,6]. In such diodes the reverse leakage current is diminished but also the forward current is lower comparing to an equivalent Schottky diode. We have studied how the *I-V* curves of the MIS diodes forming the varactor influence the *C-V* curves of the back-to-back connected diodes.

978-1-5090-3084-2/16 $31.00 © 2016 IEEE   231

2. Theory and experiment

For electrical current density flowing through the Schottky diode the well-know thermionic expression

$$J = A^{**}T^2 \exp(-\frac{q\varphi}{kT})[\exp(\frac{qV}{nkT})-1],$$ (1)

is valid, where A^{**} is the Richardson constant and other symbols have their usual meaning. If we calculate the current flowing for two Schottky diodes in series and connected back-to-back we have to come out from the fact that the current flowing through the diodes is the same. Attempting to write an expression for I-V curve of such a structure we find out that we can write the electrical potential as a function of the current flowing through the diodes as [4]

$$V = \frac{n_1 kT}{q}\ln(\frac{J}{J_{01}}+1) - \frac{n_2 kT}{q}\ln(-\frac{J}{J_{02}}+1) + RAJ,$$ (2)

where J_{01}, J_{02} are saturation current densities of the diodes, A is the diodes area and R is the series resistance of the circuit. In this expression different ideality factors of the contacts are taken into account.

Capacitance of such a two diodes combination is determined as a capacitance of two capacitors in series. Since the diodes have not only capacitance but also some conductance in parallel the direct current voltage is not divided on them according to the electrostatics. The voltages on single diodes are determined by the voltage drops caused by the direct current flow through the diodes.

The diodes forming the varactor are polarized in the opposite direction. If the first diode is polarized forwardly, the second one is polarized reversely and vice versa. If the polarizing voltage is close to zero the varactor has the capacitance approximately half of the capacitance of the diodes forming the varactor. If the capacitance of one of the diodes – forwardly or reversely polarized – decreases, the capacitance of the whole varactor decreases.

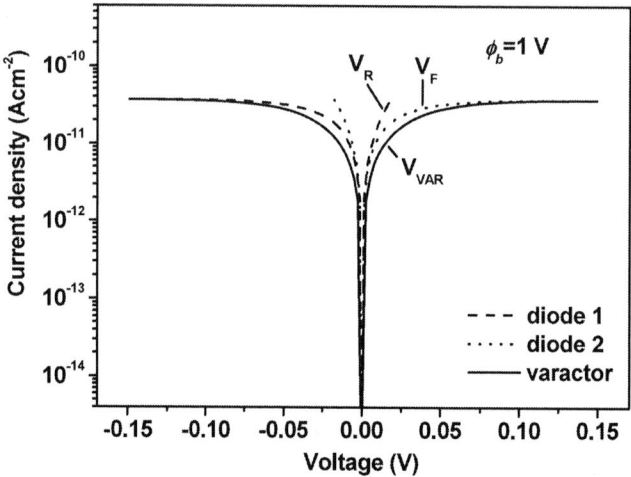

Fig. 1. Theoretical *I-V* curves of particular diodes and the varactor as a function of voltage; V_R, V_F are the voltage drops on the reversely and forwardly biased diodes and V_{VAR} is the external voltage on the varactor.

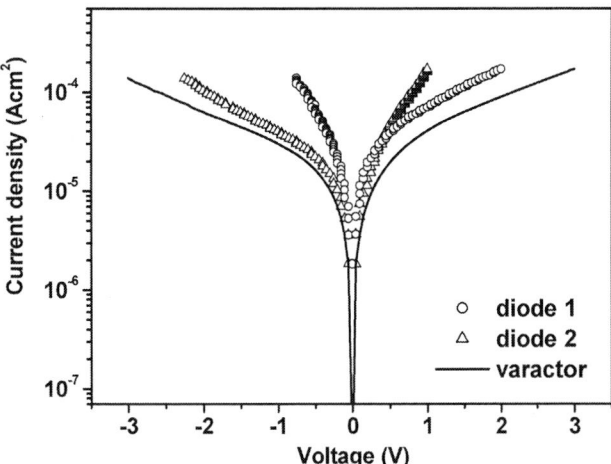

Fig. 2. Experimental *I-V* curves of MIS capacitors and the varactor as a function of voltage.

The question may arise which of the diodes is responsible for the capacitance decrease. Obviously forwardly polarized diode needs lower voltage for the capacitance decrease than the reversely biased one but the voltage drop on this diode is generally much lower. The problem can be solved theoretically as well as experimentally. Results of the theoretical calculation for the Schottky diodes with the barrier height $\varphi = 1$ V and ideality factor $n = 1$ are shown in Fig. 1. It is seen that the voltage drop on the forwardly biased diode is negligible comparing to the voltage drop on the reversely biased diode and for external voltage larger than ~ 0.02 V does not increase at all. That means that practically whole voltage is on the reversely polarized diode. It further means that the decrease of the capacitance of the varactor is the consequence of the capacitance decrease of the reversely polarized diode in spite of the fact that this diode needs higher voltage for the capacitance drop.

This is the theoretical result for ideal Schottky diodes. For real Schottky diodes and

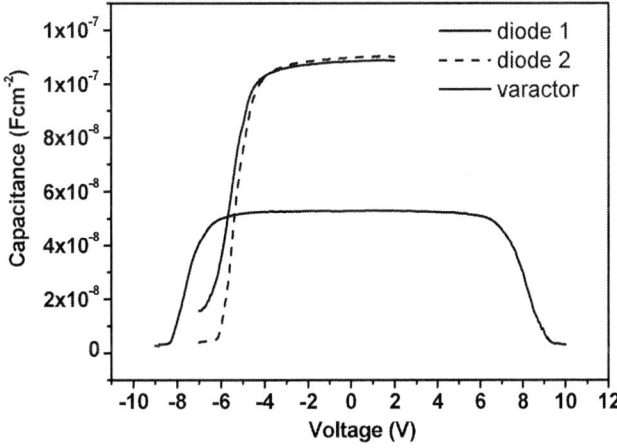

Fig. 3. Capacitance-voltage curves for the single diodes and the back-to-back connected diodes.

MIS structures, the *I-V* curve does not obviously obey the theoretical assumptions. The reverse leakage current is diminished because of the dielectric layer but forward current is diminished, too. The ration between the forward and the leakage current is then changed.

We prepare MIS diodes on AlGaN/GaN heterostructure with 20 nm thick AlGaN layer and 15 nm thick Al_2O_3 layer. The dielectric layer was prepared by atomic layer deposition method (ALD) at 100 °C. Fig. 2 shows the relation between the forward and reverse currents on experimental sample. The ratio between the forward and reverse current is much lower than theoretically predicted for the Schottky diode. The voltage drop on the forwardly biased diode reaches even ~ 1 V which may be enough voltage in certain cases for the capacitance decrease of the forwardly polarized diode. According to these results, for the capacitance decrease of the varactor may, in principle, be responsible also the forwardly polarized diode.

In Fig. 3 capacitance curves of the single diodes and back-to-back diodes are shown. The single diodes have slightly different *C-V* curves probably as a result of the technological process. It is seen that the difference between the voltage where the capacitance of the single diode and the back-to-back connected diodes decrease is approximately 2 V. This corresponds to Fig. 2 where the total voltage of 3 V is divided between the diodes approximately as 2 V potential drop on the reverse biased diode and 1 V on the forward biased one. In 8 V of total bias the appropriate voltage drops increase for both diodes.

3. Conclusion

We have shown that the voltage where the capacitance of the back-to-back connected diodes changes from high to low depends on ratio between voltage drops on the particular diode diodes. These voltage drops polarize particular diodes and determine the total voltage necessary to switch the diode capacitance from high to low.

Acknowledgement

The author is thankful for the financial support received during work on this study from the Slovak Grant Agency for Science under Contract No. 2/0167/13, Agency for Research and Development APVV-14-0613 and from the Structural Funds of the European Union by means of the Agency of the Ministry of Education, Science, Research and Sport of the Slovak Republic in the project "CENTE II" ITMS code 26240120019.

References

[1] L. B. Chang, A. Das, R. M. Lin, S. Maikap, M. J. Jeng, S. T. Chou, *Appl. Phys. Lett.* **98**, 222106, 2011.

[2] M. Marso, M. Wolter, P. Javorka, A. Fox, P. Kordoš, *Electron Lett.* **37**, 1476, 2001.

[3] M. Marso, A. Fox, G. Heidelberger, P. Kordoš, H. Lüth, *IEEE Electron Device Lett.* **27**, 945, 2006.

[4] J. Osvald, *Phys. Status Solidi* **212**, 2754, 2015.

[5] V. Adivarahan; M. Gaevski; W. H. Sun; H. Fatima; A. Koudymov; S. Saygi; G. Simin; J. Yang; M. A. Khan; A. Tarakji; M. S. Shur; R. Gaska, *IEEE Electron. Dev. Let.* **24**, 541, 2003.

[6] K. Balachander, S. Arulkumaran, T. Egawa, Y. Sano, K. Baskar, *Mat. Sci. Engn. B* **119**, 36, 2005.

ASDAM 2016, The 11th International Conference on Advanced Semiconductor
Devices And Microsystems, November 13-16, 2016, Smolenice, Slovakia

Evaluation of TIGBT field stop layer generated by helium implantation

Radim Pechal[*], Lubomír Dorňák, Juraj Vavro, Adam Kozelský, Adam Klimsza

ON Semiconductor, 1. máje 2230, 756 61 Rožnov pod Radhoštěm, Czech Republic
[*]e-mail: radim.pechal@onsemi.com

A helium implantation is a common technique used for electron and hole lifetime control in semiconductor devices. Siemieniec [1] shows that suitable annealing conditions after the helium implantation lead to an increased conductivity of N-type silicon substrate in the affected area. Based on these observations a possibility of using helium implantation for generation of the TIGBT field stop layer was evaluated. Required concentration profiles were prepared for test wafers, however, a decreased conductivity in the helium implantation area was observed when using production wafers with standard process flow. The root cause of this discrepancy is diffusion of nitrogen and hydrogen atoms into the silicon during previous diffusion operations and an interaction of these atoms with the implanted helium.

1. Introduction

A trench insulated gate bipolar transistor (TIGBT) is a common semiconductor device, which is typically used as an electronic switch in power applications. A structure of a new TIGBT device during its development is optimized with respect to several parameters: break down voltage, Vceon, turn-off power losses, etc. Those parameters are interdependent. Power losses can be improved by drift region thickness reduction but TIGBT with thinner drift region has lower break down voltage. This decrease can be compensated by including an additional layer called field stop layer (FS). This structure has been published by Laska [2].

One way in which FS layer can be generated is implantation of phosphorus into back-side of the wafer after a wafer thinning process. The implanted layer is then activated by a furnace annealing. An active dose is limited by a maximum furnace annealing temperature, which is typically limited by the melting temperature of a metallization present on a front side of the wafer. Those limitations can be avoided by using different elements like hydrogen or helium, which has lower level of energy required for their activation in silicon matrix.

The results of field stop layer generated by multiple hydrogen implantations were published by Niedernostheide [3]. This solution is covered by a patent [4]. Results of TIGBT with FS layer generated by helium have not been published yet.

An electrical characterization of helium implantation into the silicon wafer has been studied by Siemieniec [1]. Based on a deep level transient spectroscopy three different energy levels were identified. Two levels are near the middle of a band gap (E_C-E_T = 0.43 eV, E_T-E_V = 0.35 eV) and one level is close to a conduction band (E_C-E_T = 0.17 eV). The concentrations of those three levels depend on the annealing conditions. The levels near to the middle of the bang gap are dominant for the low temperature annealing (~300°C). An effective dopant concentration for N-type substrate decreased for these conditions. The level close to the conduction band is dominant for higher temperature annealing (~400°C and higher) contributing to the increasing effective N-type doping concentration.

978-1-5090-3084-2/16 $31.00 © 2016 IEEE

2.   TIGBT field stop layer with helium implantation

A standard ON Semiconductor 1200 V FSII TIGBT [5] was used for evaluation. This TIGBT was compared to the same device with additional field stop layer generated by a helium (He) implantation. Helium was implanted with energy of 1 MeV and dose $1 \cdot 10^{13}$ cm^{-2}. The activation annealing temperature was 450°C for the duration of 30 min in nitrogen ambient. For the TIGBT with the He implantation, we expect higher break-down voltage.

The break-down voltage for TIGBT with the additional helium field stop and the standard TIGBT were similar. But the high Ice leakage (collector – emitter current) was observed (Fig. 1). A resistivity profile of TIGBT wafer measured by spreading resistance profiling technique (SRP) [6] was compared with a profile prepared on the test wafer. Both profiles are different significantly (Fig. 2).

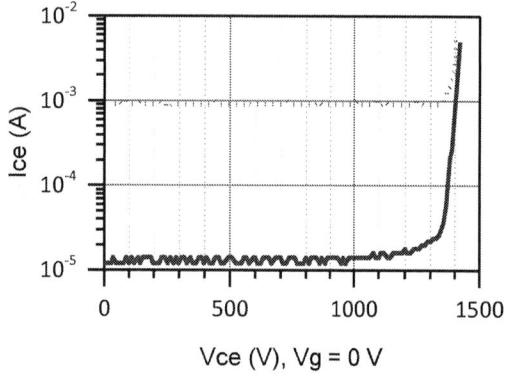

Standard 1200 V TIGBT

Standard 1200 V TIGBT with He implantation

Fig. 1: Comparison of the Ice leakage for the standard 1200 V TIGBT with and without the He implantation.

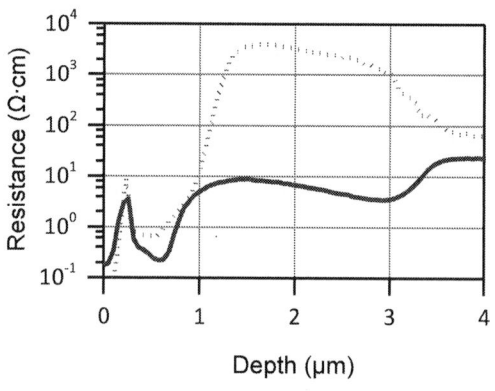

Standard 1200 V TIGBT

Standard 1200 V TIGBT with He implantation

Fig. 2: Back-side SRP profile of the standard 1200 V FSII TIGBT with the He implantation compared with a profile prepared on the test wafer.

3.   Test wafers experiment

For the evaluation of the difference of the SRP profile between TIGBT wafer and test wafer (Fig. 2) several test wafers were prepared (summarized in table 1) and the SRP profiles were measured. In the first experiment test wafer 1 was implanted with standard phosphorus and boron (P+B) back-side implantation used for standard 1200 V TIGBT, wafer 2 was processed with the He implantation and wafer 3 was processed with a combination of the P+B back-side implantation and the He implantation. After the implantations were finished all wafers were annealed at 450°C for 30 min. The resulting resistance profiles are shown in figure 3. The resistance profile of wafer 2 decreased due to He implantation. This observation is in agreement with Siemieniecs [1]. The resistance profile for wafer 3 is a parallel combination of resistance profiles for wafer 1 and 2. This profile was expected.

Wafer 4 was processed with similar process conditions with an additional annealing between the B+P back-side implantation and the He implantation. The resistance profile in the area of He implantation did not decrease but increased (Fig. 2).

The test wafers (wafer 5, 6 and 7) without P+B back-side implantation as well as various annealing conditions preceeding the He implantation were prepared to evaluate this phenomenon. The resistance profiles of those wafers are comparable with wafer 4. All wafers with the annealing before the He implantation have higher resistance in the area affected by the He implantation (Fig. 3).

Table 1: Summary of the test wafers used for the experiments and their process conditions.

Wafer	1st step: P+B Back-side implantation	2nd step: Annealing	3rd step: He implantation	4th step: Annealing
1	Yes	No	No	450°C 30 min
2	No	No	1 MeV 1E+13 cm^{-2}	450°C 30 min
3	Yes	No	1 MeV 1E+13 cm^{-2}	450°C 30 min
4	Yes	450°C 30 min N$_2$	1 MeV 1E+13 cm^{-2}	450°C 30 min
5	No	450°C 30 min N$_2$	1 MeV 1E+13 cm^{-2}	450°C 30 min
6	No	900°C 10 min H$_2$+O$_2$	1 MeV 1E+13 cm^{-2}	450°C 30 min
7	No	1000°C 45 sec N$_2$+O$_2$	1 MeV 1E+13 cm^{-2}	450°C 30 min

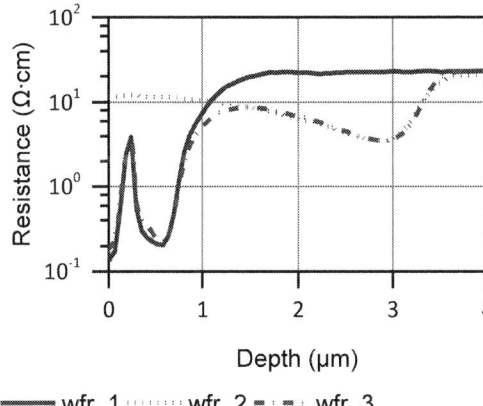

Fig. 3: Comparison of the SRP profiles of wafer 1 with the P+B back-side implantation, wafer 2 with the He implantation and wafer 3 with the combination of these implantations.

Fig. 4: Comparison of the SRP profiles of wafer 1 with the P+B back-side implantation, wafer 2 with the He implantation and wafer 4 with the combination of these implantations and the annealing before the He implantation.

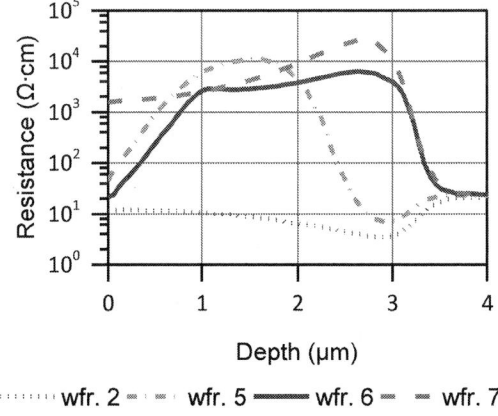

Fig. 5: Comparison of SRP profiles of wafer 2 without annealing before He implantation and the test wafers with the annealing before the He implantation.

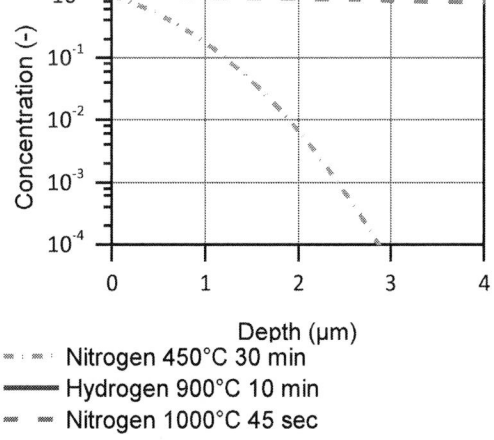

Fig. 6: Theoretical concentration profiles of the impurities after the annealing process.

978-1-5090-3084-2/16 $31.00 © 2016 IEEE   237

4. Interaction between the He implantation profile and the wafer process history

The He implantation resistivity profile is very sensitive to impurities or defects in the silicon. It has been demonstrated by Privitera [7], who used the He implantation as a marker for evaluation of vacancy and interstitial diffusion in silicon.

The test wafers with the annealing before the He implantation were affected by an analogous mechanism. The annealing before the He implantation was implemented in a specific furnace ambient (nitrogen, hydrogen…). Gas atoms diffused into the silicon during the process. Concentration profiles (Fig. 6) of the impurities were calculated by following formula:

$$C(Depth) \approx \mathrm{erfc}\left(\frac{Depth}{2\sqrt{Dt}}\right), \tag{1}$$

where t is an annealing time and D is a diffusion coefficient ($D_H = 6.94 \cdot 10^{-5}$ cm^{2}s^{-1} for hydrogen at 900°C [8]). Nitrogen diffusion in silicon is a complex problem and several mechanisms have been described. If nitrogen anomalous diffusion model [9] ($D_{N\ 450°C}$ =1.51·10^{-12} cm^{2}s^{-1} for nitrogen at 450°C, $D_{N\ 1000°C}$ =2.17·10^{-8} cm^{2}s^{-1} for nitrogen at 1000°C) is used, the layer in 2 μm depth and more is affected by the hydrogen and nitrogen diffusion.

The occurrence of the impurities generated by the gas atoms diffusion into the silicon during the activation annealing is the root cause of the resistivity increase in the area affected by He implantation.

5. Conclusions

The resistance profile induced by He implantation can be significantly affected by the impurities or the defects in silicon during the He activation. Those impurities are generated by gas atoms diffusion during the annealing process preceding the He implantation.

The field stop layer generated by the He implantation is limited by the process before the He implantation. During the front-side process, a TIGBT wafer is annealed at a high temperature in an oxide atmosphere and oxygen concentration in the field stop layer reaches levels that do not allow a successful application of helium field stop layer in this process.

Acknowledgement

This work has been supported by R&D grant TH01010419 awarded by the Technology Agency of the Czech Republic.

References

[1] R. Siemieniec et al., Compensation and doping effects in heavily helium-radiated silicon for power device applications, *Microelectronics Journal* 37.3: 204-212, 2006.

[2] T. Laska et al., The field stop IGBT (FS IGBT)-a new power device concept with a great improvement potential, *ISPO'2000: international symposium on power semiconductor devices and IC's*, 2000.

[3] F-J. Niedernostheide et al., Tailoring of field-stop layers in power devices by hydrogen-related donor formation, *28th International Symposium on Power Semiconductor Devices and ICs (ISPSD)*, 2016.

[4] H-J. Schulze et al., *Semiconductor device with a field stop zone*, U.S. Patent No. 7,538,412, 2009.

[5] ON Semiconductor, *IGBT - Field Stop II*, NGTB40N120FL2WG datasheet, Rev. 6. April, 2015.

[6] D. K. Schroder, *Semiconductor material and device characterization*, John Wiley & Sons, 2006.

[7] V. Privitera et al., Room-temperature migration and interaction of ion beam generated defects in crystalline silicon, *Applied Physics Letters* 68.24: 3422-3424, 1996.

[8] *Semiconductor Technology Handbook*, Technology Associates, 1985.

[9] V. V. Voronkov and R. J. Falster., Nitrogen diffusion and interaction with oxygen in Si, *Solid State Phenomena*, Vol. 95. Trans Tech Publications, 2004.

ASDAM 2016, The 11th International Conference on Advanced Semiconductor Devices And Microsystems, November 13-16, 2016, Smolenice, Slovakia

Incremental Control Techniques for Layout Modification

Patrik VACULA[1,2], Vlastimil KOTĚ[1,2], Adam KUBAČÁK[1], Milan LŽÍČAŘ[1],
Radek ZELENÝ[1], Miroslav HUSÁK[2] and Jiří JAKOVENKO[2],

[1]STMicroelectronics, IBC Building, Pobřežní 620/3, Prague, Czech Republic
[2]Department of Microelectronics, Faculty of Electrical Engineering, Czech Technical University in Prague, Technická 2, Prague 6, 16627, Czech Republic

e-mail: patrik.vacula@st.com, vlastimil.kote@st.com, adam.kubacak@st.com,
milan.lzicar@st.com, radek.zeleny@st.com, husak@fel.cvut.cz, jakovenk@fel.cvut.cz

During creation of analog layout in IC CAD environment layout engineers spend a lot of time by modifying objects in a database. By applying new control concept and targeting modern control approaches our solution unifies and simplifies control of any layout object to speed-up work. Discussed new control techniques are compatible with IC CAD environment and both current control devices as keyboard/mouse and new gesture tracking devices. New control concept is very intuitive and improves productivity of analog layout.

1. Introduction

In process of an analog layout creation, the layout objects properties has to be often modified. In the past, considerable effort has been devoted to the creation of a user-friendly interface [1]. A general concept for modification of selected layout objects is through complicated forms in window. This general concept is working well but consist from many steps. Because modifications are done very often, our improvement of the general control concept speeds up the analog layout creation.

2. New Incremental Control Concept

The principles of the new control concept for layout object modifications are as follows:

- Replacement of numeric value typing by incremental approach.
- Suppression of pop-up Edit Properties windows for most frequently operations.
- Display of changed values in a status bar.
- Application of the same control bindkeys on different layout objects.
- Application of different bindkey modifiers on different layout object properties.
- Possibility of using classic and modern control methods for incremental operations.

This new control concept is called as the Incremental Control, where an increment value is applied directly instead of typing a numeric value. The modification flow is simplified to scrolling of a mouse wheel, which is shown in Fig. 1 for wire object. The exact width value is displayed in the status bar, which allows user to have all the modifications under control.

The Incremental Control allows dimension modifications of basic layout objects such as wires, pins with their labels, rectangles, or rows and columns of vias and also allows modifications of aspect ratio of rectangles. With this control concept it is also possible to modify a number of gate fingers (NGF) of metal-oxide-semiconductor field effect transistors

(MOSFETs). The Incremental Control applied on MOSFET allows to quickly browse among all possible variants which shorten the time needed to find an optimal MOSFET shape with the relevant NGF. The Incremental Control applied on MOSFET does not change a length and a width of a transistor channel but only adjusts the NGF. Other modifiable layout objects are capacitors where theirs aspect ratios can be changed. If a requested value of a layout object property is very far from current value an adaptive increment or a predefined list of allowed values is used to speed up transformations and improves user experience.

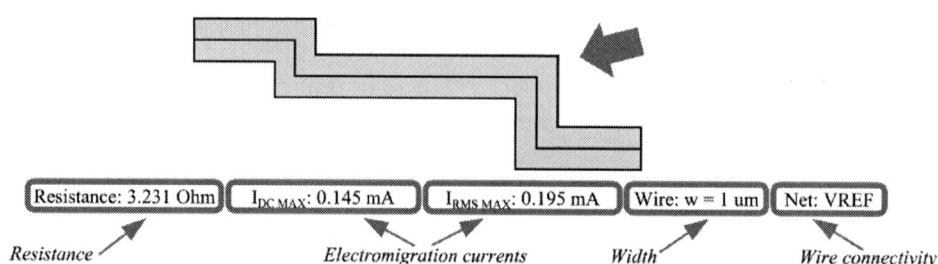

Fig. 1: Two steps for the wire width modification

2.1 Status Bar for Effective Incremental Control

To prevent doing compromise between the Incremental Control simplicity and precision of modifications, the status bar is used. Thus, the user is informed about a current value of a modified property. For example for a wire the status bar displays not only actual width but even more useful electrical quantities such as resistance, maximum electromigration current and connectivity. The status bar is shown in Fig. 2.

Fig. 2: The selected wire electrical and physical properties displayed in the status bar

2.2 Gesture Control with Incremental Control

This discussed control technique is compatible with the IC CAD environment and with current control devices such as keyboard, mouse. Moreover this control technique is also compatible with new gesture tracking systems similar as in [2] and [3]. The Incremental Control is usually mapped to mouse wheel but has been experimentally mapped to gesture recognition camera Creative Senz3D using [4]. Based on experience precision improvement of tested device must be put in place to be suitable for real analog layout application.

2.3 SKILL Implementation of Incremental Control

Virtuoso Layout Suite is universal environment for physical implementation of integrated circuits [5]. The Cadence CAD (Computer aided design) environment supports the programming language Cadence SKILL, which allows implementation of new features as described in [6]. A flow chart describing the Incremental Control implementation in this programming language is shown in Fig. 3.

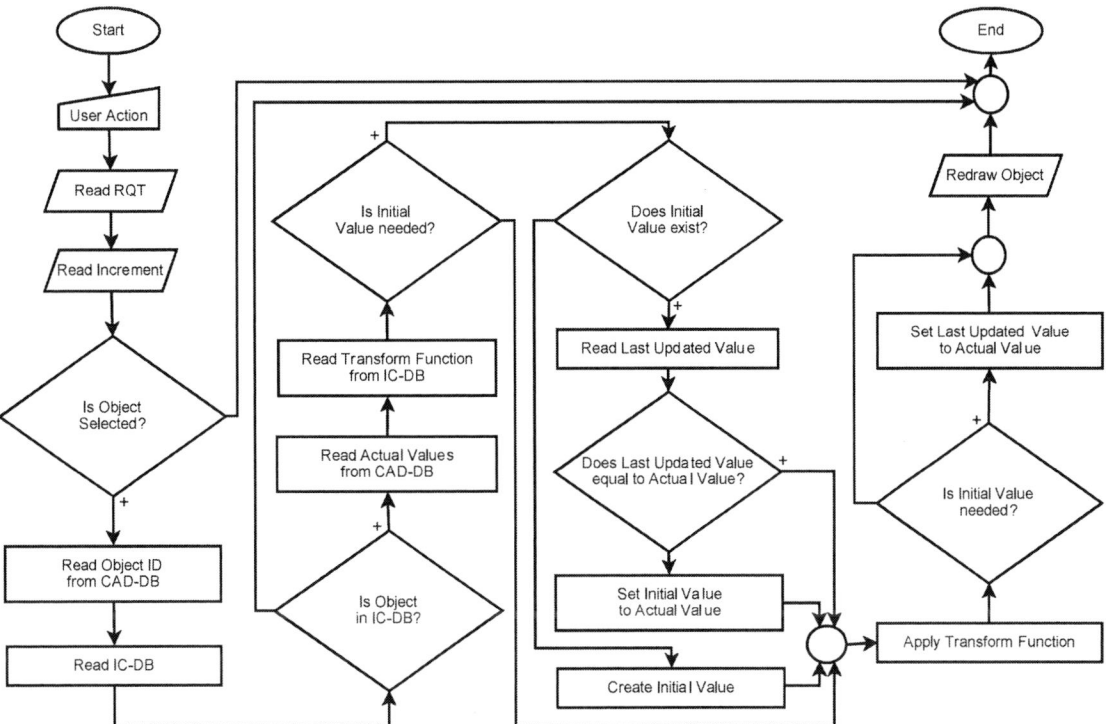

Fig. 3: The flow chart of the Incremental Control

The program starts when the user selects a request type (RQT) by pressing a modifier and performs an increment by the mouse wheel. After the request type and the increment detection, the check of selection is performed. When objects are selected, the program detects object IDs from the CAD database (CAD-DB) and reads the Incremental Control Database (IC-DB) which contains a list of changeable properties of each object type and determines the way of all possible modifications. For case of continuous values to be entered, the IC-DB also contains a list of predefined values which the program reads incrementally.

During layout object modification, an actual value of a changeable property can be slightly different than desired value in order to align layout objects to design grid. After repeated modifications, the actual value can shift from the desired value and an error can occur. Hence, an initial value of the changeable property is introduced. It allows that the actual value to be always calculated from the original correct value. Hereby the maximal possible accuracy of incremental changes done by the program is ensured. If the initial value is required and not yet set, the program stores the initial value as a parameter of the layout object and together with this also creates a record of the last updated value, which serves as a decisive criterion, whether the layout object has been changed in different way by this program. When the initial value is set, then is read and compared with the actual value. If these values equal, the layout object has not been modified by an external intervention and the initial value is untouched. If these values are different, the external intervention has been performed and the initial value is set to the actual value. So, the program respects the external intervention.

After these steps, update of a current state is executed based on obtained parameters and the function which has been read from the IC-DB. When the layout object needs to have the initial value, the last updated value is set to the actual value. So, the last updated value is the parameter of the layout object and it allows its value to be used in next run of this program. By cyclic applications of this program, the layout object is controlled incrementally.

2.4 Layout Productivity Gain

The Incremental Control brings an intuitive common control concept which reduces the number of actions needed to reach optimal result. To quantify layout productivity gain, the average time of standard control modification t_{SC} and the average time of Incremental Control modification t_{IC} to reach target value of layout objects with random initial value are introduced. Relative difference of t_{IC} and t_{SC} has been used for calculation of average layout productivity gain for each layout object type. Based on measurement which is noted in Tab. 1, time saving is in range of 23% to 66%. In Tab. 1, there is evident that higher efficiency is reached when different objects are modified simultaneously.

Tab. 1: Layout productivity measurement results

Layout object	t_{SC} (s)	t_{IC} (s)	$(t_{IC} - t_{SC}) / t_{SC}$ (%)	Layout object	t_{SC} (s)	t_{IC} (s)	$(t_{IC} - t_{SC}) / t_{SC}$ (%)
via	5.6	2.0	-64.7	pin + label	15.1	5.0	-66.7
wire	4.6	3.5	-23.4	rectangle	5.2	3.2	-38.5
pin	7.0	4.5	-36.1	transistor	5.6	3.7	-33.6
label	5.8	3.5	-39.9	capacitor	5.4	3.5	-35.2

3.   Conclusion

The Incremental Control is new control concept which is very intuitive and simplifies interaction with the IC CAD environment. The Incremental Control is implemented in Cadence Virtuoso to be used for modification of layout objects such as wires, pins, labels, vias, rectangles, MOSFETs and capacitors. Main applied ideas are typing removal and incremental approach. Using the Incremental Control, the productivity of analog layout creation has been improved in range of 23% to 66%. Moreover, this type of interaction seems to be more native for all backend designers who use it. By using gesture control camera mapped to Incremental Control bindkeys, layout objects have been changed. Based on our experience, the gesture tracking device needs an additional improvement of precision.

Acknowledgement

This work is part of the CTU SGS grant no. SGS14/195/OHK3/3T/13, (MiNa).

References

[1]   N. Munro, *ECSTASY – A Control System CAD Environment,* Proceedings of the International Conference on Control, 1988. CONTROL 88, Oxford, United Kingdom, April 13–15, pp. 76–80, 1988.

[2]   W. H. Chen, Y. H. Lin and S. J. Yang, *A generic framework for the design of visual-based gesture control interface,* Proceedings of the 5th IEEE Conference on Industrial Electronics and Applications (ICIEA), Taichung, Taiwan, June 15–17, pp. 1522–1525, 2010.

[3]   D. Ionescu, B. Ionescu, C. Gadea and S. Islam, *An intelligent gesture interface for controlling TV sets and set-top boxes,* Proceedings of the 6th IEEE International Symposium on Applied Computational Intelligence and Informatics (SACI), Timisoara, Romania, May 19–21, pp. 159–164, 2011.

[4]   *Intel® Perceptual Computing SDK – Human Interface Guidelines,* Intel Corporation, Revision 3.0, February 25, 2013, [cited 10. 3. 2016], available from: https://software.intel.com/sites/default/   files/article/401008/perc-humaninterfaceguidelines.pdf.

[5]   *Virtuoso® Layout Suite XL User Guide,* Cadence Design Systems, Product Version 6.1.6, June, 2015.

[6]   *Cadence® User Interface SKILL Functions Reference,* Cadence Design Systems, Product Version 6.1.6, August, 2015.

ASDAM 2016, The 11th International Conference on Advanced Semiconductor
Devices And Microsystems, November 13-16, 2016, Smolenice, Slovakia

The influence of ozone pre-treatment in HfO₂-based resistive switching memory structures

P. Benko[1], M. Mikolášek[1], L. Harmatha[1], K. Fröhlich[2]

[1]Slovak University of Technology, Faculty of Electrical Engineering and Information
Technology, Institute of Electronics and Photonics, Ilkovičova 3,
812 19 Bratislava, Slovakia
[2]Slovak Academy of Sciences, Institute of Electrical Engineering, Dúbravská cesta 9,
841 04 Bratislava, Slovakia
e-mail: peter.benko1@stuba.sk

The paper examines the influence of ozone pre-treatment of the bottom TiN electrode in a HfO₂-based resistive switching metal-insulator-metal structure. The HfO₂ layers with thicknesses 5.7, 4 and 3 nm were prepared by ozone assisted atomic layer deposition. Selected samples were pre-treated by ozone-only cycles prior to HfO₂ deposition. Stable and reproducible bipolar resistive switching was obtained with all structures. The resistivity ratio at a read-out voltage of +0.2 V was above 100 for Pt/HfO₂(5.7 nm)/TiN structure with ozone pre-treatment, Pt/HfO₂(4 nm)/TiN and Pt/HfO₂(3 nm)/TiN without ozone pre-treatment. Moreover, ozone pre-treated structures exhibited a lower value of the forming voltage.

1. Introduction

In recent years, the phenomenon of resistive switching (RS) in thin dielectric films and its application in non-volatile memories has been studied intensely. The random access memories based on the principle of resistive switching, referred to ReRAM or RRAM, are considered as good candidates of a new generation of fast high capacity non-volatile memories. RRAM memories combine the properties of dynamic RAMs (fast response) and FLESH memories (non-volatility) and are believed to be a candidate to replace the current NAND FLASH memories. The advantage of RRAM memories is first of all the simplicity of the elementary memory cell. This consists of a metal-insulator-metal (MIM) structure allowing a high scalability and good compatibility with the complementary MIS technology. Resistive switching is repeated and reproducible switching between two defined states: a low-resistance state (LRS) representing the ON state, and a high-resistance state (HRS) representing the OFF state. Usually, formation of a memory cell (thus a soft dielectric breakdown) is performed as a first step after which the memory cell is in the LRS. Subsequent switching from LRS to HRS or vice versa is achieved by applying an external voltage. There are two types of switching modes. First, the bipolar resistive switching (BRS) mode requires opposite electrical polarity during the SET and RESET processes and, second, the unipolar resistive switching mode (URS) requires the same electrical polarity. RS effect has been widely investigated in numerous high permittivity oxide materials including binary transition metal oxides such as TiO₂, HfO₂, ZnO, NiO, Ta₂O₅, CuO and perovskite oxides such as SrTiO₃, LaSrMnO₃ and PrCaMnO₃ [1-5]. In previous years, HfO₂ has been adopted as a gate dielectric in advanced Si technology. For this reason, it is advantageous to develop an HfO₂-based RRAM technology to leverage its compatibility with Si [3, 4, 5].

2. Experiment

The $Pt/HfO_2/TiN$ structures were prepared on a Si substrate. Bottom 70 nm thick TiN electrodes were prepared by reactive sputtering. Thin HfO_2 films were deposited by atomic layer deposition (ALD) at 300°C using the ozone assisted process in the Beneq TFS 200 equipment. The precursor was tetrakis (ethylmethylamino) hafnium, $Hf [N(CH_3)(C_2H_5)]_4$ that was heated at 70°C. Pure nitrogen was used as a carrier gas, while oxygen was employed as a reactant gas. The following time sequence was used for the ALD process: Hf-precursor dosing 1.5 s, purging 10 s, ozone supply 0.4 s, purging 10 s. A growth of 0.07 nm per cycle was obtained. Selected samples were in situ pre-treated by 30 ozone-only cycles prior to HfO_2 deposition in order to alter the chemical state of the thin bottom electrodes. Electron beam vacuum evaporation through a shadow mask was used to deposit the top 30 nm thick Pt electrodes with sizes 1×10^4 μm^2. The electrodes were capped with 30 nm thick Au layer. The MIM structures were electrically characterized using a Keithley 2612A source meter instrument. The bottom electrodes were grounded and the top ones biased during the measurement.

Tab. 1 Summary of samples parameters

Sample	HfO_2 thickness	O_3 pre-treatment	Resistivity ratio	Forming voltage
Hfl78-9	5.7 nm	no	above 60	~ -4 V
Hfl78-4		yes	above 100	~ -1.8 V
Hfl79-9	4 nm	no	above 100	~ -2.9 V
Hfl79-4		yes	above 40	~ -1.2 V
Hfl80-9	3 nm	no	above 100	~ -2.6 V
Hfl80-4		yes	above 20	~ -1.2 V

3. Results and discussion

Figures 1 and 2 show the bipolar resistive switching characteristics for all samples. The thickness of the HfO_2 films was between 3 and 6 nm. In the first step, all samples were formed by applying a negative voltage with current compliance (CC) in the range from 10^{-4} to 10^{-2} A. Resistive switching cycles were measured with different current limitations: 3 mA for samples Hfl78-9 (Fig. 1(a)) and Hfl80-4 (Fig. 2(c)), 1 mA for sample Hfl79-9 (Fig. 1(b)), 2 mA for samples Hfl80-9 (Fig. 1(c)) and Hfl78-4 (Fig. 2(a)), and 5 mA for sample Hfl79-4 (Fig. 2(b)). All samples exhibit fairly stable cycles of bipolar resistive switching. The SET transition occurred as a rapid change close to −0.8 V for all samples. On the other hand, the RESET transition is a gradual resistance increase which starts close to +1 V for all samples. The values of the resistance ratio (RR) defined as HRS/LRS resistance ratio were determined at a read-out voltage of +0.2 V (Tab. 1). Desired values of RR, more than 10, were obtained for all samples. Excellent RR values above 100 were obtained for samples Hfl78-4, Hfl79-9 and Hfl80-9. Figure 3 shows the forming *I-V* curves for single samples. The forming voltage is approx. −4 V for sample Hfl78-9, −2.9 V for sample Hfl79-9, and −2.6 V for sample Hfl80-9. As assumed, the magnitude of the forming voltage is lower with a decreasing thickness of the HfO_2 layer. In the case of ozone treated samples, a marked decrease of the forming voltage was observed to a value of approx. −1.8 V for sample Hfl78-4, and to −1.2 V for samples Hfl79-4 and Hfl80-4.

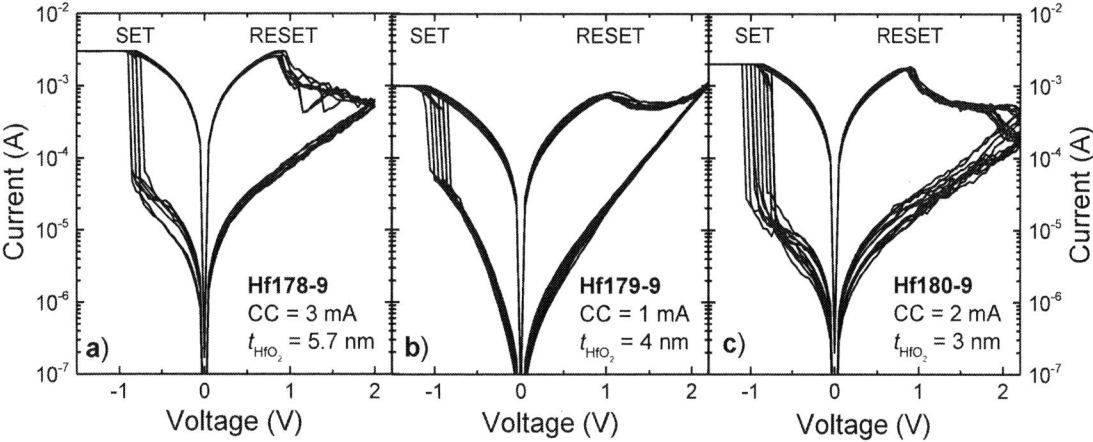

Fig. 1 Resistive switching cycles of the Pt/HfO$_2$/TiN structure without ozone pre-treatment for different thicknesses of the HfO$_2$ layer.

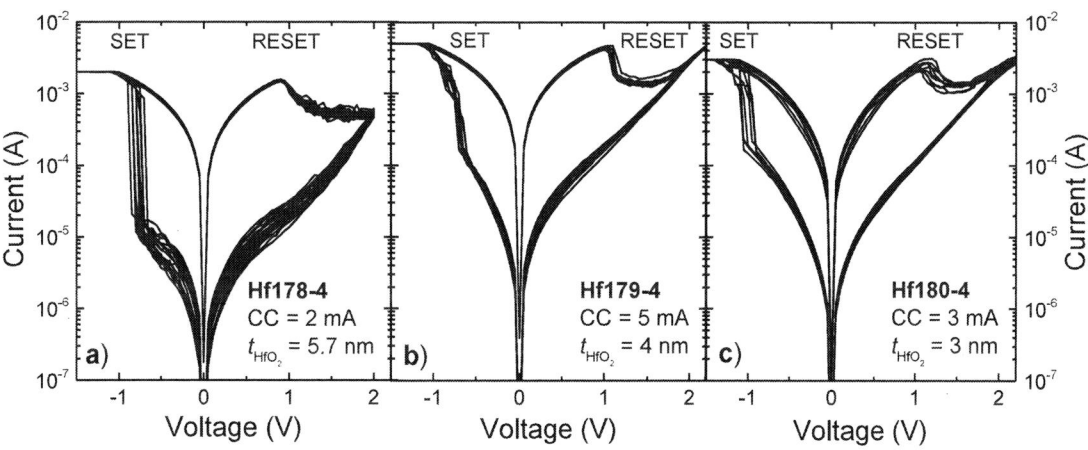

Fig. 2 Resistive switching cycles of the Pt/HfO$_2$/TiN structure with ozone pre-treatment for different thicknesses of the HfO$_2$ layer.

The value of the forming voltage for ozone treated samples (mainly for those with 3 and 4 nm thick HfO$_2$ layers) is close to the value of the SET voltage (post-forming operation) which is approx. –0.8 V for all samples. During structure forming, a soft breakdown occurs due to a high field intensity across the dielectric. In the case of an inappropriately set current compliance, in combination with a high value of the electric field (higher values of the forming voltage), the so-called current overshoot may occur, which results in a difference between the first and next BRS cycles. Also, the RESET voltage of this first reset sweep may be typically higher compared with a reset during stable BRS operation [6]. Thus a lower forming voltage may achieve a less dramatic current overshoot during structure forming. On the other hand, one can clearly see that in samples treated by ozone a marked rise is observed of the current flowing through the pristine (non-formed) cell (Fig. 3). Parameters of all samples are summarized in Tab. 1.

978-1-5090-3084-2/16 $31.00 © 2016 IEEE

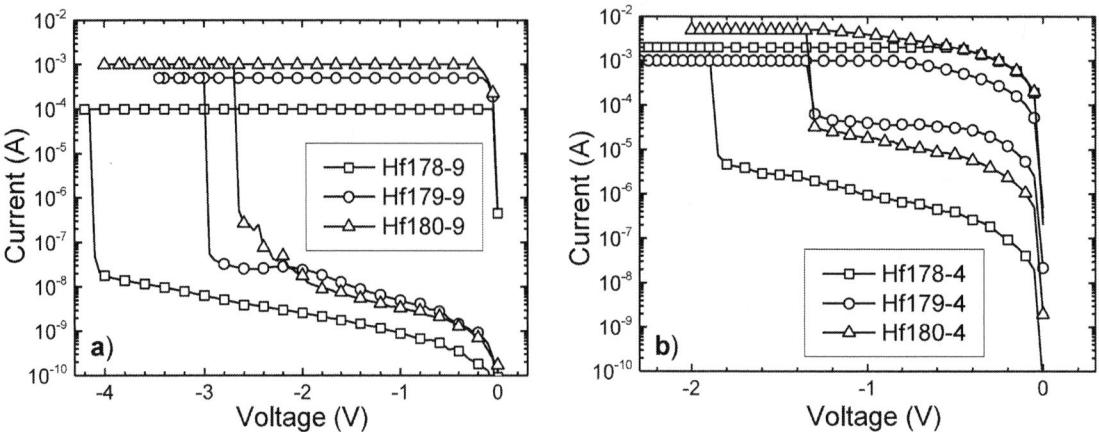

Fig. 3 Forming *IV* characteristic of the Pt/HfO$_2$/TiN structures a) without and b) with ozone pre-treatment for different thicknesses of the HfO$_2$ layer. Different values of current compliance were used in the forming process.

4. Conclusion

We have investigated the influence of ozone pre-treatment in Pt/HfO$_2$/TiN resistive switching structures with 5.7 nm, 4 nm and 3 nm HfO$_2$ layers deposited by ozone assisted ALD. All samples exhibit stable bipolar resistive switching. The RR at the read-out voltage +0.2 V was above 100 for samples Hf178-4, Hf179-9 and Hf180-9. By using the ozone treatment we have managed to lower the forming voltage. The samples treated by ozone exhibited lower forming voltages, roughly by one half, in comparison with non-treated samples with the same thickness of the HfO$_2$ layer.

Acknowledgement

This work was financially supported by a grant of the Scientific Grant Agency of the Ministry of Education of the Slovak Republic and of the Slovak Academy of Sciences No. VEGA-1/0651/16 and by a grant of the Slovak Research and Development Agency APVV-0509-10.

References

[1]   K. L. Lin, T. H. Hou, J. Shieh, *J. Appl. Phys.* **109**, 084104:1-7, 2011.
[2]   Ch. Yoshida, K. Tsunoda, H. Noshiro, et.al., *Appl. Phys. Lett.* **91**, 223510:1-3, 2007.
[3]   S. Brivio, J. Frascali, S. Spiga, *Appl. Phys. Lett.* **107**, 023504:1-5, 2015.
[4]   J.W. Yoon, J.H. Yoon, J.H. Lee, Ch.S. Hwang., *Nanoscale* **6**, 6668-6678, 2014.
[5]   P. Jančovič, B. Hudec, E. Dobročka, *Appl. Surf. Sci.* **312**, 112-116, 2014.
[6]   B. Hudec, I-T. Wang, W-L. Lai, et.al., *J. Phys. D: Appl. Phys.* **49**, 215102:1-9, 2016.

AUTHOR INDEX

Adam	R.	73
Ahmeda	K.	181, 203
Al-Nashash	H.	61
Andok	R.	133, 137
Arbet	J.	141, 215
Babchenko	O.	157, 227
Badura	M.	149
Benbakhti	B.	203
Benčurová	A.	85, 133, 137
Benko	P.	161, 165, 243
Bielak	K.	149
Bíly	A.	109
Bird	J.	89
Blaho	M.	177
Boháček	P.	23, 141, 215, 219
Cesur	E.	211
Davydova	M.	193
Dawidowski	W.	33
Delage	S. L.	145
Di Benedetto	L.	19, 169, 199
Donoval	D.	49, 129, 165, 173
Donoval	M.	129, 173
Dorňák	L.	235
Dowben	P.	89
Dubecký	F.	141, 219
Duffy	S. J.	203
Ďurina	P.	137
Ďurišová	J.	33
Dzuba	J.	157, 223, 227
Faramehr	S.	181, 203
Florovič	M.	185
Foit	J.	101
Fox	A.	65, 73
Fröhlich	K.	1, 9, 45, 189, 207, 243
Gašo	P.	33, 53
Gombia	E.	219
Goraus	M.	33, 53
Grančič	B.	137
Gregušová	D.	1, 9, 177, 185, 189, 207
Grosser	A.	5
Grützmacher	D.	65, 69, 73, 81
Gucmann	F.	1, 9, 207

Hardtdegen	H.	27, 65, 69, 73, 77, 81
Harmatha	L.	45, 145, 149, 243
Hashizume	T.	1
Haščik	Š.	9, 85, 177, 207
Hentschel	R.	5
Horvát	F.	227
Hospodková	A.	41, 57, 93
Hotový	I.	85, 121, 153, 161
Hronec	P.	37
Hrubčín	L.	141, 215
Hubík	P.	219
Hučko	B.	227
Hulicius	E.	41, 57, 93
Huran	J.	215
Husák	M.	109, 113, 117, 239
Hušeková	K.	9, 207
Chimento	F.	13
Chvála	A.	49, 129, 165, 173, 177
Igič	P.	181, 203
Ižák	T.	157, 193
Jagelka	M.	129, 173
Jahn	A.	5
Jakovenko	J.	239
Jandura	D.	53
Janicek	V.	97
Jom	P.	153
Kalisch	H.	5
Kalna	K.	181, 203
Kindl	D.	219
Kleinová	A.	215
Klimsza	A.	235
Kobzev	A. P.	215
Kočan	M.	65
Koleva	E.	137
Kordoš	P.	65, 73, 77, 185
Kósa	A.	145, 149
Kostic	I.	85, 133, 137
Kotě	V.	239
Kováč	J.	33, 37, 145, 149, 185
Kováč	T.	227

Kováč, jr.	J.	33, 37		Racko	J.	45
Kováčová	S.	125		Rezek	B.	193
Kozelský	A.	235		Ritomský	A.	133
Kromka	A.	157, 193		Rossberg	D.	85
Kroutil	J.	113, 117		Rubino	A.	169, 199
Kubačák	A.	239		Řehaček	V.	45, 85, 121, 161
Kuldová	K.	41		Sasinková	V.	215
Kúš	P.	137		Sciana	B.	33, 149
Kuzmík	J.	1, 9, 177, 189, 207		Sedlačková	K.	23, 141
Lalinský	T.	157, 223, 227		Seifertová	A.	177
Laposa	A.	113, 117		Sekáčová	M.	141, 215, 219
Licciardo	G. D.	19, 169, 199		Sharma	N.	89
Lüth	H.	65, 69, 73, 77		Schmult	S.	5
Lžičař	M.	239		Schubert	J.	65
Marek	J.	49, 129, 165, 173, 177		Schuck	M.	73
Marshall	A.	89		Sio	R.	117
Marso	M.	65, 69, 73, 77, 81		Skuratov	V. A.	141
				Sobolewski	R.	73
Merkel	U.	5		Sofer	Z.	65, 69, 77, 81,
Mikiolajick	T.	5		Soltani	A.	203
Mikolášek	M.	45, 49, 85, 153, 161, 215, 243		Spiess	L.	85
				Stoklas	R.	189
Mikulics	M.	27, 65, 69, 73, 77, 81		Stuchliková	Ľ.	129, 145, 149, 173
Moers	J.	81		Suslik	L.	33, 53
Nagy	L.	177		Szabó	O.	125
Narayanan	M.	61		Szarvas	M.	227
Nawaz	M.	13		Szobolovszky	R.	145
Nečas	V.	23, 141		Šagátová	A.	23, 141
Nemec	P.	85		Šatka	A.	177
Novak	J.	101		Škriniarová	J.	37, 133, 185
Novotný	I.	125		Šoltýs	J.	177
Öksüz	D. I.	211		Ťapajna	M.	1, 9, 45, 177, 207
Osvald	J.	231				
Oswald	J.	41, 57		Tlaczala	M.	33, 149
Pangrác	J.	41, 57, 93		Trellenkamp	St.	69, 81
Pechal	R.	235		Truchlý	M.	137
Petrus	M.	145		Tvarožek	V.	125
Plecenik	A.	137		Ubochi	B. C.	181
Pohorelec	O.	9, 207		Uherek	F.	37
Predanocy	M.	85, 121, 153, 161		Vacula	P.	239
				Válik	L.	1, 9, 207
Príbytný	P.	49, 129, 165, 173		Vanko	G.	157, 219, 223, 227
Priesol	J.	177		Vavro	J.	235
Puci	F.	105		Vescan	A.	5
Pudiš	D.	33, 53		Vojs	M.	157

Vutova	K.	133, 137
Wachowiak	A.	5
Wille	A.	5
Winden	A.	69
Yarimbiyik	A. M.	211
Zaťko	B.	23, 141, 219
Zborowska-Lindert	I.	33
Zehetner	J.	223
Zelený	R.	239
Ziková	M.	41, 57, 93

IEEE
445 Hoes Lane
Piscataway, NJ 08854-4141

ISBN 978-1-5090-3084-2